HUMBOLDT'S PERSONAL NARRATIVE

NARRATIVE

VOLUME 2.

ISBN: 978-1-78139-331-4

Cover: *The River of Light* by Frederic Edwin Church.

Printed on acid-free ANSI archival quality paper.

HUMBOLDT'S PERSONAL NARRATIVE

PERSONAL NARRATIVE OF TRAVELS TO THE EQUINOCTIAL
REGIONS OF AMERICA DURING THE YEARS 1799-1804

BY

ALEXANDER VON HUMBOLDT

AND

AIMÉ BONPLAND

TRANSLATED FROM THE FRENCH OF
ALEXANDER VON HUMBOLDT
AND EDITED BY
THOMASINA ROSS.

IN THREE VOLUMES

VOLUME 2.

Contents

Personal Narrative

of a

Journey

to the

Equinoctial Regions

of the

New Continent.

Volume 2.

CHAPTER 2.16.

LAKE OF TACARIGUA. HOT SPRINGS OF MARIARA. TOWN OF NUEVA VALENCIA DEL REY. DESCENT TOWARDS THE COASTS OF PORTO CABELLO.

The valleys of Aragua form a narrow basin between granitic and calcareous mountains of unequal height. On the north, they are separated by the Sierra Mariara from the sea-coast; and towards the south, the chain of Guacimo and Yusma serves them as a rampart against the heated air of the steppes. Groups of hills, high enough to determine the course of the waters, close this basin on the east and west like transverse dykes. We find these hills between the Tuy and La Victoria, as well as on the road from Valencia to Nirgua, and at the mountains of Torito.[*] From this extraordinary configuration of the land, the little rivers of the valleys of Aragua form a peculiar system, and direct their course towards a basin closed on all sides. These rivers do not bear their waters to the ocean; they are collected in a lake; and subject to the peculiar influence of evaporation, they lose

---

[*] The lofty mountains of Los Teques, where the Tuy takes its source, may be looked upon as the eastern boundary of the valleys of Aragua. The level of the ground continues, in fact, to rise from La Victoria to the Hacienda de Tuy; but the river Tuy, turning southward in the direction of the sierras of Guairaima and Tiara has found an issue on the east; and it is more natural to consider as the limits of the basin of Aragua a line drawn through the sources of the streams flowing into the lake of Valencia. The charts and sections I have traced of the road from Caracas to Nueva Valencia, and from Porto Cabello to Villa de Cura, exhibit the whole of these geological relations.

themselves, if we may use the expression, in the atmosphere. On the existence of rivers and lakes, the fertility of the soil and the produce of cultivation in these valleys depend. The aspect of the spot, and the experience of half a century, have proved that the level of the waters is not invariable; the waste by evaporation, and the increase from the waters running into the lake, do not uninterruptedly balance each other. The lake being elevated one thousand feet above the neighbouring steppes of Calabozo, and one thousand three hundred and thirty-two feet above the level of the ocean, it has been suspected that there are subterranean communications and filtrations. The appearance of new islands, and the gradual retreat of the waters, have led to the belief that the lake may perhaps, in time, become entirely dry. An assemblage of physical circumstances so remarkable was well fitted to fix my attention on those valleys where the wild beauty of nature is embellished by agricultural industry, and the arts of rising civilization.

The lake of Valencia, called Tacarigua by the Indians, exceeds in magnitude the lake of Neufchatel in Switzerland; but its general form has more resemblance to the lake of Geneva, which is nearly at the same height above the level of the sea. As the slope of the ground in the valleys of Aragua tends towards the south and the west, that part of the basin still covered with water is the nearest to the southern chain of the mountains of Guigue, of Yusma, and of Guacimo, which stretch towards the high savannahs of Ocumare. The opposite banks of the lake of Valencia display a singular contrast; those on the south are desert, and almost uninhabited, and a screen of high mountains gives them a gloomy and monotonous aspect. The northern shore on the contrary, is cheerful, pastoral, and decked with the rich cultivation of the sugar-cane, coffee-tree, and cotton. Paths bordered with cestrums, azedaracs, and other shrubs always in flower, cross the plain, and join the scattered farms. Every house is surrounded by clumps of trees. The ceiba with its large yellow flowers[*] gives a peculiar character to the landscape, mingling its branches with those of the purple erythrina. This mixture of vivid vegetable colours contrasts finely with the uniform tint of an unclouded sky. In the season of drought, where the burning soil is covered with an undulating vapour, artificial irrigations preserve verdure and promote fertility. Here and there the granite rock pierces through the cultivated ground. Enormous stony masses rise abruptly in the midst of the valley. Bare and forked, they nourish a few succulent plants, which prepare mould for future ages. Often on the summit of these lonely hills may be seen a fig-tree or a clusia with fleshy leaves, which has fixed its roots in the rock, and towers over the landscape. With their dead and withered branches, these trees look like signals erected on a steep cliff. The form of these mounts unfolds the secret of their ancient origin; for when the whole of this valley was filled with water, and the waves beat at the foot of the peaks of Mariara (the Devil's Nook[†] and the chain of the coast, these rocky hills were shoals or islets.

These features of a rich landscape, these contrasts between the two banks of the lake of Valencia, often reminded me of the Pays de Vaud, where the soil, everywhere cultivated, and everywhere fertile, offers the husbandman, the shepherd, and the vine-dresser, the secure fruit of their labours, while, on the opposite side, Chablais presents only a mountainous and half-desert country. In these distant climes surrounded by exotic productions, I loved to recall to mind the enchanting descriptions with which the aspect of the Leman lake and the rocks of La Meillerie inspired a great writer. Now, while in the centre of civilized Europe, I endeavour in my turn to paint the scenes of the New World, I do not imagine I present the reader with clearer images, or more precise ideas, by comparing our landscapes with those of the equinoctial regions. It cannot be too often repeated that nature, in every zone, whether wild or

---

[*] Carnes tollendas, Bombax hibiscifolius.
[†] El Rincon del Diablo.)

cultivated, smiling or majestic, has an individual character. The impressions which she excites are infinitely varied, like the emotions produced by works of genius, according to the age in which they were conceived, and the diversity of language from which they in part derive their charm. We must limit our comparisons merely to dimensions and external form. We may institute a parallel between the colossal summit of Mont Blanc and the Himalaya Mountains; the cascades of the Pyrenees and those of the Cordilleras: but these comparisons, useful with respect to science, fail to convey an idea of the characteristics of nature in the temperate and torrid zones. On the banks of a lake, in a vast forest, at the foot of summits covered with eternal snow, it is not the mere magnitude of the objects which excites our admiration. That which speaks to the soul, which causes such profound and varied emotions, escapes our measurements as it does the forms of language. Those who feel powerfully the charms of nature cannot venture on comparing one with another, scenes totally different in character.

But it is not alone the picturesque beauties of the lake of Valencia that have given celebrity to its banks. This basin presents several other phenomena, and suggests questions, the solution of which is interesting alike to physical science and to the well-being of the inhabitants. What are the causes of the diminution of the waters of the lake? Is this diminution more rapid now than in former ages? Can we presume that an equilibrium between the waters flowing in and the waters lost will be shortly re-established, or may we apprehend that the lake will entirely disappear?

According to astronomical observations made at La Victoria, Hacienda de Cura, Nueva Valencia, and Guigue, the length of the lake in its present state from Cagua to Guayos, is ten leagues, or twenty-eight thousand eight hundred toises. Its breadth is very unequal. If we judge from the latitudes of the mouth of the Rio Cura and the village of Guigue, it nowhere surpasses 2.3 leagues, or six thousand five hundred toises; most commonly it is but four or five miles. The dimensions, as deduced from my observations are much less than those hitherto adopted by the natives. It might be thought that, to form a precise idea of the progressive diminution of the waters, it would be sufficient to compare the present dimensions of the lake with those attributed to it by ancient chroniclers; by Oviedo for instance, in his History of the Province of Venezuela, published about the year 1723. This writer in his emphatic style, assigns to "this inland sea, this monstruoso cuerpo de la laguna de Valencia"[*], fourteen leagues in length and six in breadth. He affirms that at a small distance from the shore the lead finds no bottom; and that large floating islands cover the surface of the waters, which are constantly agitated by the winds. No importance can be attached to estimates which, without being founded on any measurement, are expressed in leagues (leguas) reckoned in the colonies at three thousand, five thousand, and six thousand six hundred and fifty varas.[†] Oviedo, who must so often have passed over the valleys of Aragua, asserts that the town of Nueva Valencia del Rey was built in 1555, at the distance of half a league from the lake; and that the proportion between the length of the lake and its breadth, is as seven to three. At present, the town of Valencia is separated from the lake by level ground of more than two thousand seven hundred toises (which Oviedo would no doubt have estimated as a space of a league and a half); and the length

---

[*] "Enormous body of the lake of Valencia."

[†] Seamen being the first, and for a long time the only, persons who introduced into the Spanish colonies any precise ideas on the astronomical position and distances of places, the legua nautica of 6650 varas, or of 2854 toises (20 in a degree), was originally used in Mexico and throughout South America; but this legua nautica has been gradually reduced to one-half or one-third, on account of the slowness of travelling across steep mountains, or dry and burning plains. The common people measure only time directly; and then, by arbitrary hypotheses, infer from the time the space of ground travelled over. In the course of my geographical researches, I have had frequent opportunities of examining the real value of these leagues, by comparing the itinerary distances between points lying under the same meridian with the difference of latitudes.

of the basin of the lake is to its breadth as 10 to 2.3, or as 7 to 1.6. The appearance of the soil between Valencia and Guigue, the little hills rising abruptly in the plain east of the Cano de Cambury, some of which (el Islote and la Isla de la Negra or Caratapona) have even preserved the name of islands, sufficiently prove that the waters have retired considerably since the time of Oviedo. With respect to the change in the general form of the lake, it appears to me improbable that in the seventeenth century its breadth was nearly the half of its length. The situation of the granite mountains of Mariara and of Guigue, the slope of the ground which rises more rapidly towards the north and south than towards the east and west, are alike repugnant to this supposition.

In treating the long-discussed question of the diminution of the waters, I conceive we must distinguish between the different periods at which the sinking of their level has taken place. Wherever we examine the valleys of rivers, or the basins of lakes, we see the ancient shore at great distances. No doubt seems now to be entertained, that our rivers and lakes have undergone immense diminutions; but many geological facts remind us also, that these great changes in the distribution of the waters have preceded all historical times; and that for many thousand years most lakes have attained a permanent equilibrium between the produce of the water flowing in, and that of evaporation and filtration. Whenever we find this equilibrium broken, it will be well rather to examine whether the rupture be not owing to causes merely local, and of very recent date, than to admit an uninterrupted diminution of the water. This reasoning is conformable to the more circumspect method of modern science. At a time when the physical history of the world, traced by the genius of some eloquent writers, borrowed all its charms from the fictions of imagination, the phenomenon of which we are treating would have been adduced as a new proof of the contrast these writers sought to establish between the two continents. To demonstrate that America rose later than Asia and Europe from the bosom of the waters, the lake of

Tacarigua would have been described as one of those interior basins which have not yet become dry by the effects of slow and gradual evaporation. I have no doubt that, in very remote times, the whole valley, from the foot of the mountains of Cocuyza to those of Torito and Nirgua, and from La Sierra de Mariara to the chain of Guigue, of Guacimo, and La Palma, was filled with water. Everywhere the form of the promontories, and their steep declivities, seem to indicate the shore of an alpine lake, similar to those of Styria and Tyrol. The same little helicites, the same valvatae, which now live in the lake of Valencia, are found in layers of three or four feet thick as far inland as Turmero and La Concesion near La Victoria. These facts undoubtedly prove a retreat of the waters; but nothing indicates that this retreat has continued from a very remote period to our days. The valleys of Aragua are among the portions of Venezuela most anciently peopled; and yet there is no mention in Oviedo, or any other old chronicler, of a sensible diminution of the lake. Must we suppose, that this phenomenon escaped their observation, at a time when the Indians far exceeded the white population, and when the banks of the lake were less inhabited? Within half a century, and particularly within these thirty years, the natural desiccation of this great basin has excited general attention. We find vast tracts of land which were formerly inundated, now dry, and already cultivated with plantains, sugar-canes, or cotton. Wherever a hut is erected on the bank of the lake, we see the shore receding from year to year. We discover islands, which, in consequence of the retreat of the waters, are just beginning to be joined to the continent, as for instance the rocky island of Culebra, in the direction of Guigue; other islands already form promontories, as the Morro, between Guigue and Nueva Valencia, and La Cabrera, south-east of Mariara; others again are now rising in the islands themselves like scattered hills. Among these last, so easily recognised at a distance, some are only a quarter of a mile, others a league from the present shore. I may cite as the most remarkable three granite islands, thirty or forty toises high, on the road from the Hacienda de Cura to

Aguas Calientes; and at the western extremity of the lake, the Serrito de Don Pedro, Islote, and Caratapona. On visiting two islands entirely surrounded by water, we found in the midst of brushwood, on small flats (four, six, and even eight toises height above the surface of the lake,) fine sand mixed with helicites, anciently deposited by the waters. (Isla de Cura and Cabo Blanco. The promontory of Cabrera has been connected with the shore ever since the year 1750 or 1760 by a little valley, which bears the name of Portachuelo.) In each of these islands may be perceived the most certain traces of the gradual sinking of the waters. But still farther (and this accident is regarded by the inhabitants as a marvellous phenomenon) in 1796 three new islands appeared to the east of the island Caiguira, in the same direction as the islands Burro, Otama, and Zorro. These new islands, called by the people Los nuevos Penones, or Los Aparecidos,[*] form a kind of banks with surfaces quite flat. They rose, in 1800, more than a foot above the mean level of the water.

It has already been observed that the lake of Valencia, like the lakes of the valley of Mexico, forms the centre of a little system of rivers, none of which have any communication with the ocean. These rivers, most of which deserve only the name of torrents, or brooks,[†] are twelve or fourteen in number. The inhabitants, little acquainted with the effects of evaporation, have long imagined that the lake has a subterranean outlet, by which a quantity of water runs out equal to that which flows in by the rivers. Some suppose that this outlet communicates with grottos, supposed to be at great depth; others believe that the water flows through an oblique channel into the basin of the ocean. These bold hypotheses on the communication between

---

[*] Los Nuevos Penones, the New Rocks. Los Aparecidos, the Unexpectedly-appeared.

[†] The following are their names: Rios de Aragua, Turmero, Maracay, Tapatapa, Agnes Calientes, Mariara, Cura, Guacara, Guataparo, Valencia, Cano Grande de Cambury, etc.

two neighbouring basins have presented themselves in every zone to the imagination of the ignorant, as well as to that of the learned; for the latter, without confessing it, sometimes repeat popular opinions in scientific language. We hear of subterranean gulfs and outlets in the New World, as on the shores of the Caspian sea, though the lake of Tacarigua is two hundred and twenty-two toises higher, and the Caspian sea fifty-four toises lower, than the sea; and though it is well known, that fluids find the same level, when they communicate by a lateral channel.

The changes which the destruction of forests, the clearing of plains, and the cultivation of indigo, have produced within half a century in the quantity of water flowing in on the one hand, and on the other the evaporation of the soil, and the dryness of the atmosphere, present causes sufficiently powerful to explain the progressive diminution of the lake of Valencia. I cannot concur in the opinion of M. Depons[*] (who visited these countries since I was there) "that to set the mind at rest, and for the honour of science," a subterranean issue must be admitted. By felling the trees which cover the tops and the sides of mountains, men in every climate prepare at once two calamities for future generations; want of fuel and scarcity of water. Trees, by the nature of their perspiration, and the radiation from their leaves in a sky without clouds, surround themselves with an atmosphere constantly cold and misty. They affect the copiousness of springs, not, as was long believed, by a peculiar attraction for the vapours diffused through the air, but because, by sheltering the soil from the direct action of the sun, they diminish the evaporation of water produced by rain. When forests are destroyed, as they are everywhere in America by the European planters, with imprudent precipitancy, the springs are entirely

---

[*] In his Voyage a la Terre Ferme M. Depons says, "The small extent of the surface of the lake renders impossible the supposition that evaporation alone, however considerable within the tropics, could remove as much water as the rivers furnish." In the sequel, the author himself seems to abandon what he terms "this occult case, the hypothesis of an aperture."

dried up, or become less abundant. The beds of the rivers, remaining dry during a part of the year, are converted into torrents whenever great rains fall on the heights. As the sward and moss disappear with the brushwood from the sides of the mountains, the waters falling in rain are no longer impeded in their course; and instead of slowly augmenting the level of the rivers by progressive filtrations, they furrow, during heavy showers, the sides of the hills, bearing down the loosened soil, and forming sudden and destructive inundations. Hence it results, that the clearing of forests, the want of permanent springs, and the existence of torrents, are three phenomena closely connected together. Countries situated in opposite hemispheres, as, for example, Lombardy bordered by the Alps, and Lower Peru inclosed between the Pacific and the Cordillera of the Andes, afford striking proofs of the justness of this assertion.

Till the middle of the last century, the mountains round the valleys of Aragua were covered with forests. Great trees of the families of mimosa, ceiba, and the fig-tree, shaded and spread coolness along the banks of the lake. The plain, then thinly inhabited, was filled with brushwood, interspersed with trunks of scattered trees and parasite plants, enveloped with a thick sward, less capable of emitting radiant caloric than the soil that is cultivated and consequently not sheltered from the rays of the sun. With the destruction of the trees, and the increase of the cultivation of sugar, indigo, and cotton, the springs, and all the natural supplies of the lake of Valencia, have diminished from year to year. It is difficult to form a just idea of the enormous quantity of evaporation which takes place under the torrid zone, in a valley surrounded with steep declivities, where a regular breeze and descending currents of air are felt towards evening, and the bottom of which is flat, and looks as if levelled by the waters. It has been remarked, that the heat which prevails throughout the year at Cura, Guacara, Nueva Valencia, and on the borders of the lake, is the same as that felt at midsummer in Naples and Sicily. The mean annual temperature of the valleys of Aragua is nearly 25.5°; my hygrometrical observations of the month of February, taking the mean of day and night, gave 71.4° of the hair hygrometer. As the words great drought and great humidity have no determinate signification, and air that would be called very dry in the lower regions of the tropics would be regarded as humid in Europe, we can judge of these relations between climates only by comparing spots situated in the same zone. Now at Cumana, where it sometimes does not rain during a whole year, and where I had the means of collecting a great number of hygrometric observations made at different hours of the day and night, the mean humidity of the air is 86°; corresponding to the mean temperature of 27.7°. Taking into account the influence of the rainy months, that is to say, estimating the difference observed in other parts of South America between the mean humidity of the dry months and that of the whole year; an annual mean humidity is obtained, for the valleys of Aragua, at farthest of 74°, the temperature being 25.5°. In this air, so hot, and at the same time so little humid, the quantity of water evaporated is enormous. The theory of Dalton estimates, under the conditions just stated, for the thickness of the sheet of water evaporated in an hour's time, 0.36 mill., or 3.8 lines in twenty-four hours. Assuming for the temperate zone, for instance at Paris, the mean temperature to be 10.6°, and the mean humidity 82°, we find, according to the same formulae, 0.10 mill., an hour, and 1 line for twenty-four hours. If we prefer substituting for the uncertainty of these theoretical deductions the direct results of observation, we may recollect that in Paris, and at Montmorency, the mean annual evaporation was found by Sedileau and Cotte, to be from 32 in. 1 line to 38 in. 4 lines. Two able engineers in the south of France, Messrs. Clausade and Pin, found, that in subtracting the effects of filtrations, the waters of the canal of Languedoc, and the basin of Saint Ferreol lose every year from 0.758 met. to 0.812 met., or from 336 to 360 lines. M. de Prony found nearly similar results in the Pontine marshes. The whole of these experiments, made in the latitudes of 41 and 49°, and at 10.5 and 16° of mean temperature, indicate a mean evaporation of one line, or one and three-tenths a day. In the torrid zone, in the West India Islands for instance, the effect of evaporation is three times as much, according to Le Gaux, and double according to Cassan. At Cumana, in a place where the atmosphere is far more loaded with humidity than in the valley of Aragua, I have often seen evaporate during twelve hours, in the sun, 8.8 mill., in the shade 3.4 mill.; and I believe, that the annual produce of evaporation in the rivers near Cumana is not less than one hundred and thirty inches. Experiments of this kind are extremely delicate, but what I have stated will suffice to demonstrate how great must be the quantity of vapour that rises from the lake of Valencia, and from the surrounding country, the waters of which flow into the lake. I shall have occasion elsewhere to resume this subject; for, in a work which displays the great laws of nature in different zones, we must endeavour to solve the problem of the mean tension of the vapours contained in the atmosphere in different latitudes, and at different heights above the surface of the ocean.

A great number of local circumstances cause the produce of evaporation to vary; it changes in proportion as more or less shade covers the basin of the waters, with their state of motion or repose, with their depth, and the nature and colour of their bottom; but in general evaporation depends only on three circumstances, the temperature, the tension of the vapours contained in the atmosphere, and the resistance which the air, more or less dense, more or less agitated, opposes to the diffusion of vapour. The quantity of water that evaporates in a given spot, everything else being equal, is proportionate to the difference between the quantity of vapour which the ambient air can contain when saturated, and the quantity which it actually contains. Hence it follows that the evaporation is not so great in the torrid zone as might be expected from the enormous augmentation of temperature; because, in those ardent climates, the air is habitually very humid.

Since the increase of agricultural industry in the valleys of Aragua, the little rivers which run into the lake of Valencia can no longer be regarded as positive supplies during the six months succeeding December. They remain dried up in the lower part of their course, because the planters of indigo, coffee, and sugar-canes, have made frequent drainings (azequias), in order to water the ground by trenches. We may observe also, that a pretty considerable river, the Rio Pao, which rises at the entrance of the Llanos, at the foot of the range of hills called La Galera, heretofore mingled its waters with those of the lake, by uniting with the Cano de Cambury, on the road from the town of Nueva Valencia to Guigue. The course of this river was from south to north. At the end of the seventeenth century, the proprietor of a neighbouring plantation dug at the back of the hill a new bed for the Rio Pao. He turned the river; and, after having employed part of the water for the irrigation of his fields, he caused the rest to flow at a venture southward, following the declivity of the Llanos. In this new southern direction the Rio Pao, mingled with three other rivers, the Tinaco, the Guanarito, and the Chilua, falls into the Portuguesa, which is a branch of the Apure. It is a remarkable phenomenon, that by a particular position of the ground, and the lowering of the ridge of division to south-west, the Rio Pao separates itself from the little system of interior rivers to which it originally belonged, and for a century past has communicated, through the channel of the Apure and the Orinoco, with the ocean. What has been here effected on a small scale by the hand of man, nature often performs, either by progressively elevating the level of the soil, or by those falls of the ground occasioned by violent earthquakes. It is probable, that in the lapse of ages, several rivers of Soudan, and of New Holland, which are now lost in the sands, or in inland basins, will open for themselves a course to the shores of the ocean. We cannot at least doubt, that in both continents there are systems of interior rivers, which may be considered as not entirely developed; and which communicate with each other, either in the time of great risings, or by permanent bifurcations.

13

The Rio Pao has scooped itself out a bed so deep and broad, that in the season of rains, when the Cano Grande de Cambury inundates all the land to the north-west of Guigue, the waters of this Cano, and those of the lake of Valencia, flow back into the Rio Pao itself; so that this river, instead of adding water to the lake, tends rather to carry it away. We see something similar in North America, where geographers have represented on their maps an imaginary chain of mountains, between the great lakes of Canada and the country of the Miamis. At the time of floods, the waters flowing into the lakes communicate with those which run into the Mississippi; and it is practicable to proceed by boats from the sources of the river St. Mary to the Wabash, as well as from the Chicago to the Illinois. These analogous facts appear to me well worthy of the attention of hydrographers.

The land that surrounds the lake of Valencia being entirely flat and even, a diminution of a few inches in the level of the water exposes to view a vast extent of ground covered with fertile mud and organic remains.[*] In proportion as the lake retires, cultivation advances towards the new shore. These natural desiccations, so important to agriculture, have been considerable during the last ten years, in which America has suffered from great droughts. Instead of marking the sinuosities of the present banks of the lake, I have advised the rich landholders in these countries to fix columns of granite in the basin itself, in order to observe from year to year the mean height of the waters. The Marquis del Toro has undertaken to put this design into execution, employing the fine granite of the Sierra de Mariara, and establishing limnometers, on a bottom of gneiss rock, so common in the lake of Valencia.

It is impossible to anticipate the limits, more or less narrow, to which this basin of water will one day be confined, when an equilibrium between the streams flowing in and the produce of evaporation and filtration, shall be completely established. The

---

[*] This I observed daily in the Lake of Mexico.

idea very generally spread, that the lake will soon entirely disappear, seems to me chimerical. If in consequence of great earthquakes, or other causes equally mysterious, ten very humid years should succeed to long droughts; if the mountains should again become clothed with forests, and great trees overshadow the shore and the plains of Aragua, we should more probably see the volume of the waters augment, and menace that beautiful cultivation which now trenches on the basin of the lake.

While some of the cultivators of the valleys of Aragua fear the total disappearance of the lake, and others its return to the banks it has deserted, we hear the question gravely discussed at Caracas, whether it would not be advisable, in order to give greater extent to agriculture, to conduct the waters of the lake into the Llanos, by digging a canal towards the Rio Pao. The possibility* of this enterprise cannot be denied, particularly by having recourse to tunnels, or subterranean canals. The progressive retreat of the waters has given birth to the beautiful and luxuriant plains of Maracay, Cura, Mocundo, Guigue, and Santa Cruz del Escoval, planted with tobacco, sugar-canes, coffee, indigo, and cacao; but how can it be doubted for a moment that the lake alone spreads fertility over this country? If

---

* The dividing ridge, namely, that which divides the waters between the valleys of Aragua and the Llanos, lowers so much towards the west of Guigue, as we have already observed, that there are ravines which conduct the waters of the Cano de Cambury, the Rio Valencia, and the Guataparo, in the time of floods, to the Rio Pao; but it would be easier to open a navigable canal from the lake of Valencia to the Orinoco, by the Pao, the Portuguesa, and the Apure, than to dig a draining canal level with the bottom of the lake. This bottom, according to the sounding, and my barometric measurements, is 40 toises less than 222, or 182 above the surface of the ocean. On the road from Guigue to the Llanos, by the table-land of La Villa de Cura, I found, to the south of the dividing ridge, and on its southern declivity, no point of level corresponding to the 182 toises, except near San Juan. The absolute height of this village is 194 toises. But, I repeat that, farther towards the west, in the country between the Cano de Cambury and the sources of the Rio Pao, which I was not able to visit, the point of level of the bottom of the lake is much further north.

deprived of the enormous mass of vapour which the surface of the waters sends forth daily into the atmosphere, the valleys of Aragua would become as dry and barren as the surrounding mountains.

The mean depth of the lake is from twelve to fifteen fathoms; the deepest parts are not, as is generally admitted, eighty, but thirty-five or forty deep. Such is the result of soundings made with the greatest care by Don Antonio Manzano. When we reflect on the vast depths of all the lakes of Switzerland, which, notwithstanding their position in high valleys, almost reach the level of the Mediterranean, it appears surprising that greater cavities are not found at the bottom of the lake of Valencia, which is also an Alpine lake. The deepest places are between the rocky island of Burro and the point of Cana Fistula, and opposite the high mountains of Mariara. But in general the southern part of the lake is deeper than the northern: nor must we forget that, if all the shores be now low, the southern part of the basin is the nearest to a chain of mountains with abrupt declivities; and we know that even the sea is generally deepest where the coast is elevated, rocky, or perpendicular.

The temperature of the lake at the surface during my abode in the valleys of Aragua, in the month of February, was constantly from 23 to 23.7°, consequently a little below the mean temperature of the air. This may be from the effect of evaporation, which carries off caloric from the air and the water; or because a great mass of water does not follow with an equal rapidity the changes in the temperature of the atmosphere, and the lake receives streams which rise from several cold springs in the neighbouring mountains. I have to regret that, notwithstanding its small depth, I could not determine the temperature of the water at thirty or forty fathoms. I was not provided with the thermometrical sounding apparatus which I had used in the Alpine lakes of Salzburg, and in the Caribbean Sea. The experiments of Saussure prove that, on both sides of the Alps, the lakes which are from one hundred and ninety to two hundred and seventy-four toises of absolute elevation[*] have, in the middle of winter, at nine hundred, at six hundred, and sometimes even at one hundred and fifty feet of depth, a uniform temperature from 4.3 to 6°: but these experiments have not yet been repeated in lakes situated under the torrid zone. The strata of cold water in Switzerland are of an enormous thickness. They have been found so near the surface in the lakes of Geneva and Bienne, that the decrement of heat in the water was one centesimal degree for ten or fifteen feet; that is to say, eight times more rapid than in the ocean, and forty-eight times more rapid than in the atmosphere. In the temperate zone, where the heat of the atmosphere sinks to the freezing point, and far lower, the bottom of a lake, even were it not surrounded by glaciers and mountains covered with eternal snow, must contain particles of water which, having during winter acquired at the surface the maximum of their density, between 3.4 and 4.4°, have consequently fallen to the greatest depth. Other particles, the temperature of which is +0.5°, far from placing themselves below the stratum at 4°, can only find their hydrostatic equilibrium above that stratum. They will descend lower only when their temperature is augmented 3 or 4° by the contact of strata less cold. If water in cooling continued to condense uniformly to the freezing point, there would be found, in very deep lakes and basins having no communication with each other (whatever the latitude of the place), a stratum of water, the temperature of which would be nearly equal to the maximum of refrigeration above the freezing point, which the lower regions of the ambient atmosphere annually attain. Hence it is probable, that, in the plains of the torrid zone, or in the valleys but little elevated, the mean heat of which is from 25.5 to 27°, the temperature of the bottom of the lakes can never be below 21 or 22°. If in the same zone the ocean contain at depths of seven or eight hundred fathoms, water the temperature of which is at 7°,

---

[*] This is the difference between the absolute elevations of the lakes of Geneva and Thun.

that is to say, twelve or thirteen degrees colder than the maximum of the heat[*] of the equinoctial atmosphere over the sea, I think it must be considered as a direct proof of a submarine current, carrying the waters of the pole towards the equator. We will not here solve the delicate problem, as to the manner in which, within the tropics and in the temperate zone, (for example, in the Caribbean Sea and in the lakes of Switzerland,) these inferior strata of water, cooled to 4 or 7°, act upon the temperature of the stony strata of the globe which they cover; and how these same strata, the primitive temperature of which is, within the tropics, 27°, and at the lake of Geneva 10°, react upon the half-frozen waters at the bottom of the lakes, and of the equinoctial ocean. These questions are of the highest importance, both with regard to the economy of animals that live habitually at the bottom of fresh and salt waters, and to the theory of the distribution of heat in lands surrounded by vast and deep seas.

The lake of Valencia is full of islands, which embellish the scenery by the picturesque form of their rocks, and the beauty of the vegetation with which they are covered: an advantage which this tropical lake possesses over those of the Alps. The islands are fifteen in number, distributed in three groups;[†] without reckoning Morro and Cabrera, which are already joined to the shore. They are partly cultivated, and extremely fertile on account of the vapours that rise from the lake. Burro, the largest

---

[*] It is almost superfluous to observe that I am considering here only that part of the atmosphere lying on the ocean between 10° north and 10° south latitude. Towards the northern limits of the torrid zone, in latitude 23°, whither the north winds bring with an extreme rapidity the cold air of Canada, the thermometer falls at sea as low as 16°, and even lower.

[†] The position of these islands is as follows: northward, near the shore, the Isla de Cura; on the south-east, Burro, Horno, Otama, Sorro, Caiguira, Nuevos Penones, or the Aparecidos; on the north-west, Cabo Blanco, or Isla de Aves, and Chamberg; on the south-west, Brucha and Culebra. In the centre of the lake rise, like shoals or small detached rocks, Vagre, Fraile, Penasco, and Pan de Azucar.

of these islands, is two miles in length, and is inhabited by some families of mestizos, who rear goats. These simple people seldom visit the shore of Mocundo. To them the lake appears of immense extent; they have plantains, cassava, milk, and a little fish. A hut constructed of reeds; hammocks woven from the cotton which the neighbouring fields produce; a large stone on which the fire is made; the ligneous fruit of the tutuma (the calabash) in which they draw water, constitute their domestic establishment. An old mestizo who offered us some goat's milk had a beautiful daughter. We learned from our guide, that solitude had rendered him as mistrustful as he might perhaps have been made by the society of men. The day before our arrival, some hunters had visited the island. They were overtaken by the shades of night; and preferred sleeping in the open air to returning to Mocundo. This news spread alarm throughout the island. The father obliged the young girl to climb up a very lofty zamang or acacia, which grew in the plain at some distance from the hut, while he stretched himself at the foot of the tree, and did not permit his daughter to descend till the hunters had departed.

The lake is in general well stocked with fish; though it furnishes only three kinds, the flesh of which is soft and insipid, the guavina, the vagre, and the sardina. The two last descend into the lake with the streams that flow into it. The guavina, of which I made a drawing on the spot, is 20 inches long and 3.5 broad. It is perhaps a new species of the genus erythrina of Gronovius. It has large silvery scales edged with green. This fish is extremely voracious, and destroys other kinds. The fishermen assured us that a small crocodile, the bava,[*] which often approached us when we were bathing, contributes also to the destruction of the fish. We never could succeed in procuring this

---

[*] The bava, or bavilla, is very common at Bordones, near Cumana. See volume 1. The name of bava, baveuse, has misled M. Depons; he takes this reptile for a fish of our seas, the Blennius pholis. Voyage a la Terre Ferme. The Blennius pholis, smooth blenny, is called by the French baveuse (slaverer), in Spanish, baba.

reptile so as to examine it closely: it generally attains only three or four feet in length. It is said to be very harmless; its habits however, as well as its form, much resemble those of the alligator (Crocodilus acutus). It swims in such a manner as to show only the point of its snout, and the extremity of its tail; and places itself at mid-day on the bare beach. It is certainly neither a monitor (the real monitors living only in the old continent,) nor the sauvegarde of Seba (Lacerta teguixin,) which dives and does not swim. It is somewhat remarkable that the lake of Valencia, and the whole system of small rivers flowing into it, have no large alligators, though this dangerous animal abounds a few leagues off in the streams which flow either into the Apure or the Orinoco, or immediately into the Caribbean Sea between Porto Cabello and La Guayra.

In the islands that rise like bastions in the midst of the waters, and wherever the rocky bottom of the lake is visible, I recognised a uniform direction in the strata of gneiss. This direction is nearly that of the chains of mountains on the north and south of the lake. In the hills of Cabo Blanco there are found among the gneiss, angular masses of opaque quartz, slightly translucid on the edges, and varying from grey to deep black. This quartz passes sometimes into hornstein, and sometimes into kieselschiefer (schistose jasper). I do not think it constitutes a vein. The waters of the lake* decompose the gneiss by erosion in a very extraordinary manner.

The island of Chamberg is remarkable for its height. It is a rock of gneiss, with two summits in the form of a saddle, and

---

* The water of the lake is not salt, as is asserted at Caracas. It may be drunk without being filtered. On evaporation it leaves a very small residuum of carbonate of lime, and perhaps a little nitrate of potash. It is surprising that an inland lake should not be richer in alkaline and earthy salts, acquired from the neighbouring soils. I have found parts of it porous, almost cellular, and split in the form of cauliflowers, fixed on gneiss perfectly compact. Perhaps the action ceases with the movement of the waves, and the alternate contact of air and water.

raised two hundred feet above the surface of the water. The slope of this rock is barren, and affords only nourishment for a few plants of clusia with large white flowers. But the view of the lake and of the richly cultivated neighbouring valleys is beautiful, and their aspect is wonderful after sunset, when thousands of aquatic birds, herons, flamingoes, and wild ducks cross the lake to roost in the islands, and the broad zone of mountains which surrounds the horizon is covered with fire. The inhabitants, as we have already mentioned, burn the meadows in order to produce fresher and finer grass. Gramineous plants abound, especially at the summit of the chain; and those vast conflagrations extend sometimes the length of a thousand toises, and appear like streams of lava overflowing the ridge of the mountains. When reposing on the banks of the lake to enjoy the soft freshness of the air in one of those beautiful evenings peculiar to the tropics, it is delightful to contemplate in the waves as they beat the shore, the reflection of the red fires that illumine the horizon.

Among the plants which grow on the rocky islands of the lake of Valencia, many have been believed to be peculiar to those spots, because till now they have not been discovered elsewhere. Such are the papaw-trees of the lake; and the tomato[*] of the island of Cura. The latter differs from our Solanum lycopersicum; the fruit is round and small, but has a fine flavour; it is now cultivated at La Victoria, at Nueva Valencia, and everywhere in the valleys of Aragua. The papaw-tree of the lake (papaya de la laguna) abounds also in the island of Cura and at Cabo Blanco; its trunk shoots higher than that of the common papaw (Carica papaya), but its fruit is only half as large, perfectly spherical, without projecting ribs, and four or five inches in diameter. When cut open it is found quite filled with seeds, and without those hollow places which occur constantly in the common papaw. The taste of this fruit, of which I have

---

[*] The tomatoes are cultivated, as well as the papaw-tree of the lake, in the Botanical Garden of Berlin, to which I had sent some seeds.

often eaten, is extremely sweet.[*] I know not whether it be a variety of the Carica microcarpa, described by Jacquin.

The environs of the lake are insalubrious only in times of great drought, when the waters in their retreat leave a muddy sediment exposed to the rays of the sun. The banks, shaded by tufts of Coccoloba barbadensis, and decorated with fine liliaceous plants,[†] remind us, by the appearance of the aquatic vegetation, of the marshy shores of our lakes in Europe. We find there, pondweed (potamogeton), chara, and cats'-tail three feet high, which it is difficult not to confound with the Typha angustifolia of our marshes. It is only after a careful examination, that we recognise each of these plants for distinct species,[‡] peculiar to the new continent. How many plants of the straits of Magellan, of Chile, and the Cordilleras of Quito have formerly been confounded with the productions of the northern temperate zone, owing to their analogy in form and appearance.

The inhabitants of the valleys of Aragua often inquire why the southern shore of the lake, particularly the south-west part towards los Aguacotis, is generally more shaded, and exhibits fresher verdure than the northern side. We saw, in the month of February, many trees stripped of their foliage, near the Hacienda de Cura, at Mocundo, and at Guacara; while to the south-east of Valencia everything presaged the approach of the rains. I believe that in the early part of the year, when the sun has southern declination, the hills around Valencia, Guacara, and Cura are scorched by the heat of the solar rays, while the southern shore receives, along with the breeze when it enters the valley by the Abra de Porto Cabello, an atmosphere which has crossed the lake, and is loaded with aqueous vapour. On this southern shore, near Guaruto, are situated the finest plantations of tobacco in the whole province.

---

[*] The people of the country attribute to it an astringent quality, and call it tapaculo.

[†] Pancratium undulatum, Amaryllis nervosa.

[‡] Potamogeton tenuifolium, Chara compressa, Typha tenuifolia.

Among the rivers flowing into the lake of Valencia some owe their origin to thermal springs, and deserve particular attention. These springs gush out at three points of the granitic Cordillera of the coast; near Onoto, between Turmero and Maracay; near Mariara, north-east of the Hacienda de Cura; and near Las Trincheras, on the road from Nueva Valencia to Porto Cabello. I could examine with care only the physical and geological relations of the thermal waters of Mariara and Las Trincheras. In going up the small river Cura towards its source, the mountains of Mariara are seen advancing into the plain in the form of a vast amphitheatre, composed of perpendicular rocks, crowned by peaks with rugged summits. The central point of the amphitheatre bears the strange name of the Devil's Nook (Rincon del Diablo). The range stretching to the east is called El Chaparro; that to the west, Las Viruelas. These ruin-like rocks command the plain; they are composed of a coarse-grained granite, nearly porphyritic, the yellowish white feldspar crystals of which are more than an inch and a half long. Mica is rare in them, and is of a fine silvery lustre. Nothing can be more picturesque and solemn than the aspect of this group of mountains, half covered with vegetation. The Peak of Calavera, which unites the Rincon del Diablo to the Chaparro, is visible from afar. In it the granite is separated by perpendicular fissures into prismatic masses. It would seem as if the primitive rock were crowned with columns of basalt. In the rainy season, a considerable sheet of water rushes down like a cascade from these cliffs. The mountains connected on the east with the Rincon del Diablo, are much less lofty, and contain, like the promontory of La Cabrera, and the little detached hills in the plain, gneiss and mica-slate, including garnets.

In these lower mountains, two or three miles north-east of Mariara, we find the ravine of hot waters called Quebrada de Aguas Calientes. This ravine, running north-west 75°, contains several small basins. Of these the two uppermost, which have no communication with each other, are only eight inches in diameter; the three lower, from two to three feet. Their depth varies from three to fifteen inches. The temperature of these different funnels (pozos) is from 56 to 59°; and what is remarkable, the lower funnels are hotter than the upper, though the difference of the level is only seven or eight inches. The hot waters, collected together, form a little rivulet, called the Rio de Aguas Calientes, which, thirty feet lower, has a temperature of only 48°. In seasons of great drought, the time at which we visited the ravine, the whole body of the thermal waters forms a section of only twenty-six square inches. This is considerably augmented in the rainy season; the rivulet is then transformed into a torrent, and its heat diminishes for it appears that the hot springs themselves are subject only to imperceptible variations. All these springs are slightly impregnated with sulphuretted hydrogen gas. The fetid smell, peculiar to this gas, can be perceived only by approaching very near the springs. In one of these wells only, the temperature of which is 56.2°, bubbles of air are evolved at nearly regular intervals of two or three minutes. I observed that these bubbles constantly rose from the same points, which are four in number; and that it was not possible to change the places from which the gas is emitted, by stirring the bottom of the basin with a stick. These places correspond no doubt to holes or fissures on the gneiss; and indeed when the bubbles rise from one of the apertures, the emission of gas follows instantly from the other three. I could not succeed in inflaming the small quantities of gas that rise above the thermal waters, or those I collected in a glass phial held over the springs, an operation that excited in me a nausea, caused less by the smell of the gas, than by the excessive heat prevailing in this ravine. Is this sulphuretted hydrogen mixed with a great proportion of carbonic acid or atmospheric air? I am doubtful of the first of these mixtures, though so common in thermal waters; for example at Aix la Chapelle, Enghien, and Bareges. The gas collected in the tube of Fontana's eudiometer had been shaken for a long time with water. The small basins are covered with a light film of sulphur, deposited by the sulphuretted hydrogen in its slow combustion in contact with the atmospheric oxygen. A few plants near the springs were encrusted with sulphur. This deposit is scarcely visible when the water of Mariara is suffered to cool in an open vessel; no doubt because the quantity of disengaged gas is very small, and is not renewed. The water, when cold, gives no precipitate with a solution of nitrate of copper; it is destitute of flavour, and very drinkable. If it contain any saline substances, for example, the sulphates of soda or magnesia, their quantities must be very insignificant. Being almost destitute of chemical tests,[*] we contented ourselves with filling at the spring two bottles, which were sent, along with the nourishing milk of the tree called palo de vaca, to MM. Fourcroy and Vauquelin, by the way of Porto Cabello and the Havannah. This purity in hot waters issuing immediately from granite mountains is in Europe, as well as in the New Continent, a most curious phenomenon.[†] How can we explain the origin of the sulphuretted hydrogen? It cannot proceed from the decomposition of sulphurets of iron, or pyritic strata. Is it owing to sulphurets of calcium, of magnesium, or other earthy metalloids, contained in the interior of our planet, under its rocky and oxidated crust?

In the ravine of the hot waters of Mariara, amidst little funnels, the temperature of which rises from 56 to 59°, two species of aquatic plants vegetate; the one is membranaceous, and contains bubbles of air; the other has parallel fibres. The first much resembles the Ulva labyrinthiformis of Vandelli, which the thermal waters of Europe furnish. At the island of Amsterdam, tufts of lycopodium and marchantia have been seen

---

[*] A small case, containing acetate of lead, nitrate of silver, alcohol, prussiate of potash, etc., had been left by mistake at Cumana. I evaporated some of the water of Mariara, and it yielded only a very small residuum, which, digested with nitric acid, appeared to contain only a little silica and extractive vegetable matter.

[†] Warm springs equally pure are found issuing from the granites of Portugal, and those of Cantal. In Italy, the Pisciarelli of the lake Agnano have a temperature equal to 93°. Are these pure waters produced by condensed vapours?

in places where the heat of the soil was far greater: such is the effect of an habitual stimulus on the organs of plants. The waters of Mariara contain no aquatic insects. Frogs are found in them, which, being probably chased by serpents, have leaped into the funnels, and there perished.

South of the ravine, in the plain extending towards the shore of the lake, another sulphureous spring gushes out, less hot and less impregnated with gas. The crevice whence this water issues is six toises higher than the funnel just described. The thermometer did not rise in the crevice above 42°. The water is collected in a basin surrounded by large trees; it is nearly circular, from fifteen to eighteen feet diameter, and three feet deep. The slaves throw themselves into this bath at the end of the day, when covered with dust, after having worked in the neighbouring fields of indigo and sugar-cane. Though the water of this bath (bano) is habitually from 12 to 14° hotter than the air, the negroes call it refreshing; because in the torrid zone this term is used for whatever restores strength, calms the irritation of the nerves, or causes a feeling of comfort. We ourselves experienced the salutary effects of the bath. Having slung our hammocks on the trees round the basin, we passed a whole day in this charming spot, which abounds in plants. We found near the bano of Mariara the volador, or gyrocarpus. The winged fruits of this large tree turn like a fly-wheel, when they fall from the stalk. On shaking the branches of the volador, we saw the air filled with its fruits, the simultaneous fall of which presents the most singular spectacle. The two membranaceous and striated wings are turned so as to meet the air, in falling, at an angle of 45°. Fortunately the fruits we gathered were at their maturity. We sent some to Europe, and they have germinated in the gardens of Berlin, Paris, and Malmaison. The numerous plants of the volador, now seen in hot-houses, owe their origin to the only tree of the kind found near Mariara. The geographical distribution of the different species of gyrocarpus, which Mr.

Brown considers as one of the laurineae, is very singular. Jacquin saw one species near Carthagena in America.[*] This is the same which we met with again in Mexico, near Zumpango, on the road from Acapulco to the capital.[†] Another species, which grows on the mountains of Coromandel,[‡] has been described by Roxburgh; the third and fourth[§] grow in the southern hemisphere, on the coasts of Australia.

After getting out of the bath, while, half-wrapped in a sheet, we were drying ourselves in the sun, according to the custom of the country, a little man of the mulatto race approached us. After bowing gravely, he made us a long speech on the virtues of the waters of Mariara, adverting to the numbers of invalids by whom they have been visited for some years past, and to the favourable situation of the springs, between the two towns Valencia and Caracas. He showed us his house, a little hut covered with palm-leaves, situated in an enclosure at a small distance, on the bank of a rivulet, communicating with the bath. He assured us that we should there find all the conveniences of life; nails to suspend our hammocks, ox-leather to stretch over benches made of reeds, earthern vases always filled with cool water, and what, after the bath, would be most salutary of all, those great lizards (iguanas), the flesh of which is known to be a refreshing aliment. We judged from his harangue, that this good man took us for invalids, who had come to stay near the spring. His counsels and offers of hospitality were not altogether disinterested. He styled himself the inspector of the waters, and the pulpero[**] of the place. Accordingly all his obliging attentions to us ceased as soon as he heard that we had come merely to satisfy our

---

[*] The Gyrocarpus Jacquini of Gartner, or Gyrocarpus americanus of Willdenouw.

[†] The natives of Mexico called it quitlacoctli. I saw some of its young leaves with three and five lobes; the full-grown leaves are in the form of a heart, and always with three lobes. We never met with the volador in flower.

[‡] This is the Gyrocarpus asiaticus of Willdenouw.

[§] Gyrocarpus sphenopterus, and G. rugosus.

[**] Proprietor of a pulperia, or little shop where refreshments are sold.

curiosity; or as they express it in the Spanish colonies, those lands of idleness, para ver, no mas, to see, and nothing more. The waters of Mariara are used with success in rheumatic swellings, and affections of the skin. As the waters are but very feebly impregnated with sulphuretted hydrogen, it is necessary to bathe at the spot where the springs issue. Farther on, these same waters are employed for the irrigation of fields of indigo. A wealthy landed proprietor of Mariara, Don Domingo Tovar, had formed the project of erecting a bathing-house, and an establishment which would furnish visitors with better resources than lizard's flesh for food, and leather stretched on a bench for their repose.

On the 21st of February, in the evening, we set out from the beautiful Hacienda de Cura for Guacara and Nueva Valencia. We preferred travelling by night, on account of the excessive heat of the day. We passed by the hamlet of Punta Zamuro, at the foot of the high mountains of Las Viruelas. The road is bordered with large zamang-trees, or mimosas, the trunks of which rise to sixty feet high. Their branches, nearly horizontal, meet at more than one hundred and fifty feet distance. I have nowhere seen a vault of verdure more beautiful and luxuriant. The night was gloomy: the Rincon del Diablo with its denticulated rocks appeared from time to time at a distance, illumined by the burning of the savannahs, or wrapped in ruddy smoke. At the spot where the bushes were thickest, our horses were frightened by the yell of an animal that seemed to follow us closely. It was a large jaguar, which had roamed for three years among these mountains. He had constantly escaped the pursuits of the boldest hunters, and had carried off horses and mules from the midst of enclosures; but, having no want of food, had not yet attacked men. The negro who conducted us uttered wild cries, expecting by these means to frighten the tiger; but his efforts were ineffectual. The jaguar, like the wolf of Europe, follows travellers even when he will not attack them; the wolf in the open fields and in unsheltered places, the jaguar skirting the road and appearing only at intervals between the bushes.

We passed the day on the 23rd in the house of the Marquis de Toro, at the village of Guacara, a very considerable Indian community. An avenue of carolineas leads from Guacara to Mocundo. It was the first time I had seen in the open air this majestic plant, which forms one of the principal ornaments of the extensive conservatories of Schonbrunn.[*] Mocundo is a rich plantation of sugar-canes, belonging to the family of Toro. We there find, what is so rare in that country, a garden, artificial clumps of trees, and on the border of the water, upon a rock of gneiss, a pavilion with a mirador, or belvidere. The view is delightful over the western part of the lake, the surrounding mountains, and a forest of palm-trees that separates Guacara from the city of Nueva Valencia. The fields of sugar-cane, from the soft verdure of the young reeds, resemble a vast meadow. Everything denotes abundance; but it is at the price of the liberty of the cultivators. At Mocundo, with two hundred and thirty negroes, seventy-seven tablones, or cane-fields, are cultivated, each of which, ten thousand varas square,[†] yields a net profit of two hundred or two hundred and forty piastres a-year. The creole cane and the cane of Otaheite[‡] are planted in the month of April, the first at four, the second at five feet distance. The cane ripens in fourteen months. It flowers in the month of October, if the plant be sufficiently vigorous; but the top is cut off before the panicle unfolds. In all the monocotyledonous plants (for example, the maguey cultivated at Mexico for extracting pulque, the wine-yielding palm-tree, and the sugar-cane), the flowering alters the quality of the juices. The preparation of sugar, the boiling, and the claying, are very imperfect in Terra Firma,

---

[*] Every tree of the Carolinea princeps at Schonbrunn has sprung from seeds collected from one single tree of enormous size, near Chacao, east of Caracas.

[†] A tablon, equal to 1849 square toises, contains nearly an acre and one-fifth: a legal acre has 1344 square toises, and 1.95 legal acre is equal to one hectare.

[‡] In the island of Palma, where in the latitude of 29° the sugar-cane is said to be cultivated as high as 140 toises above the level of the Atlantic, the Otaheite cane requires more heat than the Creole cane.

because it is made only for home consumption; and for wholesale, papelon is preferred to sugar, either refined or raw. This papelon is an impure sugar, in the form of little loaves, of a yellow-brown colour. It contains a mixture of molasses and mucilaginous matter. The poorest man eats papelon, as in Europe he eats cheese. It is believed to have nutritive qualities. Fermented with water it yields the guarapo, the favourite beverage of the people. In the province of Caracas subcarbonate of potash is used, instead of lime, to purify the juice of the sugar-cane. The ashes of the bucare, which is the Erythrina corallodendrum, are preferred.

The sugar-cane was introduced very late, probably towards the end of the sixteenth century, from the West India Islands, into the valleys of Aragua. It was known in India, in China, and in all the islands of the Pacific, from the most remote antiquity; and it was planted at Khorassan, in Persia, as early as the fifth century of our era, in order to obtain from it solid sugar.[*] The Arabs carried this reed, so useful to the inhabitants of hot and temperate countries, to the shores of the Mediterranean. In 1306, its cultivation was yet unknown in Sicily; but was already common in the island of Cyprus, at Rhodes, and in the Morea. A hundred years after it enriched Calabria, Sicily, and the coasts of Spain. From Sicily the Infante Don Henry transported the cane to Madeira: from Madeira it passed to the Canary Islands, where it was entirely unknown; for the ferulae of Juba, quae expressae liquorem fundunt potui ucundum, are euphorbias (the Tabayba dulce), and not, as has been recently asserted,[†] sugar-canes. Twelve sugar-manufactories (ingenios de azucar) were soon established in the island of Great Canary, in that of Palma, and between Adexe, Icod, and Guarachico, in the island of Teneriffe.

---

[*] The Indian name for the sugar-cane is sharkara. Thence the word sugar.

[†] On the origin of cane-sugar, in the Journal de Pharmacie 1816 page 387. The Tabayba dulce is, according to Von Buch, the Euphorbia balsamifera, the juice of which is neither corrosive nor bitter like that of the cardon, or Euphorbia canariensis.

Negroes were employed in this cultivation, and their descendants still inhabit the grottos of Tiraxana, in the Great Canary. Since the sugar-cane has been transplanted to the West Indies, and the New World has given maize to the Canaries, the cultivation of the latter has taken the place of the cane at Teneriffe and the Great Canary. The cane is now found only in the island of Palma, near Argual and Tazacorte,[*] where it yields scarcely one thousand quintals of sugar a year. The sugar-cane of the Canaries, which Aiguilon transported to St. Domingo, was there cultivated extensively as early as 1513, or during the six or seven following years, under the auspices of the monks of St. Jerome. Negroes were employed in this cultivation from its commencement; and in 1519 representations were made to government, as in our own time, that the West India Islands would be ruined and made desert, if slaves were not conveyed thither annually from the coast of Guinea.

For some years past the culture and preparation of sugar has been much improved in Terra Firma; and, as the process of refining is prohibited by the laws at Jamaica, they reckon on the fraudulent exportation of refined sugar to the English colonies. But the consumption of the provinces of Venezuela, in papelon, and in raw sugar employed in making chocolate and sweetmeats (dulces) is so enormous, that the exportation has been hitherto entirely null. The finest plantations of sugar are in the valleys of Aragua and of the Tuy, near Pao de Zarate, between La Victoria and San Sebastian, near Guatire, Guarenas, and Caurimare. The first canes arrived in the New World from the Canary Islands; and even now Canarians, or Islenos, are placed at the head of most of the great plantations, and superintend the labours of cultivation and refining.

It is this connexion between the Canarians and the inhabitants of Venezuela, that has given rise to the introduction

---

[*] "Notice sur la Culture du Sucre dans les Isles Canariennes" by Leopold von Buch.

31

of camels into those provinces. The Marquis del Toro caused three to be brought from Lancerote. The expense of conveyance was very considerable, owing to the space which these animals occupy on board merchant-vessels, and the great quantity of water they require during a long sea-voyage. A camel, bought for thirty piastres, costs between eight and nine hundred before it reaches the coast of Caracas. We saw four of these animals at Mocundo; three of which had been bred in America. Two others had died of the bite of the coral, a venomous serpent very common on the banks of the lake. These camels have hitherto been employed only in the conveyance of the sugarcanes to the mill. The males, stronger than the females, carry from forty to fifty arrobas. A wealthy landholder in the province of Varinas, encouraged by the example of the Marquis del Toro, has allotted a sum of 15,000 piastres for the purpose of bringing fourteen or fifteen camels at once from the Canary Islands. It is presumed these beasts of burden may be employed in the conveyance of merchandise across the burning plains of Casanare, from the Apure and Calabozo, which in the season of drought resemble the deserts of Africa. How advantageous it would have been had the Conquistadores, from the beginning of the sixteenth century, peopled America with camels, as they have peopled it with horned cattle, horses, and mules. Wherever there are immense distances to cross¦ in uninhabited lands; wherever the construction of canals becomes difficult (as in the isthmus of Panama, on the table-land of Mexico, and in the deserts that separate the kingdom of Quito from Peru, and Peru from Chile), camels would be of the highest importance, to facilitate inland commerce. It seems the more surprising, that their introduction was not encouraged by the government at the beginning of the conquest, as, long after the taking of Grenada, camels, for which the Moors had a great predilection, were still very common in the south of Spain. A Biscayan, Juan de Reinaga, carried some of these animals at his own expense to Peru. Father Acosta saw them at the foot of the Andes, about the end of the sixteenth century; but little care being taken of them, they scarcely ever bred, and the race soon became extinct. In those times of oppression and cruelty, which have been described as the era of Spanish glory, the commendatories (encomenderos) let out the Indians to travellers like beasts of burden. They were assembled by hundreds, either to carry merchandise across the Cordilleras, or to follow the armies in their expeditions of discovery and pillage. The Indians endured this service more patiently, because, owing to the almost total want of domestic animals, they had long been constrained to perform it, though in a less inhuman manner, under the government of their own chiefs. The introduction of camels attempted by Juan de Reinaga spread an alarm among the encomenderos, who were, not by law, but in fact, lords of the Indian villages. The court listened to the complaints of the encomenderos; and in consequence America was deprived of one of the means which would have most facilitated inland communication, and the exchange of productions. Now, however, there is no reason why the introduction of camels should not be attempted as a general measure. Some hundreds of these useful animals, spread over the vast surface of America, in hot and barren places, would in a few years have a powerful influence on the public prosperity. Provinces separated by steppes would then appear to be brought nearer to each other; several kinds of inland merchandize would diminish in price on the coast; and by increasing the number of camels, above all the species called hedjin, or the ship of the desert, a new life would be given to the industry and commerce of the New World.

On the evening of the 22nd we continued our journey from Mocundo by Los Guayos to the city of Nueva Valencia. We passed a little forest of palm-trees, which resembled, by their appearance, and their leaves spread like a fan, the Chamaerops humilis of the coast of Barbary. The trunk, however, rises to twenty-four and sometimes thirty feet high. It is probably a new species of the genus corypha; and is called in the country palma de sombrero, the footstalks of the leaves being employed in weaving hats resembling our straw hats. This grove of palm-trees, the withered foliage of which rustles at the least breath of air—the camels feeding in the plain—the undulating motion of the vapours on a soil scorched by the ardour of the sun, give the landscape an African aspect. The aridity of the land augments as the traveller approaches the town, after passing the western extremity of the lake. It is a clayey soil, which has been levelled and abandoned by the waters. The neighbouring hills, called Los Morros de Valencia, are composed of white tufa, a very recent limestone formation, immediately covering the gneiss. It is again found at Victoria, and on several other points along the chain of the coast. The whiteness of this tufa, which reflects the rays of the sun, contributes greatly to the excessive heat felt in this place. Everything seems smitten with sterility; scarcely are a few plants of cacao found on the banks of the Rio de Valencia; the rest of the plain is bare, and destitute of vegetation. This appearance of sterility is here attributed, as it is everywhere in the valleys of Aragua, to the cultivation of indigo; which, according to the planters, is, of all plants, that which most exhausts (cansa) the ground. The real physical causes of this phenomenon would be an interesting inquiry, since, like the effects of fallowing land, and of a rotation of crops, it is far from being sufficiently understood. I shall only observe in general, that the complaints of the increasing sterility of cultivated land become more frequent between the tropics, in proportion as they are near the period of their first breaking-up. In a region almost destitute of herbs, where every plant has a ligneous stem, and tends to raise itself as a shrub, the virgin soil remains shaded either by great trees, or by bushes; and under this tufted shade it preserves everywhere coolness and humidity. However active the vegetation of the tropics may appear, the number of roots that penetrate into the earth, is not so great in an uncultivated soil; while the plants are nearer to each other in lands subjected to cultivation, and covered with indigo, sugar-canes, or cassava. The trees and shrubs, loaded with branches and leaves, draw a great part of their nourishment from the ambient air; and the virgin soil augments its fertility by the decomposition of the vegetable substances which progressively accumulate. It is not so in the fields covered with indigo, or other herbaceous plants; where the rays of the sun penetrate freely into the earth, and by the accelerated combustion of the hydrurets of carbon and other acidifiable principles, destroy the germs of fecundity. These effects strike the imagination of the planters the more forcibly, as in lands newly inhabited they compare the fertility of a soil which has been abandoned to itself during thousands of years, with the produce of ploughed fields. The Spanish colonies on the continent, and the great islands of Porto-Rico and Cuba, possess remarkable advantages with respect to the produce of agriculture over the lesser West India islands. The former, from their extent, the variety of their scenery, and their small relative population, still bear all the characters of a new soil; while at Barbadoes, Tobago, St. Lucia, the Virgin Islands, and the French part of St. Domingo, it may be perceived that long cultivation has begun to exhaust the soil. If in the valleys of Aragua, instead of abandoning the indigo grounds, and leaving them fallow, they were covered during several years, not with corn, but with other alimentary plants and forage; if among these plants such as belong to different families were preferred, and which shade the soil by their large leaves, the amelioration of the fields would be gradually accomplished, and they would be restored to a part of their former fertility.

The city of Nueva Valencia occupies a considerable extent of ground, but its population scarcely amounts to six or seven thousand souls. The streets are very broad, the market place, (plaza mayor,) is of vast dimensions; and, the houses being low, the disproportion between the population of the town, and the space that it occupies, is still greater than at Caracas. Many of the whites, (especially the poorest,) forsake their houses, and live the greater part of the year in their little plantations of indigo and cotton, where they can venture to work with their own hands; which, according to the inveterate prejudices of that country, would be a disgrace to them in the town.

Nueva Valencia, founded in 1555 under the government of Villacinda, by Alonzo Diaz Moreno, is twelve years older than Caracas. Valencia was at first only a dependency of Burburata; but this latter town is nothing now but a place of embarkation for mules. It is regretted, and perhaps justly, that Valencia has not become the capital of the country. Its situation in a plain, on the banks of a lake, recalls to mind the position of Mexico. When we reflect on the easy communication afforded by the valleys of Aragua with the Llanos and the rivers that flow into the Orinoco; when we recognize the possibility of opening an inland navigation, by the Rio Pao and the Portuguesa, as far as the mouths of the Orinoco, the Cassiquiare, and the Amazon, it may be conceived that the capital of the vast provinces of Venezuela would have been better placed near the fine harbour of Porto Cabello, beneath a pure and serene sky, than near the unsheltered road of La Guayra, in a temperate but constantly foggy valley. Near the kingdom of New Grenada, and situate between the fertile corn-lands of La Victoria and Barquesimeto, the city of Valencia ought to have prospered; but, notwithstanding these advantages, it has been unable to maintain the contest with Caracas.

Only those who have seen the myriads of ants, that infest the countries within the torrid zone, can form an idea of the destruction and the sinking of the ground occasioned by these insects. They abound to such a degree on the site of Valencia, that their excavations resemble subterranean canals, which are filled with water in the time of the rains, and become very dangerous to the buildings. Here recourse has not been had to the extraordinary means employed at the beginning of the sixteenth century in the island of St. Domingo, when troops of ants ravaged the fine plains of La Vega, and the rich possessions of the order of St. Francis. The monks, after having in vain burnt the larvae of the ants, and had recourse to fumigations, advised the inhabitants to choose by lot a saint, who would act as a mediator against the plague of the ants.[*] The honour of the choice fell on St. Saturnin; and the ants disappeared as soon as the first festival of this saint was celebrated. Incredulity has made great progress since the time of the conquest; and it was only on the back of the Cordilleras that I found a small chapel, destined, according to its inscription, for prayers to be addressed to Heaven for the destruction of the termites.

Valencia affords some historical remembrances; but these, like everything connected with the colonies, have no remote date, and recall to mind either civil discords or sanguinary conflicts with the savages. Lopez de Aguirre, whose crimes and adventures form some of the most dramatic episodes of the history of the conquest, proceeded in 1561, from Peru, by the river Amazon to the island of Margareta; and thence, by the port of Burburata, into the valleys of Aragua. On his entrance into Valencia, which proudly entitles itself the City of the King, he proclaimed the independence of country, and the deposition of Philip II. The inhabitants withdrew to the islands of the lake of Tacarigua, taking with them all the boats from the shore, to be more secure in their retreat. In consequence of this stratagem, Aguirre could exercise his cruelties only on his own people. From Valencia he addressed to the king of Spain, a remarkable letter, in which he boasts alternately of his crimes and his piety; at the same time giving advice to the king on the government of the colonies, and the system of missions. Surrounded by savage Indians, navigating on a great sea of fresh water, as he calls the Amazon, he is alarmed at the heresies of Martin Luther, and the increasing influence of schismatics in Europe.[†]

---

[*] Un abogado contra los harmigos.

[†] The following are some remarkable passages in the letter from Aguirre to the king of Spain.

"King Philip, native of Spain, son of Charles the Invincible! I, Lopez de Aguirre, thy vassal, an old Christian, of poor but noble parents, and a native of the town of Onate in Biscay, passed over young to Peru, to labour lance in hand. I rendered thee great services in the conquest of India. I fought for thy glory, without demanding pay of thy officers, as is proved by the books of thy treasury. I firmly believe, Christian King and Lord, that, very ungrateful to me and my companions, all those who write to thee from this land [America], deceive thee much, because thou seest things from too far off. I recommend to thee to be more just toward the good vassals whom thou hast in this country: for I and mine, weary of the cruelties and injustice which thy viceroys, thy governors, and thy judges, exercise in thy name, are resolved to obey thee no more. We regard ourselves no longer as Spaniards. We wage a cruel war against thee, because we will not endure the oppression of thy ministers; who, to give places to their nephews and their children, dispose of our lives, our reputation, and our fortune. I am lame in the left foot from two shots of an arquebuss, which I received in the valley of Coquimbo, fighting under the orders of thy marshal, Alonzo de Alvarado, against Francis Hernandez Giron, then a rebel, as I am at present, and shall be always; for since thy viceroy, the Marquis de Canete, a cowardly, ambitious, and effeminate man, has hanged our most valiant warriors, I care no more for thy pardon than for the books of Martin Luther. It is not well in thee, King of Spain, to be ungrateful toward thy vassals; for it was whilst thy father, the emperor Charles, remained quietly in Castile, that they procured for thee so many kingdoms and vast countries. Remember, King Philip, that thou hast no right to draw revenues from these provinces, the conquest of which has been without danger to thee, but inasmuch as thou recompensest those who have rendered thee such great services. I am certain that few kings go to heaven. Therefore we regard ourselves as very happy to be here in the Indies, preserving in all their purity the commandments of God, and of the Roman Church; and we intend, though sinners during life, to become one day martyrs to the glory of God. On going out of the river Amazon, we landed in an island called La Margareta. We there received news from Spain of the great faction and machination (maquina) of the Lutherans. This news alarmed us extremely; we found among us one of that faction; his name was Monteverde. I had him cut to pieces, as was just: for, believe me, Senor, wherever I am, people live according to the law. But the corruption of morals among the monks is so great in this land that it is necessary to chastise it severely. There is not an ecclesiastic here who does not think himself higher than the governor of a province. I beg of thee, great King, not to believe what the monks tell thee down yonder in Spain. They are always talking of the sacrifices they make, as well as of the hard and bitter life they are forced to lead in America: while they occupy the richest lands, and the Indians hunt and fish for them every day. If they shed tears before thy throne, it is that thou mayest send them hither to govern provinces. Dost thou know what sort of life they lead here? Given up to luxury, acquiring possessions, selling the

Lopez de Aguirre, or as he is still called by the common people, the Tyrant, was killed at Barquesimeto, after having been abandoned by his own men. At the moment when he fell, he plunged a dagger into the bosom of his only daughter, "that she might not have to blush before the Spaniards at the name of the daughter of a traitor." The soul of the tyrant (such is the belief of the natives) wanders in the savannahs, like a flame that flies the approach of men.[*]

---

sacraments, being at once ambitious, violent, and gluttonous; such is the life they lead in America. The faith of the Indians suffer by such bad examples. If thou dost not change all this, O King of Spain, thy government will not be stable.

"What a misfortune that the Emperor, thy father, should have conquered Germany at such a price, and spent, on that conquest, the money we procured for him in these very Indies! In the year 1559 the Marquis de Canete sent to the Amazon, Pedro de Ursua, a Navarrese, or rather a Frenchman: we sailed on the largest rivers of Peru till we came to a gulf of fresh water. We had already gone three hundred leagues when we killed that bad and ambitious captain. We chose a caballero of Seville, Fernando de Guzman, for king: and we swore fealty to him, as is done to thyself. I was named quarter-master-general: and because I did not consent to all he willed, he wanted to kill me. But I killed this new king, the captain of his guards, his lieutenant-general, his chaplain, a woman, a knight of the order of Rhodes, two ensigns, and five or six domestics of the pretended king. I then resolved to punish thy ministers and thy auditors (counsellors of the audiencia). I named captains and sergeants: these again wanted to kill me, but I had them all hanged. In the midst of these adventures we navigated for eleven months, till we reached the mouth of the river. We sailed more than fifteen hundred leagues. God knows how we got through that great mass of water. I advise thee, O great King, never to send Spanish fleets into that accursed river. God preserve thee in his holy keeping."

This letter was given by Aguirre to the vicar of the island of Margareta, Pedro de Contreras, in order to be transmitted to King Philip II. Fray Pedro Simon, Provincial of the Franciscans in New Grenada, saw several manuscript copies of it both in America and in Spain. It was printed, for the first time, in 1723, in the History of the Province of Venezuela, by Oviedo, volume 1 page 206. Complaints no less violent, on the conduct of the monks of the 16th century, were addressed directly to the pope by the Milanese traveller, Girolamo Benzoni.

[*] See volume 1 chapter 1.4.

The second historical event connected with the name of Valencia is the great incursion made by the Caribs of the Orinoco in 1578 and 1580. That cannibal horde went up the banks of the Guarico, crossing the plains or llanos. They were happily repulsed by the valour of Garcia Gonzales, one of the captains whose names are still most revered in those provinces. It is gratifying to recollect, that the descendants of those very Caribs now live in the missions as peaceable husbandmen, and that no savage nation of Guiana dares to cross the plains which separate the region of the forests from that of cultivated land. The Cordillera of the coast is intersected by several ravines, very uniformly directed from south-east to north-west. This phenomenon is general from the Quebrada of Tocume, between Petares and Caracas, as far as Porto Cabello. It would seem as if the impulsion had everywhere come from the south-east; and this fact is the more striking, as the strata of gneiss and mica-slate in the Cordillera of the coast are generally directed from the south-west to the north-east. Most of these ravines penetrate into the mountains at their southern declivity, without crossing them entirely. But there is an opening (abra) on the meridian of Nueva Valencia, which leads towards the coast, and by which a cooling sea-breeze penetrates every evening into the valleys of Aragua. This breeze rises regularly two or three hours after sunset.

By this abra, the farm of Barbula, and an eastern branch of the ravine, a new road is being constructed from Valencia to Porto Cabello. It will be so short, that it will require only four hours to reach the port; and the traveller will be able to go and return in the same day from the coast to the valleys of Aragua. In order to examine this road, we set out on the 26th of February in the evening for the farm of Barbula.

On the morning of the 27th we visited the hot springs of La Trinchera, three leagues from Valencia. The ravine is very large, and the descent almost continual from the banks of the lake to the sea-coast. La Trinchera takes its name from some fortifications of earth, thrown up in 1677 by the French buccaneers, who sacked the town of Valencia. The hot springs (and this is a remarkable geological fact,) do not issue on the south side of the mountains, like those of Mariara, Onoto, and the Brigantine; but they issue from the chain itself almost at its northern declivity. They are much more abundant than any we had till then seen, forming a rivulet which, in times of the greatest drought, is two feet deep and eighteen wide. The temperature of the water, measured with great care, was 90.3° of the centigrade thermometer. Next to the springs of Urijino, in Japan, which are asserted to be pure water at 100° of temperature, the waters of the Trinchera of Porto Cabello appear to be the hottest in the world. We breakfasted near the spring; eggs plunged into the water were boiled in less than four minutes. These waters, strongly charged with sulphuretted hydrogen, gush out from the back of a hill rising one hundred and fifty feet above the bottom of the ravine, and tending from south-south-east to north-north-west. The rock from which the springs gush, is a real coarse-grained granite, resembling that of the Rincon del Diablo, in the mountains of Mariara. Wherever the waters evaporate in the air, they form sediments and incrustations of carbonate of lime; possibly they traverse strata of primitive limestone, so common in the mica-slate and gneiss of the coasts of Caracas. We were surprised at the luxuriant vegetation that surrounds the basin; mimosas with slender pinnate leaves, clusias, and fig-trees, have pushed their roots into the bottom of a pool, the temperature of which is 85°; and the branches of these trees extended over the surface of the water, at two or three inches distance. The foliage of the mimosas, though constantly enveloped in the hot vapours, displayed the most beautiful verdure. An arum, with a woody stem, and with large sagittate leaves, rose in the very middle of a pool the temperature of which was 70°. Plants of the same species vegetate in other parts of those mountains at the brink of torrents, the temperature of which is not 18°. What is still more singular, forty feet distant from the point whence the springs gush out at a temperature of 90°, other springs are found

41

perfectly cold. They all follow for some time a parallel direction; and the natives showed us that, by digging a hole between the two rivulets, they could procure a bath of any given temperature they pleased. It seems remarkable, that in the hottest as well as the coldest climates, people display the same predilection for heat. On the introduction of Christianity into Iceland, the inhabitants would be baptized only in the hot springs of Hecla: and in the torrid zone, in the plains, as well as on the Cordilleras, the natives flock from all parts to the thermal waters. The sick, who come to La Trinchera to use vapour-baths, form a sort of frame-work over the spring with branches of trees and very slender reeds. They stretch themselves naked on this frame, which appeared to me to possess little strength, and to be dangerous of access. The Rio de Aguas Calientes runs towards the north-east, and becomes, near the coast, a considerable river, swarming with great crocodiles, and contributing, by its inundations, to the insalubrity of the shore.

We descended towards Porto Cabello, having constantly the river of hot water on our right. The road is extremely picturesque, and the waters roll down on the shelves of rock. We might have fancied we were gazing on the cascades of the Reuss, that flows down Mount St. Gothard; but what a contrast in the vigour and richness of the vegetation! The white trunks of the cecropia rise majestically amid bignonias and melastomas. They do not disappear till we are within a hundred toises above the level of the ocean. A small thorny palm-tree extends also to this limit; the slender pinnate leaves of which look as if they had been curled toward the edges. This tree is very common in these mountains; but not having seen either its fruit or its flowers, we are ignorant whether it be the piritu palm-tree of the Caribbees, or the Cocos aculeata of Jacquin.

The rock on this road presents a geological phenomenon, the more remarkable as the existence of real stratified granite has long been disputed. Between La Trinchera and the Hato de Cambury a coarse-grained granite appears, which, from the disposition of the spangles of mica, collected in small groups, scarcely admits of confounding with gneiss, or with rocks of a schistose texture. This granite, divided into ledges of two or three feet thick, is directed 52° north-east, and slopes to the north-west regularly at an angle of from 30 or 40°. The feldspar, crystallized in prisms with four unequal sides, about an inch long, passes through every variety of tint from a flesh-red to yellowish white. The mica, united in hexagonal plates, is black, and sometimes green. The quartz predominates in the mass; and is generally of a milky white. I observed neither hornblende, black schorl, nor rutile titanite, in this granite. In some ledges we recognised round masses, of a blackish gray, very quartzose, and almost destitute of mica. They are from one to two inches diameter; and are found in every zone, in all granite mountains. These are not imbedded fragments, as at Greiffenstein in Saxony, but aggregations of particles which seem to have been subjected to partial attractions. I could not follow the line of junction of the gneiss and granitic formations. According to angles taken in the valleys of Aragua, the gneiss appears to descend below the granite, which must consequently be of a more recent formation. The appearance of a stratified granite excited my attention the more, because, having had the direction of the mines of Fichtelberg in Franconia for several years, I was accustomed to see granites divided into ledges of three or four feet thick, but little inclined, and forming masses like towers, or old ruins, at the summit of the highest mountains.[*]

The heat became stifling as we approached the coast. A reddish vapour veiled the horizon. It was near sunset, and the

---

[*] At Ochsenkopf, at Rudolphstein, at Epprechtstein, at Luxburg, and at Schneeberg. The dip of the strata of these granites of Fichtelberg is generally only from 6 to 10°, rarely (at Schneeberg) 18°. According to the dips I observed in the neighbouring strata of gneiss and mica-slate, I should think that the granite of Fichtelberg is very ancient, and serves as a basis for other formations; but the strata of grunstein, and the disseminated tin-ore which it contains, may lead us to doubt its great antiquity, from the analogy of the granites of Saxony containing tin.

breeze was not yet stirring. We rested in the lonely farms known under the names of the Hato de Cambury and the house of the Canarian (Casa del Isleno). The river of hot water, along the banks of which we passed, became deeper. A crocodile, more than nine feet long, lay dead on the strand. We wished to examine its teeth, and the inside of its mouth; but having been exposed to the sun for several weeks, it exhaled a smell so fetid that we were obliged to relinquish our design and remount our horses. When we arrived at the level of the sea, the road turned eastward, and crossed a barren shore a league and a half broad, resembling that of Cumana. We there found some scattered cactuses, a sesuvium, a few plants of Coccoloba uvifera, and along the coast some avicennias and mangroves. We forded the Guayguaza and the Rio Estevan, which, by their frequent overflowing, form great pools of stagnant water. Small rocks of meandrites, madrepores, and other corals, either ramified or with a rounded surface, rise in this vast plain, and seem to attest the recent retreat of the sea. But these masses, which are the habitations of polypi, are only fragments imbedded in a breccia with a calcareous cement. I say a breccia, because we must not confound the fresh and white corallites of this very recent littoral formation, with the corallites blended in the mass of transition-rocks, grauwacke, and black limestone. We were astonished to find in this uninhabited spot a large Parkinsonia aculeata loaded with flowers. Our botanical works indicate this tree as peculiar to the New World; but during five years we saw it only twice in a wild state, once in the plains of the Rio Guayguaza, and once in the llanos of Cumana, thirty leagues from the coast, near la Villa del Pao, but there was reason to believe that this latter place had once been a conuco, or cultivated enclosure. Everywhere else on the continent of America we saw the Parkinsonia, like the Plumeria, only in the gardens of the Indians.

At Porto Cabello, as at La Guayra, it is disputed whether the port lies east or west of the town, with which the communications are the most frequent. The inhabitants believe that Porto Cabello is north-north-west of Nueva Valencia; and my observations give a longitude of three or four minutes more towards the west.

We were received with the utmost kindness in the house of a French physician, M. Juliac, who had studied medicine at Montpelier. His small house contained a collection of things the most various, but which were all calculated to interest travellers. We found works of literature and natural history; notes on meteorology; skins of the jaguar and of large aquatic serpents; live animals, monkeys, armadilloes, and birds. Our host was principal surgeon to the royal hospital of Porto Cabello, and was celebrated in the country for his skilful treatment of the yellow fever. During a period of seven years he had seen six or eight thousand persons enter the hospitals, attacked by this cruel malady. He had observed the ravages that the epidemic caused in Admiral Ariztizabal's fleet, in 1793. That fleet lost nearly a third of its men; for the sailors were almost all unseasoned Europeans, and held unrestrained intercourse with the shore. M. Juliac had heretofore treated the sick as was commonly practised in Terra Firma, and in the island, by bleeding, aperient medicines, and acid drinks. In this treatment no attempt was made to raise the vital powers by the action of stimulants, so that, in attempting to allay the fever, the languor and debility were augmented. In the hospitals, where the sick were crowded, the mortality was often thirty-three per cent among the white Creoles; and sixty-five in a hundred among the Europeans recently disembarked. Since a stimulant treatment, the use of opium, of benzoin, and of alcoholic draughts, has been substituted for the old debilitating method, the mortality has considerably diminished. It was believed to be reduced to twenty in a hundred among Europeans, and ten among Creoles;[*] even when black vomiting, and

---

[*] I have treated in another work of the proportions of mortality in the yellow fever. (Nouvelle Espagne volume 2 pages 777, 785, and 867.) At Cadiz the average mortality was, in 1800, twenty per cent; at Seville, in 1801, it amounted to sixty per cent. At Vera Cruz the mortality does not exceed

haemorrhage from the nose, ears, and gums, indicated a high degree of exacerbation in the malady. I relate faithfully what was then given as the general result of observation: but I think, in these numerical comparisons, it must not be forgotten, that, notwithstanding appearances, the epidemics of several successive years do not resemble each other; and that, in order to decide on the use of fortifying or debilitating remedies, (if indeed this difference exist in an absolute sense,) we must distinguish between the various periods of the malady.

The climate of Porto Cabello is less ardent than that of La Guayra. The breeze there is stronger, more frequent, and more regular. The houses do not lean against rocks that absorb the rays of the sun during the day, and emit caloric at night, and the air can circulate more freely between the coast and the mountains of Ilaria. The causes of the insalubrity of the atmosphere must be sought in the shores that extend to the east, as far as the eye can reach, towards the Punta de Tucasos, near the fine port of Chichiribiche. There are situated the salt-works; and there, at the beginning of the rainy season, tertian fevers prevail, and easily degenerate into asthenic fevers. It is affirmed that the mestizoes who are employed in the salt-works are more tawny, and have a yellower skin, when they have suffered several successive years from those fevers, which are called the malady of the coast. The poor fishermen, who dwell on this shore, are of opinion that it is not the inundations of the sea, and the retreat of the salt-water, which render the lands covered with mangroves so unhealthful;[*] they believe that the insalubrity of the air is owing to the fresh water, that is, to the overflowings of

---

twelve or fifteen per cent, when the sick can be properly attended. In the civil hospitals of Paris the number of deaths, one year with another, is from fourteen to eighteen per cent; but it is asserted that a great number of patients enter the hospitals almost dying, or at very advanced time of life.

[*] In the West India Islands all the dreadful maladies which prevail during the wintry season, have been for a long time attributed to the south winds. These winds convey the emanations of the mouths of the Orinoco and of the small rivers of Terra Firma toward the high latitudes.

the Guayguaza and Estevan, the swell of which is so great and sudden in the months of October and November. The banks of the Rio Estevan have been less insalubrious since little plantations of maize and plantains have been established; and, by raising and hardening the ground, the river has been confined within narrower limits. A plan is formed of giving another issue to the Rio San Estevan, and thus to render the environs of Porto Cabello more wholesome. A canal is to lead the waters toward that part of the coast which is opposite the island of Guayguaza.

The salt-works of Porto Cabello somewhat resemble those of the peninsula of Araya, near Cumana. The earth, however, which they lixivate by collecting the rain-water into small basins, contains less salt. It is questioned here, as at Cumana, whether the ground be impregnated with saline particles because it has been for ages covered at intervals with sea-water evaporated by the heat of the sun, or whether the soil be muriatiferous, as in a mine very poor in native salt. I had not leisure to examine this plain with the same attention as the peninsula of Araya. Besides, does not this problem reduce itself to the simple question, whether the salt be owing to new or very ancient inundations? The labouring at the salt-works of Porto Cabello being extremely unhealthy, the poorest men alone engage in it. They collect the salt in little stores, and afterwards sell it to the shopkeepers in the town.

During our abode at Porto Cabello, the current on the coast, generally directed towards the west,[*] ran from west to east. This upward current (corriente por arriba), is very frequent during two or three months of the year, from September to November. It is believed to be owing to some north-west winds that have

---

[*] The wrecks of the Spanish ships, burnt at the island of Trinidad, at the time of its occupation by the English in 1797, were carried by the general or rotary current to Punta Brava, near Porto Cabello. This general current toward the east, from the coasts of Paria to the isthmus of Panama and the western extremity of the island of Cuba, was the subject of a violent dispute between Don Diego Columbus, Oviedo, and the pilot Andres, in the sixteenth century.

47

blown between Jamaica and Cape St. Antony in the island of Cuba.

The military defence of the coasts of Terra Firma rests on six points: the castle of San Antonio at Cumana; the Morro of Nueva Barcelona; the fortifications of La Guayra, (mounting one hundred and thirty-four guns); Porto Cabello; fort San Carlos, (at the mouth of the lake of Maracaybo); and Carthagena. Porto Cabello is, next to Carthagena, the most important fortified place. The town of Porto Cabello is quite modern, and the port is one of the finest in the world. Art has had scarcely anything to add to the advantages which the nature of the spot presents. A neck of land stretches first towards the north, and then towards the west. Its western extremity is opposite to a range of islands connected by bridges, and so close together that they might be taken for another neck of land. These islands are all composed of a calcareous breccia of extremely recent formation, and analagous to that on the coast of Cumana, and near the castle of Araya. It is a conglomerate, containing fragments of madrepores and other corals cemented by a limestone basis and grains of sand. We had already seen this conglomerate near the Rio Guayguaza. By a singular disposition of the ground the port resembles a basin or a little inland lake, the southern extremity of which is filled with little islands covered with mangroves. The opening of the port towards the west contributes much to the smoothness of the water.* One vessel only can enter at a time; but the largest ships of the line can anchor very near land to take in water. There is no other danger in entering the harbour than the reefs of Punta Brava, opposite which a battery of eight guns has been erected. Towards the west and south-west we see the fort, which is a regular pentagon with five bastions, the battery

---

* It is disputed at Porto Cabello whether the port takes its name from the tranquillity of its waters, "which would not move a hair (cabello)," or (which is more probable) derived from Antonio Cabello, one of the fishermen with whom the smugglers of Curacoa had formed a connexion at the period when the first hamlet was constructed on this half-desert coast.

of the reef, and the fortifications that surround the ancient town, founded on an island of a trapezoidal form. A bridge and the fortified gate of the Staccado join the old to the new town; the latter is already larger than the former, though considered only as its suburb. The bottom of the basin or lake which forms the harbour of Porto Cabello, turns behind this suburb to the south-west. It is a marshy ground filled with noisome and stagnant water. The town, which has at present nearly nine thousand inhabitants, owes its origin to an illicit commerce, attracted to these shores by the proximity of the town of Burburata, which was founded in 1549. It is only since the administration of the Biscayans, and of the company of Guipuzcoa, that Porto Cabello, which was but a hamlet, has been converted into a well-fortified town. The vessels of La Guayra, which is less a port than a bad open roadstead, come to Porto Cabello to be caulked and repaired.

The real defence of the harbour consists in the low batteries on the neck of land at Punta Brava, and on the reef; but from ignorance of this principle, a new fort, the Mirador of Solano[*] has been constructed at a great expense, on the mountains commanding the suburb towards the south. More than ten thousand mules are annually exported from Porto Cabello. It is curious enough to see these animals embarked; they are thrown down with ropes, and then hoisted on board the vessels by means of a machine resembling a crane. Ranged in two files, the mules with difficulty keep their footing during the rolling and pitching of the ship; and in order to frighten and render them more docile, a drum is beaten during a great part of the day and night. We may guess what quiet a passenger enjoys, who has the courage to embark for Jamaica in a schooner laden with mules.

We left Porto Cabello on the first of March, at sunrise. We saw with surprise the great number of boats that were laden with

---

[*] The Mirador is situate eastward of the Vigia Alta, and south-east of the battery of the salt-works and the powder-mill.

fruit to be sold at the market. It reminded me of a fine morning at Venice. The town presents in general, on the side towards the sea, a cheerful and agreeable aspect. Mountains covered with vegetation, and crowned with peaks called Las Tetas de Ilaria, which, from their outline would be taken for rocks of a trap-formation, form the background of the landscape. Near the coast all is bare, white, and strongly illumined, while the screen of mountains is clothed with trees of thick foliage that project their vast shadows upon the brown and rocky ground. On going out of the town we visited an aqueduct that had been just finished. It is five thousand varas long, and conveys the waters of the Rio Estevan by a trench to the town. This work has cost more than thirty thousand piastres; but its waters gush out in every street.

We returned from Porto Cabello to the valleys of Aragua, and stopped at the Farm of Barbula, near which, a new road to Valencia is in the course of construction. We had heard, several weeks before, of a tree, the sap of which is a nourishing milk. It is called the cow-tree; and we were assured that the negroes of the farm, who drink plentifully of this vegetable milk, consider it a wholesome aliment. All the milky juices of plants being acrid, bitter, and more or less poisonous, this account appeared to us very extraordinary; but we found by experience during our stay at Barbula, that the virtues of this tree had not been exaggerated. This fine tree rises like the broad-leaved star-apple.[*] Its oblong and pointed leaves, rough and alternate, are marked by lateral ribs, prominent at the lower surface, and parallel. Some of them are ten inches long. We did not see the flower: the fruit is somewhat fleshy, and contains one and sometimes two nuts. When incisions are made in the trunk of this tree, it yields abundance of a glutinous milk, tolerably thick, devoid of all acridity, and of an agreeable and balmy smell. It was offered to us in the shell of a calabash. We drank considerable quantities of it in the evening before we went to bed, and very early in the

---

[*] Chrysophyllum cainito.

morning, without feeling the least injurious effect. The viscosity of this milk alone renders it a little disagreeable. The negroes and the free people who work in the plantations drink it, dipping into it their bread of maize or cassava. The overseer of the farm told us that the negroes grow sensibly fatter during the season when the palo de vaca furnishes them with most milk. This juice, exposed to the air, presents at its surface (perhaps in consequence of the absorption of the atmospheric oxygen) membranes of a strongly animalized substance, yellowish, stringy, and resembling cheese. These membranes, separated from the rest of the more aqueous liquid, are elastic, almost like caoutchouc; but they undergo, in time, the same phenomena of putrefaction as gelatine. The people call the coagulum that separates by the contact of the air, cheese. This coagulum grows sour in the space of five or six days, as I observed in the small portions which I carried to Nueva Valencia. The milk contained in a stopped phial, had deposited a little coagulum; and, far from becoming fetid, it exhaled constantly a balsamic odour. The fresh juice mixed with cold water was scarcely coagulated at all; but on the contact of nitric acid the separation of the viscous membranes took place. We sent two bottles of this milk to M. Fourcroy at Paris: in one it was in its natural state, and in the other, mixed with a certain quantity of carbonate of soda. The French consul residing in the island of St. Thomas, undertook to convey them to him.

The extraordinary tree of which we have been speaking appears to be peculiar to the Cordillera of the coast, particularly from Barbula to the lake of Maracaybo. Some stocks of it exist near the village of San Mateo; and, according to M. Bredemeyer, whose travels have so much enriched the fine conservatories of Schonbrunn and Vienna, in the valley of Caucagua, three days journey east of Caracas. This naturalist found, like us, that the vegetable milk of the palo de vaco had an agreeable taste and an aromatic smell. At Caucagua, the natives call the tree that furnishes this nourishing juice, the milk-tree (arbol del leche). They profess to recognize, from the thickness and colour of the foliage, the trunks that yield the most juice; as the herdsman distinguishes, from external signs, a good milch-cow. No botanist has hitherto known the existence of this plant. It seems, according to M. Kunth, to belong to the sapota family. Long after my return to Europe, I found in the Description of the East Indies by Laet, a Dutch traveller, a passage that seems to have some relation to the cow-tree. "There exist trees," says Laet,[*] "in the province of Cumana, the sap of which much resembles curdled milk, and affords a salubrious nourishment."

Amidst the great number of curious phenomena which I have observed in the course of my travels, I confess there are few that have made so powerful an impression on me as the aspect of the cow-tree. Whatever relates to milk or to corn, inspires an interest which is not merely that of the physical knowledge of things, but is connected with another order of ideas and sentiments. We can scarcely conceive how the human race could exist without farinaceous substances, and without that nourishing juice which the breast of the mother contains, and which is appropriated to the long feebleness of the infant. The amylaceous matter of corn, the object of religious veneration among so many nations, ancient and modern, is diffused in the seeds, and deposited in the roots of vegetables; milk, which serves as an aliment, appears to us exclusively the produce of animal organization. Such are the impressions we have received in our earliest infancy: such is also the source of that astonishment created by the aspect of the tree just described. It is not here the solemn shades of forests, the majestic course of rivers, the mountains wrapped in eternal

---

[*] "Inter arbores quae sponte hic passim nascuntur, memorantur a scriptoribus Hispanis quaedam quae lacteum quemdam liquorem fundunt, qui durus admodum evadit instar gummi, et suavem odorem de se fundit; aliae quae liquorem quemdam edunt, instar lactis coagulati, qui in cibis ab ipsis usurpatur sine noxa." (Among the trees growing here, it is remarked by Spanish writers that there are some which pour out a milky juice which soon grows solid, like gum, affording a pleasant odour; and also others that give out a liquid which coagulates like cheese, and which they eat at meals without any ill effects). Descriptio Indiarum Occidentalium, lib. 18.

snow, that excite our emotion. A few drops of vegetable juice recall to our minds all the powerfulness and the fecundity of nature. On the barren flank of a rock grows a tree with coriaceous and dry leaves. Its large woody roots can scarcely penetrate into the stone. For several months of the year not a single shower moistens its foliage. Its branches appear dead and dried; but when the trunk is pierced there flows from it a sweet and nourishing milk. It is at the rising of the sun that this vegetable fountain is most abundant. The negroes and natives are then seen hastening from all quarters, furnished with large bowls to receive the milk, which grows yellow, and thickens at its surface. Some empty their bowls under the tree itself; others carry the juice home to their children.

In examining the physical properties of animal and vegetable products, science displays them as closely linked together; but it strips them of what is marvellous, and perhaps, therefore, of a part of their charms. Nothing appears isolated; the chemical principles that were believed to be peculiar to animals are found in plants; a common chain links together all organic nature.

Long before chemists had recognized small portions of wax in the pollen of flowers, the varnish of leaves, and the whitish dust of our plums and grapes, the inhabitants of the Andes of Quindiu made tapers with the thick layer of wax that covers the trunk of a palm-tree.[*] It is but a few years since we discovered, in Europe, caseum, the basis of cheese, in the emulsion of almonds; yet for ages past, in the mountains of the coast of Venezuela, the milk of a tree, and the cheese separated from that vegetable milk, have been considered as a salutary aliment. How are we to account for this singular course in the development of knowledge? How have the unlearned inhabitants of one hemisphere become cognizant of a fact which, in the other, so long escaped the sagacity of the scientific? It is because a small number of elements and principles differently combined are

---

[*] Coroxylon andicola.

spread through several families of plants; it is because the genera and species of these natural families are not equally distributed in the torrid, the frigid, and the temperate zones; it is that tribes, excited by want, and deriving almost all their subsistence from the vegetable kingdom, discover nutritive principles, farinaceous and alimentary substances, wherever nature has deposited them in the sap, the bark, the roots, or the fruits of vegetables. That amylaceous fecula which the seeds of the cereal plants furnish in all its purity, is found united with an acrid and sometimes even poisonous juice, in the roots of the arums, the Tacca pinnatifida, and the Jatropha manihot. The savage of America, like the savage of the South Sea islands, has learned to dulcify the fecula, by pressing and separating it from its juice. In the milk of plants, and in the milky emulsions, matter extremely nourishing, albumen, caseum, and sugar, are found mixed with caoutchouc and with deleterious and caustic principles, such as morphine and hydrocyanic acid.[*] These mixtures vary not only in the different families, but also in the species which belong to the same genus. Sometimes it is morphine or the narcotic principle, that characterises the vegetable milk, as in some papaverous plants; sometimes it is caoutchouc, as in the hevea and the castilloa; sometimes albumen and caseum, as in the cow-tree.

The lactescent plants belong chiefly to the three families of the euphorbiaceae, the urticeae, and the apocineae.[†] Since, on examining the distribution of vegetable forms over the globe, we find that those three families are more numerous in species in the low regions of the tropics, we must thence conclude, that a very elevated temperature contributes to the elaboration of the milky

---

[*] Opium contains morphine, caoutchouc, etc.

[†] After these three great families follow the papaveraceae, the chicoraceae, the lobeliaceae, the campanulaceae, the sapoteae, and the cucurbitaceae. The hydrocyanic acid is peculiar to the group of rosaceo-amygdalaceae. In the monocotyledonous plants there is no milky juice; but the perisperm of the palms, which yields such sweet and agreeable milky emulsions, contains, no doubt, caseum. Of what nature is the milk of mushrooms?

juices, to the formation of caoutchouc, albumen, and caseous matter. The sap of the palo de vaca furnishes unquestionably the most striking example of a vegetable milk in which the acrid and deleterious principle is not united with albumen, caseum, and caoutchouc: the genera euphorbia and asclepias, however, though generally known for their caustic properties, already present us with a few species, the juice of which is sweet and harmless. Such are the Tabayba dulce of the Canary Islands, which we have already mentioned,[*] and the Asclepias lactifera of Ceylon. Burman relates that, in the latter country, when cow's milk is wanting, the milk of this asclepias is used; and that the ailments commonly prepared with animal milk are boiled with its leaves. It may be possible, as Decandolle has well observed, that the natives employ only the juice that flows from the young plant, at a period when the acrid principle is not yet developed. In fact, the first shoots of the apocyneous plants are eaten in several countries.

I have endeavoured by these comparisons to bring into consideration, under a more general point of view, the milky juices that circulate in vegetables; and the milky emulsions that the fruits of the amygdalaceous plants and palms yield. I may be permitted to add the result of some experiments which I attempted to make on the juice of the Carica papaya during my stay in the valleys of Aragua, though I was then almost destitute of chemical tests. The juice has been since examined by Vauquelin, and this celebrated chemist has very clearly recognized the albumen and caseous matter; he compares the milky sap to a substance strongly animalized—to the blood of animals; but his researches were confined to a fermented juice and a coagulum of a fetid smell, formed during the passage from the Mauritius to France. He has expressed a wish that some traveller would examine the milk of the papaw-tree just as it flows from the stem or the fruit.

---

[*] Euphorbia balsamifera. The milky juice of the Cactus mamillaris is equally sweet.

The younger the fruit of the carica, the more milk it yields: it is even found in the germen scarcely fecundated. In proportion as the fruit ripens, the milk becomes less abundant, and more aqueous. Less of that animal matter which is coagulable by acids and by the absorption of atmospheric oxygen, is found in it. As the whole fruit is viscous,[*] it might be supposed that, as it grows larger, the coagulable matter is deposed in the organs, and forms a part of the pulp, or the fleshy substance. When nitric acid, diluted with four parts of water, is added drop by drop to the milk expressed from a very young fruit, a very extraordinary phenomenon appears. At the centre of each drop a gelatinous pellicle is formed, divided by greyish streaks. These streaks are simply the juice rendered more aqueous, owing to the contact of the acid having deprived it of the albumen. At the same time, the centre of the pellicles becomes opaque, and of the colour of the yolk of an egg; they enlarge as if by the prolongation of divergent fibres. The whole liquid assumes at first the appearance of an agate with milky clouds; and it seems as if organic membranes were forming under the eye of the observer. When the coagulum extends to the whole mass, the yellow spots again disappear. By agitation it becomes granulous like soft cheese.[†] The yellow colour reappears on adding a few more

---

[*] The same viscosity is also remarked in the fresh milk of the palo de vaca. It is no doubt occasioned by the caoutchouc, which is not yet separated, and which forms one mass with the albumen and the caseum, as the butter and the caseum in animal milk. The juice of a euphorbiaceous plant (Sapium aucuparium), which also yields caoutchouc, is so glutinous that it is used to catch parrots.

[†] The substance which falls down in grumous and filamentous clots is not pure caoutchouc, but perhaps a mixture of this substance with caseum and albumen. Acids precipitate the caoutchouc from the milky juice of the euphorbiums, fig-trees, and hevea; they precipitate the caseum from the milk of animals. A white coagulum was formed in phials closely stopped, containing the milk of the hevea, and preserved among our collections, during our journey to the Orinoco. It is perhaps the development of a vegetable acid which then furnishes oxygen to the albumen. The formation of the coagulum of the hevea, or of real caoutchouc, is nevertheless much more rapid in

drops of nitric acid. The acid acts in this instance as the oxygen of the atmosphere at a temperature from 27 to 35°; for the white coagulum grows yellow in two or three minutes, when exposed to the sun. After a few hours the yellow colour turns to brown, no doubt because the carbon is set more free progressively as the hydrogen, with which it was combined, is burnt. The coagulum formed by the acid becomes viscous, and acquires that smell of wax which I have observed in treating muscular flesh and mushrooms (morels) with nitric acid. According to the fine experiments of Mr. Hatchett, the albumen may be supposed to pass partly to the state of gelatine. The coagulum of the papaw-tree, when newly prepared, being thrown into water, softens, dissolves in part, and gives a yellowish tint to the fluid. The milk, placed in contact with water only, forms also membranes. In an instant a tremulous jelly is precipitated, resembling starch. This phenomenon is particularly striking if the water employed be heated to 40 or 60°. The jelly condenses in proportion as more water is poured upon it. It preserves a long time its whiteness, only growing yellow by the contact of a few drops of nitric acid. Guided by the experiments of Fourcroy and Vauquelin on the juice of the hevea, I mixed a solution of carbonate of soda with the milk of the papaw. No clot is formed, even when pure water is poured on a mixture of the milk with the alkaline solution. The membranes appear only when, by adding an acid, the soda is neutralized, and the acid is in excess. I made the coagulum formed by nitric acid, the juice of lemons, or hot water, likewise disappear by mixing it with carbonate of soda. The sap again becomes milky and liquid, as in its primitive state; but this experiment succeeds only when the coagulum has been recently formed.

---

contact with the air. The absorption of atmospheric oxygen is not in the least necessary to the production of butter which exists already formed in the milk of animals; but I believe it cannot be doubted that, in the milk of plants, this absorption produces the pellicles of caoutchouc, of coagulated albumen, and of caseum, which are successively formed in vessels exposed to the open air.

On comparing the milky juices of the papaw, the cow-tree, and the hevea, there appears a striking analogy between the juices which abound in caseous matter, and those in which caoutchouc prevails. All the white and newly prepared caoutchouc, as well as the waterproof cloaks, manufactured in Spanish America by placing a layer of milk of hevea between two pieces of cloth, exhale an animal and nauseating smell. This seems to indicate that the caoutchouc, in coagulating, carries with it the caseum, which is perhaps only an altered albumen.

The produce of the bread-fruit tree can no more be considered as bread than plantains before the state of maturity, or the tuberous and amylaceous roots of the cassava, the dioscorea, the Convolvulus batatas, and the potato. The milk of the cow-tree contains, on the contrary, a caseous matter, like the milk of mammiferous animals. Advancing to more general considerations, we may regard, with M. Gay-Lussac, the caoutchouc as the oily part—the butter of vegetable milk. We find in the milk of plants caseum and caoutchouc; in the milk of animals, caseum and butter. The proportions of the two albuminous and oily principles differ in the various species of animals and of lactescent plants. In these last they are most frequently mixed with other substances hurtful as food; but of which the separation might perhaps be obtained by chemical processes. A vegetable milk becomes nourishing when it is destitute of acrid and narcotic principles; and abounds less in caoutchouc than in caseous matter.[*]

---

[*] The milk of the lactescent agarics has not been separately analysed; it contains an acrid principle in the Agaricus piperatus, and in other species it is sweet and harmless. The experiments of MM. Braconnot, Bouillon-Lagrange, and Vauquelin (Annales de Chimie, volume 46, volume 51, volume 79, volume 80, volume 85, have pointed out a great quantity of albumen in the substance of the Agaricus deliciosus, an edible mushroom. It is this albumen contained in their juice which renders them so hard when boiled. It has been proved that morels (Morchella esculenta) can be converted into sebaceous and adipocerous matter, capable of being used in the fabrication of soap. (De Candolle, sur les Proprietes medicinales des Plantes.) Saccharine matter has

Whilst the palo de vaca manifests the immense fecundity and the bounty of nature in the torrid zone, it also reminds us of the numerous causes which favour in those fine climates the careless indolence of man. Mungo Park has made known the butter-tree of Bambarra, which M. De Candolle suspects to be of the family of sapotas, as well as our milk-tree. The plantain, the sago-tree, and the mauritia of the Orinoco, are as much bread-trees as the rema of the South Sea. The fruits of the crescentia and the lecythis serve as vessels for containing food, while the spathes of the palms, and the bark of trees, furnish caps and garments without a seam. The knots, or rather the interior cells of the trunks of bamboos, supply ladders, and facilitate in a thousand ways the construction of a hut, and the fabrication of chairs, beds, and other articles of furniture that compose the wealth of a

---

also been found in mushrooms by Gunther. It is in the family of the fungi, more especially in the clavariae, phalli, helvetiae, the merulii, and the small gymnopae which display themselves in a few hours after a storm of rain, that organic nature produces with most rapidity the greatest variety of chemical principles—sugar, albumen, adipocire, acetate of potash, fat, ozmazome, the aromatic principles, etc. It would be interesting to examine, besides the milk of the lactescent fungi, those species which, when cut in pieces, change their colour on the contact of atmospheric air.

Though we have referred the palo de vaca to the family of the sapotas, we have nevertheless found in it a great resemblance to some plants of the urticeous kind, especially to the fig-tree, because of its terminal stipulae in the shape of a horn; and to the brosimum, on account of the structure of its fruit. M. Kunth would even have preferred this last classification; if the description of the fruit, made on the spot, and the nature of the milk, which is acrid in the urticeae, and sweet in the sapotas, did not seem to confirm our conjecture. Bredemeyer saw, like us, the fruit, and not the flower of the cow tree. He asserts that he observed [sometimes?] two seeds, lying one against the other, as in the alligator pear-tree (Laurus persea). Perhaps this botanist had the intention of expressing the same conformation of the nucleus that Swartz indicates in the description of the brosimum—"nucleus bilobus aut bipartibilis." We have mentioned the places where this remarkable tree grows: it will be easy for botanical travellers to procure the flower of the palo de vaca and to remove the doubts which still remain, of the family to which it belongs.

savage household. In the midst of this lavish vegetation, so varied in its productions, it requires very powerful motives to excite man to labour, to rouse him from his lethargy, and to unfold his intellectual faculties.

Cacao and cotton are cultivated at Barbula. We there found, what is very rare in that country, two large cylindrical machines for separating the cotton from its seed; one put in motion by an hydraulic wheel, and the other by a wheel turned by mules. The overseer of the farm, who had constructed these machines, was a native of Merida. He was acquainted with the road that leads from Nueva Valencia, by the way of Guanare and Misagual, to Varinas; and thence by the ravine of Collejones, to the Paramo de Mucuchies and the mountains of Merida covered with eternal snows. The notions he gave us of the time requisite for going from Valencia by Varinas to the Sierra Nevada, and thence by the port of Torunos, and the Rio Santo Domingo, to San Fernando de Apure, were of infinite value to us. It can scarcely be imagined in Europe, how difficult it is to obtain accurate information in a country where the communications are so rare; and where distances are diminished or exaggerated according to the desire that may be felt to encourage the traveller, or to deter him from his purpose. I had resolved to visit the eastern extremity of the Cordilleras of New Grenada, where they lose themselves in the paramos of Timotes and Niquitao. I learned at Barbula, that this excursion would retard our arrival at the Orinoco thirty-five days. This delay appeared to us so much the longer, as the rains were expected to begin sooner than usual. We had the hope of examining afterwards a great number of mountains covered with perpetual snow, at Quito, Peru, and Mexico; and it appeared to me still more prudent to relinquish our project of visiting the mountains of Merida, since by so doing we might miss the real object of our journey, that of ascertaining by astronomical observations the point of communication between the Orinoco, the Rio Negro, and the river Amazon. We returned in consequence from Barbula to

Guacara, to take leave of the family of the Marquis del Toro, and pass three days more on the borders of the lake.

It was the carnival season, and all was gaiety. The sports in which the people indulge, and which are called carnes tollendas,[*] assume occasionally somewhat of a savage character. Some led an ass loaded with water, and, where-ever they found a window open, inundated the apartment within by means of a pump. Others carried bags filled with hairs of picapica;[†] and blew the hair, which causes a great irritation of the skin, into the faces of those who passed by.

From Guacara we returned to Nueva Valencia. We found there a few French emigrants, the only ones we saw during five years passed in the Spanish colonies. Notwithstanding the ties of blood which unite the royal families of France and Spain, even French priests were not permitted to take refuge in that part of the New World, where man with such facility finds food and shelter. Beyond the Atlantic, the United States of America afford the only asylum to misfortune. A government, strong because it is free, confiding because it is just, has nothing to fear in giving refuge to the proscribed.

We have endeavoured above to give some notions of the state of the cultivation of indigo, cotton, and sugar, in the province of Caracas. Before we quit the valley of Aragua and its neighbouring coast, it remains for us to speak of the cacao-plantations, which have at all times been considered as the principal source of the prosperity of those countries. The province of Caracas,[‡] at the end of the eighteenth century, produced annually a hundred and fifty thousand fanegas, of which a hundred thousand were consumed in Spain, and thirty

---

[*] Or "farewell to flesh." The word carnival has the same meaning, these sports being always held just before the commencement of Lent.

[†] Dolichos pruriens (cowage).

[‡] The province, not the capitania-general, consequently not including the cacao plantations of Cumana, the province of Barcelona, of Maracaybo, of Varinas, and of Spanish Guiana.

thousand in the province. Estimating a fanega of cacao at only twenty-five piastres for the price given at Cadiz, we find that the total value of the exportation of cacao, by the six ports of the Capitania General of Caracas, amounts to four million eight hundred thousand piastres. So important an object of commerce merits a careful discussion; and I flatter myself, that, from the great number of materials I have collected on all the branches of colonial agriculture, I shall be able to add something to the information published by M. Depons, in his valuable work on the provinces of Venezuela.

The tree which produces the cacao is not at present found wild in the forests of Terra Firma to the north of the Orinoco; we began to find it only beyond the cataracts of Ature and Maypure. It abounds particularly near the banks of the Ventuari, and on the Upper Orinoco, between the Padamo and the Gehette. This scarcity of wild cacao-trees in South America, north of the latitude of 6°, is a very curious phenomenon of botanical geography, and yet little known. This phenomenon appears the more surprising, as, according to the annual produce of the harvest, the number of trees in full bearing in the cacao-plantations of Caracas, Nueva Barcelona, Venezuela, Varinas, and Maracaybo, is estimated at more than sixteen millions. The wild cacao-tree has many branches, and is covered with a tufted and dark foliage. It bears a very small fruit, like that variety which the ancient Mexicans called tlalcacahuatl. Transplanted into the conucos of the Indians of Cassiquiare and the Rio Negro, the wild tree preserves for several generations that force of vegetable life, which makes it bear fruit in the fourth year; while, in the province of Caracas, the harvest begins only the sixth, seventh, or eighth year. It is later in the inland parts than on the coasts and in the valley of Guapo. We met with no tribe on the Orinoco that prepared a beverage with the seeds of the cacao-tree. The savages suck the pulp of the pod, and throw away the seeds, which are often found in heaps where they have passed the night. Though chorote, which is a very weak infusion of cacao, is considered on the coast to be a very ancient beverage, no historical fact proves that chocolate, or any preparation whatever of cacao, was known to the natives of Venezuela before the arrival of the Spaniards. It appears to me more probable that the cacao-plantations of Caracas were suggested by those of Mexico and Guatimala; and that the Spaniards inhabiting Terra Firma learned the cultivation of the cacao-tree, sheltered in its youth by the foliage of the erythrina and plantain;[*] the fabrication of cakes of chocolatl, and the use of the liquid of the same name, in course of their communications with Mexico, Guatimala, and Nicaragua.

Down to the sixteenth century travellers differed in opinion respecting the chocolatl. Benzoni plainly says that it is a drink "fitter for hogs than men."[†] The Jesuit Acosta asserts, that "the Spaniards who inhabit America are fond of chocolate to excess; but that it requires to be accustomed to that black beverage not to be disgusted at the mere sight of its froth, which swims on it like yeast on a fermented liquor." He adds, "the cacao is a prejudice (una supersticion) of the Mexicans, as the coca is a prejudice of the Peruvians." These opinions remind us of Madame de Sevigne's prediction respecting the use of coffee. Fernando Cortez and his page, the gentilhombre del gran Conquistador, whose memoirs were published by Ramusio, on the contrary, highly praise chocolate, not only as an agreeable drink, though prepared cold,[‡] but in particular as a nutritious substance. "He who has drunk one cup," says the page of Fernando Cortez, "can travel a whole day without any other food, especially in very hot climates; for chocolate is by its

---

[*] This process of the Mexican cultivators, practised on the coast of Caracas, is described in the memoirs known under the title of "Relazione di certo Gentiluomo del Signor Cortez, Conquistadore del Messico." (Ramusio, tome 2 page 134).

[†] Benzoni, Istoria del Mondo Nuovo, 1572 page 104.

[‡] Father Gili has very clearly shown, from two passages in Torquemada (Monarquia Indiana, lib. 14) that the Mexicans prepared the infusion cold, and that the Spaniards introduced the custom of preparing chocolate by boiling water with the paste of cacao.

nature cold and refreshing." We shall not subscribe to the latter part of this assertion; but we shall soon have occasion, in our voyage on the Orinoco, and our excursions towards the summit of the Cordilleras, to celebrate the salutary properties of chocolate. It is easily conveyed and readily employed: as an aliment it contains a large quantity of nutritive and stimulating particles in a small compass. It has been said with truth, that in the East, rice, gum, and ghee (clarified butter), assist man in crossing the deserts; and so, in the New World, chocolate and the flour of maize, have rendered accessible to the traveller the table-lands of the Andes, and vast uninhabited forests.

The cacao harvest is extremely variable. The tree vegetates with such vigour that flowers spring out even from the roots, wherever the earth leaves them uncovered. It suffers from the north-east winds, even when they lower the temperature only a few degrees. The heavy showers that fall irregularly after the rainy season, during the winter months, from December to March, are also very hurtful to the cacao-tree. The proprietor of a plantation of fifty thousand trees often loses the value of more than four or five thousand piastres in cacao in one hour. Great humidity is favourable to the tree only when it augments progressively, and is for a long time uninterrupted. If, in the season of drought, the leaves and the young fruit be wetted by a violent shower, the fruit falls from the stem; for it appears that the vessels which absorb water break from being rendered turgid. Besides, the cacao-harvest is one of the most uncertain, on account of the fatal effects of inclement seasons, and the great number of worms, insects, birds, and quadrupeds,[*] which devour the pod of the cacao-tree; and this branch of agriculture has the disadvantage of obliging the new planter to wait eight or ten years for the fruit of his labours, and of yielding after all an article of very difficult preservation.

---

[*] Parrots, monkeys, agoutis, squirrels, and stags.

The finest plantations of cacao are found in the province of Caracas, along the coast, between Caravalleda and the mouth of the Rio Tocuyo, in the valleys of Caucagua, Capaya, Curiepe, and Guapo; and in those of Cupira, between cape Conare and cape Unare, near Aroa, Barquesimeto, Guigue, and Uritucu. The cacao that grows on the banks of the Uritucu, at the entrance of the llanos, in the jurisdiction of San Sebastian de las Reyes, is considered to be of the finest quality. Next to the cacao of Uritucu comes that of Guigue, of Caucagua, of Capaya, and of Cupira. The merchants of Cadiz assign the first rank to the cacao of Caracas, immediately after that of Socomusco; and its price is generally from thirty to forty per cent higher than that of Guayaquil.

It is only since the middle of the seventeenth century, when the Dutch, tranquil possessors of the island of Curacoa, awakened, by their smuggling, the agricultural industry of the inhabitants of the neighbouring coasts, that cacao has become an object of exportation in the province of Caracas. We are ignorant of everything that passed in those countries before the establishment of the Biscay Company of Guipuzcoa, in 1728. No precise statistical data have reached us: we only know that the exportation of cacao from Caracas scarcely amounted, at the beginning of the eighteenth century, to thirty thousand fanegas a-year. From 1730 to 1748, the company sent to Spain eight hundred and fifty-eight thousand nine hundred and seventy-eight fanegas, which make, on an average, forty-seven thousand seven hundred fanegas a-year; the price of the fanega fell, in 1732, to forty-five piastres, when it had before kept at eighty piastres. In 1763 the cultivation had so much augmented, that the exportation rose to eighty thousand six hundred and fifty-nine fanegas.

In an official document, taken from the papers of the minister of finance, the annual produce (la cosecha) of the province of Caracas is estimated at a hundred and thirty-five thousand fanegas of cacao; thirty-three thousand of which are for home consumption, ten thousand for other Spanish colonies, seventy-seven thousand for the mother-country, fifteen thousand for the illicit commerce with the French, English, Dutch, and Danish colonies. From 1789 to 1793, the importation of cacao from Caracas into Spain was, on an average, seventy-seven thousand seven hundred and nineteen fanegas a-year, of which sixty-five thousand seven hundred and sixty-six were consumed in the country, and eleven thousand nine hundred and fifty-three exported to France, Italy, and Germany.

The late wars have had much more fatal effects on the cacao trade of Caracas than on that of Guayaquil. On account of the increase of price, less cacao of the first quality has been consumed in Europe. Instead of mixing, as was done formerly for common chocolate, one quarter of the cacao of Caracas, with three-quarters of that of Guayaquil, the latter has been employed pure in Spain. We must here remark, that a great deal of cacao of an inferior quality, such as that of Maranon, the Rio Negro, Honduras, and the island of St. Lucia, bears the name, in commerce, of Guayaquil cacao. The exportation from that port amounts only to sixty thousand fanegas; consequently it is two-thirds less than that of the ports of the Capitania-General of Caracas.

Though the plantations of cacao have augmented in the provinces of Cumana, Barcelona, and Maracaybo, in proportion as they have diminished in the province of Caracas, it is still believed that, in general, this ancient branch of agricultural industry gradually declines. In many parts coffee and cotton-trees progressively take place of the cacao, of which the lingering harvests weary the patience of the cultivator. It is also asserted, that the new plantations of cacao are less productive than the old; the trees do not acquire the same vigour, and yield later and less abundant fruit. The soil is still said to be exhausted; but probably it is rather the atmosphere that is changed by the progress of clearing and cultivation. The air that reposes on a virgin soil covered with forests is loaded with humidity and those gaseous mixtures that serve for the nutriment of plants, and arise from the decomposition of organic substances. When a country has been long subjected to cultivation, it is not the proportions between the azote and oxygen that vary. The constituent bases of the atmosphere remain unaltered; but it no longer contains, in a state of suspension, those binary and ternary mixtures of carbon, hydrogen, and nitrogen, which a virgin soil exhales, and which are regarded as a source of fecundity. The air, purer and less charged with miasmata and heterogeneous emanations, becomes at the same time drier. The elasticity of the vapours undergoes a sensible diminution. On land long cleared, and consequently little favourable to the cultivation of the cacao-tree (as, for instance, in the West India Islands), the fruit is almost as small as that of the wild cacao-tree. It is on the banks of the Upper Orinoco, after having crossed the Llanos, that we find the true country of the cacao-tree; thick forests, in which, on a virgin soil, and surrounded by an atmosphere continually humid, the trees furnish, from the fourth year, abundant crops. Wherever the soil is not exhausted, the fruit has become by cultivation larger and bitter, but also later.

On seeing the produce of cacao gradually diminish in Terra Firma, it may be inquired, whether the consumption will diminish in the same proportion in Spain, Italy, and the rest of Europe; or whether it be not probable, that by the destruction of the cacao plantations, the price will augment sufficiently to rouse anew the industry of the cultivator. This latter opinion is generally admitted by those who deplore, at Caracas, the diminution of so ancient and profitable a branch of commerce. In proportion as civilization extends towards the humid forests of the interior, the banks of the Orinoco and the Amazon, or towards the valleys that furrow the eastern declivity of the Andes, the new planters will find lands and an atmosphere equally favourable to the culture of the cacao-tree.

The Spaniards, in general, dislike a mixture of vanilla with the cacao, as irritating the nervous system; the fruit, therefore, of that orchideous plant is entirely neglected in the province of Caracas, though abundant crops of it might be gathered on the moist and feverish coast between Porto Cabello and Ocumare; especially at Turiamo, where the fruits of the Epidendrum vanilla attain the length of eleven or twelve inches. The English and the Anglo-Americans often seek to make purchases of vanilla at the port of La Guayra, but the merchants procure with difficulty a very small quantity. In the valleys that descend from the chain of the coast towards the Caribbean Sea, in the province of Truxillo, as well as in the Missions of Guiana, near the cataracts of the Orinoco, a great quantity of vanilla might be collected; the produce of which would be still more abundant, if, according to the practice of the Mexicans, the plant were disengaged, from time to time, from the creeping plants by which it is entwined and stifled.

The hot and fertile valleys of the Cordillera of the coast of Venezuela occupy a tract of land which, on the west, towards the lake of Maracaybo, displays a remarkable variety of scenery. I shall exhibit in one view, to close this chapter, the facts I have been able to collect respecting the quality of the soil and the metallic riches of the districts of Aroa, of Barquesimeto, and of Carora.

From the Sierra Nevada of Merida, and the paramos of Niquitao, Bocono, and Las Rosas,[*] which contain the valuable bark-tree, the eastern Cordillera of New Granada[†] decreases in

---

[*] Many travellers, who were monks, have asserted that the little Paramo de Las Rosas, the height of which appears to be more than 1,600 toises, is covered with rosemary, and the red and white roses of Europe grow wild there. These roses are gathered to decorate the altars in the neighbouring villages on the festivals of the church. By what accident has our Rosa centifolia become wild in this country, while we nowhere found it in the Andes of Quito and Peru? Can it really be the rose-tree of our garden?

[†] The bark exported from the port of Maracaybo does not come from the territory of Venezuela, but from the mountains of Pamplona in New Grenada,

height so rapidly, that, between the ninth and tenth degrees of latitude, it forms only a chain of little mountains, which, stretching to the north-east by the Altar and Torito, separates the rivers that join the Apure and the Orinoco from those numerous rivers that flow either into the Caribbean Sea or the lake of Maracaybo. On this dividing ridge are built the towns of Nirgua, San Felipe el Fuerte, Barquesimeto, and Tocuyo. The first three are in a very hot climate; but Tocuyo enjoys great coolness, and we heard with surprise, that, beneath so fine a sky, the inhabitants have a strong propensity to suicide. The ground rises towards the south; for Truxillo, the lake of Urao, from which carbonate of soda is extracted, and La Grita, all to the east of the Cordillera, though no farther distant, are four or five hundred toises high.

On examining the law which the primitive strata of the Cordillera of the coast follow in their dip, we believe we recognize one of the causes of the extreme humidity of the land bounded by this Cordillera and the ocean. The dip of the strata is most frequently to the north-west; so that the waters flow in that direction on the ledges of rock; and form, as we have stated above, that multitude of torrents and rivers, the inundations of which become so fatal to the health of the inhabitants, from cape Codera as far as the lake of Maracaybo.

Among the rivers which descend north-east toward the coast of Porto Cabello, and La Punta de Hicacos, the most remarkable are those of Tocuyo, Aroa, and Yaracuy. Were it not for the miasmata which infect the atmosphere, the valleys of Aroa and of Yaracuy would perhaps be more populous than those of Aragua. Navigable rivers would even give the former the advantage of facilitating the exportation of their own crops of sugar and cacao, and that of the productions of the neighbouring

---

being brought down the Rio de San Faustino, that flows into the lake of Maracaybo. (Pombo, Noticias sobre las Quinas, 1814 page 65.) Some is collected near Merida, in the ravine of Viscucucuy.

lands; as the wheat of Quibor, the cattle of Monai, and the copper of Aroa. The mines from which this copper is extracted, are in a lateral valley, opening into that of Aroa; and which is less hot, and less unhealthy, than the ravines nearer the sea. In the latter the Indians have their gold-washings, and the soil conceals rich copper-ores, which no one has yet attempted to extract. The ancient mines of Aroa, after having been long neglected, have been wrought anew by the care of Don Antonio Henriquez, whom we met at San Fernando on the borders of the Apure. The total produce of metallic copper is twelve or fifteen hundred quintals a year. This copper, known at Cadiz by the name of Caracas copper, is of excellent quality. It is even preferred to that of Sweden, and of Coquimbo in Chile. Part of the copper of Aroa is employed for making bells, which are cast on the spot. Some ores of silver have been recently discovered between Aroa and Nirgua, near Guanita, in the mountain of San Pablo. Grains of gold are found in all the mountainous lands between the Rio Yaracuy, the town of San Felipe, Nirgua, and Barquesimeto; particularly in the Rio de Santa Cruz, in which the Indian gold-gatherers have sometimes found lumps of the value of four or five piastres. Do the neighbouring rocks of mica-slate and gneiss contain veins? or is the gold disseminated here, as in the granites of Guadarama in Spain, and of the Fichtelberg in Franconia, throughout the whole mass of the rock? Possibly the waters, in filtering through it, bring together the disseminated grains of gold; in which case every attempt to work the rock would be useless. In the Savana de la Miel, near the town of Barquesimeto, a shaft has been sunk in a black shining slate resembling ampelite. The minerals extracted from this shaft, which were sent to me at Caracas, were quartz, non-auriferous pyrites, and carbonated lead, crystallized in needles of a silky lustre.

In the early times of the conquest the working of the mines of Nirgua and of Buria[*] was begun, notwithstanding the incursions of the warlike nation of the Giraharas. In this very district the accumulation of negro slaves in 1553 gave rise to an event bearing some analogy to the insurrection in St. Domingo. A negro slave excited an insurrection among the miners of the Real de San Felipe de Buria. He retired into the woods, and founded, with two hundred of his companions, a town, where he was proclaimed king. Miguel, this new king, was a friend to pomp and parade. He caused his wife Guiomar, to assume the title of queen; and, according to Oviedo, he appointed ministers and counsellors of state, officers of the royal household, and even a negro bishop. He soon after ventured to attack the neighbouring town of Nueva Segovia de Barquesimeto; but, being repulsed by Diego de Losada, he perished in the conflict. This African monarchy was succeeded at Nirgua by a republic of Zamboes, the descendants of negroes and Indians. The whole municipality (cabildo) is composed of men of colour to whom the king of Spain has given the title of "his faithful and loyal subjects, the Zamboes of Nirgua." Few families of Whites will inhabit a country where the system of government is so adverse to their pretensions; and the little town is called in derision La republica de Zambos y Mulatos.

If the hot valleys of Aroa, of Yaracuy, and of the Rio Tocuyo, celebrated for their excellent timber, be rendered feverish by luxuriance of vegetation, and extreme atmospheric humidity, it is different in the savannahs of Monai and Carora. These Llanos are separated by the mountainous tract of Tocuyo and Nirgua from the great plains of La Portuguesa and Calabozo. It is very extraordinary to see barren savannahs loaded with miasmata. No marshy ground is found there, but several phenomena indicate a disengagement of hydrogen.[†] When

---

[*] The valley of Buria, and the little river of the same name, communicate with the valley of the Rio Coxede, or the Rio de Barquesimeto.

[†] What is that luminous phenomenon known under the name of the Lantern

travellers, who are not acquainted with natural inflammable gases, are shown the Cueva del Serrito de Monai, the people of the country love to frighten them by setting fire to the gaseous combination which is constantly accumulated in the upper part of the cavern. May we attribute the insalubrity of the atmosphere to the same causes as those which operate in the plains between Tivoli and Rome, namely, disengagements of sulphuretted hydrogen?[*] Possibly, also, the mountainous lands, near the llanos of Monai, may have a baneful influence on the surrounding plains. The south-easterly winds may convey to them the putrid exhalations that rise from the ravine of Villegas, and from La Sienega de Cabra, between Carora and Carache. I am desirous of collecting every circumstance having a relation to the salubrity of the air; for, in a matter so obscure, it is only by the comparison of a great number of phenomena, that we can hope to discover the truth.

The barren yet feverish savannahs, extending from Barquesimeto to the eastern shore of the lake of Maracaybo, are partly covered with cactus; but the good silvester-cochineal,

---

(farol) of Maracaybo, which is perceived every night toward the seaside as well as in the inland parts, at Merida for example, where M. Palacios observed it during two years? The distance, greater than 40 leagues, at which the light is observed, has led to the supposition that it might be owing to the effects of a thunderstorm, or of electrical explosions which might daily take place in a pass in the mountains. It is asserted that, on approaching the farol, the rolling of thunder is heard. Others vaguely allege that it is an air-volcano, and that asphaltic soils, like those of Mena, cause these inflammable exhalations which are so constant in their appearance. The phenomenon is observed on a mountainous and uninhabited spot, on the borders of the Rio Catatumbo, near the junction with the Rio Sulia. The situation of the farol is such that, being nearly in the meridian of the opening (boca) of the lake of Maracaybo, navigators are guided by it as by a lighthouse.

[*] Don Carlos del Pozo has discovered in this district, at the bottom of the Quebrada de Moroturo, a stratum of clayey earth, black, strongly soiling the fingers, emitting a powerful smell of sulphur, and inflaming spontaneously when slightly moistened and exposed for a long time to the rays of the tropical sun. The detonation of this muddy substance is very violent.

known by the vague name of grana de Carora, comes from a more temperate region, between Carora and Truxillo, and particularly from the valley of the Rio Mucuju,[*] to the cast of Merida. The inhabitants altogether neglect this production, so much sought for in commerce.

---

[*] This little river descends from the Paramo de los Conejos, and flows into the Rio Albarregas.

CHAPTER 2.17.

MOUNTAINS WHICH SEPARATE THE VALLEYS OF ARAGUA FROM THE LLANOS OF CARACAS. VILLA DE CURA. PARAPARA. LLANOS OR STEPPES. CALABOZO.

The chain of mountains, bordering the lake of Tacarigua towards the south, forms in some sort the northern shore of the great basin of the Llanos or savannahs of Caracas. To descend from the valleys of Aragua into these savannahs, it is necessary to cross the mountains of Guigue and of Tucutunemo. From a peopled country embellished by cultivation, we plunge into a vast solitude. Accustomed to the aspect of rocks, and to the shade of valleys, the traveller beholds with astonishment these savannahs without trees, these immense plains, which seem to ascend to the horizon.

Before I trace the scenery of the Llanos, or of the region of pasturage, I will briefly describe the road we took from Nueva Valencia, by Villa de Cura and San Juan, to the little village of Ortiz, at the entrance of the steppes. We left the valleys of Aragua on the 6th of March before sunrise. We passed over a plain richly cultivated, keeping along the south-west side of the lake of Valencia, and crossing the ground left uncovered by the waters of the lake. We were never weary of admiring the fertility of the soil, covered with calabashes, water-melons, and plantains. The rising of the sun was announced by the distant noise of the howling monkeys. Approaching a group of trees, which rise in the midst of the plain, between those parts which were anciently the islets of Don Pedro and La Negra, we saw numerous bands of araguatos moving as in procession and very slowly, from one tree to another. A male was followed by a great number of females; several of the latter carrying their young on their shoulders. The howling monkeys, which live in society in different parts of America, everywhere resemble each other in their manners, though the species are not always the same. The uniformity with which the araguatos[*] perform their movements is extremely striking. Whenever the branches of neighbouring trees do not touch each other, the male who leads the party suspends himself by the callous and prehensile part of his tail; and, letting fall the rest of his body, swings himself till in one of his oscillations he reaches the neighbouring branch. The whole file performs the same movements on the same spot. It is almost superfluous to add how dubious is the assertion of Ulloa, and so many otherwise well-informed travellers, according to whom, the marimondos,[†] the araguatos, and other monkeys with a prehensile tail, form a sort of chain, in order to reach the opposite side of a river.[‡] We had opportunities, during five years, of observing thousands of these animals; and for this very reason we place no confidence in statements possibly invented by the Europeans themselves, though repeated by the Indians of the Missions, as if they had been transmitted to them by their fathers. Man, the most remote from civilization, enjoys the astonishment he excites in recounting the marvels of his country. He says he has seen what he imagines may have been seen by others. Every savage is a hunter, and the stories of hunters borrow from the imagination in proportion as the animals, of which they boast the artifices, are endowed with a high degree of intelligence. Hence arise the fictions of which foxes, monkeys, crows, and the condor of the Andes, have been the subjects in both hemispheres.

---

[*] Simia ursina.

[†] Simia belzebuth.

[‡] Ulloa has not hesitated to represent in an engraving this extraordinary feat of the monkeys with a prehensile tail.—See Viage a la America Meridional, Madrid 1748.

The araguatos are accused of sometimes abandoning their young, that they may be lighter for flight when pursued by the Indian hunters. It is said that mothers have been seen removing their young from their shoulders, and throwing them down to the foot of the tree. I am inclined to believe that a movement merely accidental has been mistaken for one premeditated. The Indians have a dislike and a predilection for certain races of monkeys; they love the viuditas, the titis, and generally all the little sagoins; while the araguatos, on account of their mournful aspect, and their uniform howling, are at once detested and abused. In reflecting on the causes that may facilitate the propagation of sound in the air during the night, I thought it important to determine with precision the distance at which, especially in damp and stormy weather, the howling of a band of araguatos is heard. I believe I obtained proof of its being distinguished at eight hundred toises distance. The monkeys which are furnished with four hands cannot make excursions in the Llanos; and it is easy, amidst vast plains covered with grass, to recognize a solitary group of trees, whence the noise proceeds, and which is inhabited by howling monkeys. Now, by approaching or withdrawing from this group of trees, the maximum of the distance may be measured, at which the howling is heard. These distances appeared to me sometimes one-third greater during the night, especially when the weather was cloudy, very hot, and humid.

The Indians pretend that when the araguatos fill the forests with their howling, there is always one that chaunts as leader of the chorus. The observation is pretty accurate. During a long interval one solitary and strong voice is generally distinguished, till its place is taken by another voice of a different pitch. We may observe from time to time the same instinct of imitation among frogs, and almost all animals which live together and exert their voices in union. The Missionaries further assert, that, when a female among the araguatos is on the point of bringing forth, the choir suspends its howlings till the moment of the birth of the young. I could not myself judge of the accuracy of this assertion; but I do not believe it to be entirely unfounded. I have observed that, when an extraordinary incident, the moans for instance of a wounded araguato, fixed the attention of the band, the howlings were for some minutes suspended. Our guides assured us gravely, that, to cure an asthma, it is sufficient to drink out of the bony drum of the hyoidal bone of the araguato. This animal having so extraordinary a volume of voice, it is supposed that its larynx must necessarily impart to the water poured into it the virtue of curing affections of the lungs. Such is the science of the vulgar, which sometimes resembles that of the ancients.

We passed the night at the village of Guigue, the latitude of which I found by observations of Canopus to be 10° 4′ 11″. The village, surrounded with the richest cultivation, is only a thousand toises distant from the lake of Tacarigua. We lodged with an old sergeant, a native of Murcia, a man of a very original character. To prove to us that he had studied among the Jesuits, he recited the history of the creation of the world in Latin. He knew the names of Augustus, Tiberius, and Diocletian; and while enjoying the agreeable coolness of the nights in an enclosure planted with bananas, he employed himself in reading all that related to the courts of the Roman emperors. He inquired of us with earnestness for a remedy for the gout, from which he suffered severely. "I know," said he, "a Zambo of Valencia, a famous curioso, who could cure me; but the Zambo would expect to be treated with attentions which I cannot pay to a man of his colour, and I prefer remaining as I am."

On leaving Guigue we began to ascend the chain of mountains, extending on the south of the lake towards Guacimo and La Palma. From the top of a table-land, at three hundred and twenty toises of elevation, we saw for the last time the valleys of Aragua. The gneiss appeared uncovered, presenting the same direction of strata, and the same dip towards the north-west. Veins of quartz, that traverse the gneiss, are auriferous; and hence the neighbouring ravine bears the name of Quebrada del

Oro. We heard with surprise at every step the name of "ravine of gold," in a country where only one single mine of copper is wrought. We travelled five leagues to the village of Maria Magdalena, and two leagues more to the Villa de Cura. It was Sunday, and at the village of Maria Magdalena the inhabitants were assembled before the church. They wanted to force our muleteers to stop and hear mass. We resolved to remain; but, after a long altercation, the muleteers pursued their way. I may observe, that this is the only dispute in which we became engaged from such a cause. Very erroneous ideas are formed in Europe of the intolerance, and even of the religious fervour of the Spanish colonists.

San Luis de Cura, or, as it is commonly called, the Villa de Cura, lies in a very barren valley, running north-west and south-east, and elevated, according to my barometrical observations, two hundred and sixty-six toises above the level of the ocean. The country, with the exception of some fruit-trees, is almost destitute of vegetation. The dryness of the plateau is the greater, because (and this circumstance is rather extraordinary in a country of primitive rocks) several rivers lose themselves in crevices in the ground. The Rio de Las Minas, north of the Villa de Cura, is lost in a rock, again appears, and then is ingulphed anew without reaching the lake of Valencia, towards which it flows. Cura resembles a village more than a town. We lodged with a family who had excited the resentment of government during the revolution at Caracas in 1797. One of the sons, after having languished in a dungeon, had been sent to the Havannah, to be imprisoned in a strong fortress. With what joy his mother heard that after our return from the Orinoco, we should visit the Havannah! She entrusted me with five piastres, "the whole fruit of her savings." I earnestly wished to return them to her; but I feared to wound her delicacy, and give pain to a mother, who felt a pleasure in the privations she imposed on herself.

All the society of the town was assembled in the evening, to admire in a magic lantern views of the great capitals of Europe.

We were shown the palace of the Tuileries, and the statue of the Elector at Berlin.

An apothecary who had been ruined by an unhappy propensity for working mines, accompanied us in our excursion to the Serro de Chacao, very rich in auriferous pyrites. We continued to descend the southern declivity of the Cordillera of the coast, in which the plains of Aragua form a longitudinal valley. We passed a part of the night of the 11th of March at the village of San Juan, remarkable for its thermal waters, and the singular form of two neighbouring mountains, called the Morros of San Juan. They form slender peaks, which rise from a wall of rocks with a very extensive base. The wall is perpendicular, and resembles the Devil's Wall, which surrounds a part of the group of mountains in the Hartz.[*] These peaks, when seen from afar in the Llanos, strike the imagination of the inhabitants of the plain, who are not accustomed to the least unequal ground, and the height of the peaks is singularly exaggerated by them. They were described to us as being in the middle of the steppes (which they in reality bound on the north) far beyond a range of hills called La Galera. Judging from angles taken at the distance of two miles, these hills are scarcely more than a hundred and fifty-six toises higher than the village of San Juan, and three hundred and fifty toises above the level of the Llanos. The thermal waters glide out at the foot of these hills, which are formed of transition-limestone. The waters are impregnated with sulphuretted hydrogen, like those of Mariara, and form a little pool or lagoon, in which the thermometer rose only to 31.3°. I found, on the night of the 9th of March, by very satisfactory observations of the stars, the latitude of Villa de Cura to be 10° 2' 47".

The Villa de Cura is celebrated in the country for the miracles of an image of the Virgin, known by the name of Nuestra Senora de los Valencianos. This image was found in a

---

[*] Die Teufels Mauer near Wernigerode in Germany.

ravine by an Indian, about the middle of the eighteenth century, when it became the object of a contest between the towns of Cura and San Sebastian de los Reyes. The vicars of the latter town asserting that the Virgin had made her first appearance on the territory of their parish, the Bishop of Caracas, in order to put an end to the scandal of this long dispute, caused the image to be placed in the archives of his bishopric, and kept it thirty years under seal. It was not restored to the inhabitants of Cura till 1802.

After having bathed in the cool and limpid water of the little river of San Juan, the bottom of which is of basaltic grunstein, we continued our journey at two in the morning, by Ortiz and Parapara, to the Mesa de Paja. The road to the Llanos being at that time infested with robbers, several travellers joined us so as to form a sort of caravan. We proceeded down hill during six or seven hours; and we skirted the Cerro de Flores, near which the road turns off, leading to the great village of San Jose de Tisnao. We passed the farms of Luque and Juncalito, to enter the valleys which, on account of the bad road, and the blue colour of the slates, bear the names of Malpaso and Piedras Azules.

This ground is the ancient shore of the great basin of the steppes, and it furnishes an interesting subject of research to the geologist. We there find trap-formations, probably more recent than the veins of diabasis near the town of Caracas, which seem to belong to the rocks of igneous formation. They are not long and narrow streams as in Auvergne, but large sheets, streams that appear like real strata. The lithoid masses here cover, if we may use the expression, the shore of the ancient interior sea; everything subject to destruction, such as the liquid dejections, and the scoriae filled with bubbles, has been carried away. These phenomena are particularly worthy of attention on account of the close affinities observed between the phonolites and the amygdaloids, which, containing pyroxenes and hornblende-grunsteins, form strata in a transition-slate. The better to convey an idea of the whole situation and superposition of these rocks, we will name the formations as they occur in a profile drawn from north to south.

We find at first, in the Sierra de Mariara, which belongs to the northern branch of the Cordillera of the coast, a coarse-grained granite; then, in the valleys of Aragua, on the borders of the lake, and in the islands, it contains, as in the southern branch of the chain of the coast, gneiss and mica-slate. These last-named rocks are auriferous in the Quebrada del Oro, near Guigue; and between Villa de Cura and the Morros de San Juan, in the mountain of Chacao. The gold is contained in pyrites, which are found sometimes disseminated almost imperceptibly in the whole mass of the gneiss,[*] and sometimes united in small veins of quartz. Most of the torrents that traverse the mountains bear along with them grains of gold. The poor inhabitants of Villa de Cura and San Juan have sometimes gained thirty piastres a-day by washing the sand; but most commonly, in spite of their industry, they do not in a week find particles of gold of the value of two piastres. Here, however, as in every place where native gold and auriferous pyrites are disseminated in the rock, or by the destruction of the rocks, are deposited in alluvial lands, the people conceive the most exaggerated ideas of the metallic riches of the soil. But the success of the workings, which depends less on the abundance of the ore in a vast space of land than on its accumulation in one point, has not justified these favourable prepossessions. The mountain of Chacao, bordered by the ravine of Tucutunemo, rises seven hundred feet above the village of San Juan. It is formed of gneiss, which, especially in the superior strata, passes into mica-slate. We saw the remains of an ancient mine, known by the name of Real de Santa Barbara. The works were directed to a stratum of cellular quartz,[†] full of

---

[*] The four metals, which are found disseminated in the granite rocks, as if they were of contemporaneous formation, are gold, tin, titanium, and cobalt.

[†] This stratum of quartz, and the gneiss in which it is contained, lie hor 8 of the Freyberg compass, and dip 70° to the south-west. At a hundred toises distance from the auriferous quartz, the gneiss resumes its ordinary situation,

81

polyhedric cavities, mixed with iron-ore, containing auriferous pyrites and small grains of gold, sometimes, it is said, visible to the naked eye. It appears that the gneiss of the Cerro de Chacao also furnishes another metallic deposit, a mixture of copper and silver-ores. This deposit has been the object of works attempted with great ignorance by some Mexican miners under the superintendance of M. Avalo. The gallery* directed to the north-east, is only twenty-five toises long. We there found some fine specimens of blue carbonated copper mingled with sulphate of barytes and quartz; but we could not ourselves judge whether the ore contained any argentiferous fahlerz, and whether it occurred in a stratum, or, as the apothecary who was our guide asserted, in real veins. This much is certain, that the attempt at working the mine cost more than twelve thousand piastres in two years. It would no doubt have been more prudent to have resumed the works on the auriferous stratum of the Real de Santa Barbara.

The zone of gneiss just mentioned is, in the coast-chain from the sea to the Villa de Cura, ten leagues broad. In this great extent of land, gneiss and mica-slate are found exclusively, and they constitute one formation.† Beyond the town of Villa de

---

hor 3 to 4, with 60° dip to the north-west. A few strata of gneiss abound in silvery mica, and contain, instead of garnets, an immense quantity of small octohedrons of pyrites. This silvery gneiss resembles that of the famous mine of Himmelsfurst, in Saxony.

* La Cueva de los Mexicanos.

† This formation, which we shall call gneiss-mica-slate, is peculiar to the chain of the coast of Caracas. Five formations must be distinguished, as MM. von Buch and Raumer have so ably demonstrated in their excellent papers on Landeck and the Riesengebirge, namely, granite, granite-gneiss, gneiss, gneiss-mica-slate, and mica-slate. Geologists whose researches have been confined to a small tract of land, having confounded these formations which nature has separated in several countries in the most distinct manner, have admitted that the gneiss and mica-slate alternate everywhere in superimposed beds, or furnish insensible transitions from one rock to the other. These transitions and alternating superpositions take place no doubt in formations of granite-gneiss and gneiss-mica-slate; but because these phenomena are observed in one region, it does not follow that in other regions we may not

Cura and the Cerro de Chacao the aspect of the country presents greater geognostic variety. There are still eight leagues of declivity from the table-land of Cura to the entry of the Llanos; and on the southern slope of the mountains of the coast, four different formations of rock cover the gneiss. We shall first give the description of the different strata, without grouping them systematically.

On the south of the Cerro de Chacao, between the ravine of Tucutunemo and Piedras Negras, the gneiss is concealed beneath a formation of serpentine, of which the composition varies in the different superimposed strata. Sometimes it is very pure, very homogeneous, of a dusky olive-green, and of a conchoidal fracture: sometimes it is veined, mixed with bluish steatite, of an unequal fracture, and containing spangles of mica. In both these states I could not discover in it either garnets, hornblende, or diallage. Advancing farther to the south (and we always passed over this ground in that direction) the green of the serpentine grows deeper, and feldspar and hornblende are recognised in it: it is difficult to determine whether it passes into diabasis or alternates with it. There is, however, no doubt of its containing veins of copper-ore.[*] At the foot of this mountain two fine springs gush out from the serpentine. Near the village of San Juan, the granular diabasis appears alone uncovered, and takes a greenish black hue. The feldspar intimately mixed with the mass, may be separated into distinct crystals. The mica is very rare, and there is no quartz. The mass assumes at the surface a yellowish crust like dolerite and basalt.

---

find very distinct circumscribed formations of granite, gneiss, and mica-slate. The same considerations may be applied to the formations of serpentine, which are sometimes isolated, and sometimes belong to the eurite, mica-slate, and grunstein.

[*] One of these veins, on which two shafts have been sunk, was directed hor. 2.1, and dipped 80° east. The strata of the serpentine, where it is stratified with some regularity, run hor. 8, and dip almost perpendicularly. I found malachite disseminated in this serpentine, where it passes into grunstein.

In the midst of this tract of trap-formation, the Morros of San Juan rise like two castles in ruins. They appear linked to the mornes of St. Sebastian, and to La Galera which bounds the Llanos like a rocky wall. The Morros of San Juan are formed of limestone of a crystalline texture; sometimes very compact, sometimes spongy, of a greenish-grey, shining, composed of small grains, and mixed with scattered spangles of mica. This limestone yields a strong effervescence with acids. I could not find in it any vestige of organized bodies. It contains in subordinate strata, masses of hardened clay of a blackish blue, and carburetted. These masses are fissile, very heavy, and loaded with iron; their streak is whitish, and they produce no effervescence with acids. They assume at their surface, by their decomposition in the air, a yellow colour. We seem to recognize in these argillaceous strata a tendency either to the transition-slates, or to the kieselschiefer (schistose jasper), which everywhere characterise the black transition-limestones. When in fragments, they might be taken at first sight for basalt or hornblende.[*] Another white limestone, compact, and containing some fragments of shells, backs the Morros de San Juan. I could not see the line of junction of these two limestones, or that of the calcareous formation and the diabasis.

The transverse valley which descends from Piedras Negras and the village of San Juan, towards Parapara and the Llanos, is filled with trap-rocks, displaying close affinity with the formation of green slates, which they cover. Sometimes we seem to see serpentine, sometimes grunstein, and sometimes dolerite and basalt. The arrangement of these problematical masses is not less extraordinary. Between San Juan, Malpaso, and Piedras Azules, they form strata parallel to each other; and dipping

---

[*] I had an opportunity of examining again, with the greatest care, the rocks of San Juan, of Chacao, of Parapara, and of Calabozo, during my stay at Mexico, where, conjointly with M. del Rio, one of the most distinguished pupils of the school of Freyberg, I formed a geognostical collection for the Colegio de Mineria of New Spain.

regularly northward at an angle of 40 or 50°, they cover even the green slates in concordant stratification. Lower down, towards Parapara and Ortiz, where the amygdaloids and phonolites are connected with the grunstein, everything assumes a basaltic aspect. Balls of grunstein heaped one upon another, form those rounded cones, which are found so frequently in the Mittelgebirge in Bohemia, near Bilin, the country of phonolites. The following is the result of my partial observations.

The grunstein, which at first alternated with strata of serpentine, or was connected with that rock by insensible transitions, is seen alone, sometimes in strata considerably inclined, and sometimes in balls with concentric strata, imbedded in strata of the same substance. It lies, near Malpaso, on green slates, steatitic, mingled with hornblende, destitute of mica and grains of quartz, dipping, like the grunsteins, 45° toward the north, and directed, like them, 75° north-west.

A great sterility prevails where these green slates predominate, no doubt on account of the magnesia they contain, which (as is proved by the magnesian-limestone of England[*]) is very hurtful to vegetation. The dip of the green slates continues the same; but by degrees the direction of their strata becomes parallel to the general direction of the primitive rocks of the chain of the coast. At Piedras Azules these slates, mingled with hornblende, cover in concordant stratification a blackish-blue slate, very fissile, and traversed by small veins of quartz. The green slates include some strata of grunstein, and even contain balls of that substance. I nowhere saw the green slates alternate with the black slates of the ravine of Piedras Azules: at the line of junction these two slates appear rather to pass one into the other, the green slates becoming of a pearl-grey in proportion as they lose their hornblende.

---

[*] Magnesian limestone is of a straw-yellow colour, and contains madrepores: it lies beneath red marl, or muriatiferous red sandstone.

Farther south, towards Parapara and Ortiz, the slates disappear. They are concealed under a trap-formation more varied in its aspect. The soil becomes more fertile; the rocky masses alternate with strata of clay, which appear to be produced by the decomposition of the grunsteins, the amygdaloids, and the phonolites.

The grunstein, which farther north was less granulous, and passed into serpentine, here assumes a very different character. It contains balls of mandelstein, or amygdaloid, eight or ten inches in diameter. These balls, sometimes a little flattened, are divided into concentric layers: this is the effect of decomposition. Their nucleus is almost as hard as basalt, and they are intermingled with little cavities, owing to bubbles of gas, filled with green earth, and crystals of pyroxene and mesotype. Their basis is greyish blue, rather soft, and showing small white spots which, by the regular form they present, I should conceive to be decomposed feldspar. M. von Buch examined with a powerful lens the species we brought. He discovered that each crystal of pyroxene, enveloped in the earthy mass, is separated from it by fissures parallel to the sides of the crystal. These fissures seem to be the effect of a contraction which the mass or basis of the mandelstein has undergone. I sometimes saw these balls of mandelstein arranged in strata, and separated from each other by beds of grunstein of ten or fourteen inches thick; sometimes (and this situation is most common) the balls of mandelstein, two or three feet in diameter, are found in heaps, and form little mounts with rounded summits, like spheroidal basalt. The clay which separates these amygdaloid concretions arises from the decomposition of their crust. They acquire by the contact of the air a very thin coating of yellow ochre.

South-west of the village of Parapara rises the little Cerro de Flores, which is discerned from afar in the steppes. Almost at its foot, and in the midst of the mandelstein tract we have just been describing, a porphyritic phonolite, a mass of compact feldspar of a greenish grey, or mountain-green, containing long crystals of vitreous feldspar, appears exposed. It is the real porphyrschiefer of Werner; and it would be difficult to distinguish, in a collection of stones, the phonolite of Parapara from that of Bilin, in Bohemia. It does not, however, here form rocks in grotesque shapes, but little hills covered with tabular blocks, large plates extremely sonorous, translucid on the edges, and wounding the hands when broken.

Such are the successions of rocks, which I described on the spot as I progressively found them, from the lake of Tacarigua to the entrance of the steppes. Few places in Europe display a geological arrangement so well worthy of being studied. We saw there in succession six formations: namely, mica-slate-gneiss, green transition-slate, black transition-limestone, serpentine and grunstein, amygdaloid (with pyroxene), and phonolite.

I must observe, in the first place, that the substance just described under the name of grunstein, in every respect resembles that which forms layers in the mica-slate of Cabo Blanco, and veins near Caracas. It differs only by containing neither quartz, garnets, nor pyrites. The close relations we observed near the Cerro de Chacao, between the grunstein and the serpentine, cannot surprise these geologists who have studied the mountains of Franconia and Silesia. Near Zobtenberg[*] a serpentine rock alternates also with gabbro. In the district of Glatz the fissures of the gabbro are filled with a steatite of a greenish white colour, and the rock which was long thought to belong to the grunsteins[†] is a close mixture of feldspar and diallage.

---

[*] Between Tampadel and Silsterwiz.
[†] In the mountains of Bareuth, in Franconia, so abundant in grunstein and serpentine, these formations are not connected together. The serpentine there belongs rather to the schistose hornblende (hornblendschiefer), as in the island of Cuba. Near Guanaxuato, in Mexico, I saw it alternating with syenite. These phenomena of serpentine rocks forming layers in eurite (weisstein), in schistose hornblende, in gabbro, and in syenite, are so much the more remarkable, as the great mass of garnetiferous serpentines, which are found in

The grunsteins of Tucutunemo, which we consider as constituting the same formation as the serpentine rock, contain veins of malachite and copper-pyrites. These same metalliferous combinations are found also in Franconia, in the grunsteins of the mountains of Steben and Lichtenberg. With respect to the green slates of Malpaso, which have all the characters of transition-slates, they are identical with those which M. von Buch has so well described, near Schonau, in Silesia. They contain beds of grunstein, like the slates of the mountains of Steben just mentioned.* The black limestone of the Morros de San Juan is also a transition-limestone. It forms perhaps a subordinate stratum in the slates of Malpaso. This situation would be analogous to what is observed in several parts of Switzerland.† The slaty zone, the centre of which is the ravine of Piedras Azules, appears divided into two formations. On some points we think we observe one passing into the other.

The grunsteins, which begin again to the south of these slates, appear to me to differ little from those found north of the ravine of Piedras Azules. I did not see there any pyroxene; but on the very spot I recognized a number of crystals in the amygdaloid, which appears so strongly linked to the grunstein that they alternate several times.

The geologist may consider his task as fulfilled when he has traced with accuracy the positions of the diverse strata; and has pointed out the analogies traceable between these positions and what has been observed in other countries. But how can he avoid

---

the mountains of gneiss and mica-slate, form little distinct mounts, masses not covered by other formations. It is not the same in the mixtures of serpentine and granulated limestone.

* On advancing into the adit for draining the Friedrich-Wilhelmstollen mine, which I caused to be begun in 1794, near Steben, and which is yet only 340 toises long, there have successively been found, in the transition-slate subordinate strata of pure and porphyritic grunstein, strata, of Lydian stone and ampelite (alaunschiefer), and strata of fine-grained grunstein. All these strata characterise the transition-slates.

† For Instance, at the Glyshorn, at the Col de Balme, etc.

being tempted to go back to the origin of so many different substances, and to inquire how far the dominion of fire has extended in the mountains that bound the great basin of the steppes? In researches on the position of rocks we have generally to complain of not sufficiently perceiving the connection between the masses, which we believe to be superimposed on one another. Here the difficulty seems to arise from the too intimate and too numerous relations observed in rocks that are thought not to belong to the same family.

The phonolite (or leucostine compacte of Cordier) is pretty generally regarded by all who have at once examined burning and extinguished volcanoes, as a flow of lithoid lava. I found no real basalt or dolerite; but the presence of pyroxene in the amygdaloid of Parapara leaves little doubt of the igneous origin of those spheroidal masses, fissured, and full of cavities. Balls of this amygdaloid are enclosed in the grunstein; and this grunstein alternates on one side with a green slate, on the other with the serpentine of Tucutunemo. Here, then, is a connexion sufficiently close established between the phonolites and the green slates, between the pyroxenic amygdaloids and the serpentines containing copper-ores, between volcanic substances and others that are included under the vague name of transition-traps. All these masses are destitute of quartz like the real trap-porphyries, or volcanic trachytes. This phenomenon is the more remarkable, as the grunsteins which are called primitive almost always contain quartz in Europe. The most general dip of the slates of Piedras Azules, of the grunsteins of Parapara, and of the pyroxenic amygdaloids embedded in strata of grunstein, does not follow the slope of the ground from north to south, but is pretty regular towards the north. The strata incline towards the chain of the coast, as substances which had not been in fusion might be supposed to do. Can we admit that so many alternating rocks, imbedded one in the other, have a common origin? The nature of the phonolites, which are lithoid lavas with a feldspar basis, and the nature of the green slates intermixed with hornblende, oppose this opinion. In this state of things we may choose between two solutions of the problem in question. In one of these solutions the phonolite of the Cerro de Flores is to be regarded as the sole volcanic production of the tract; and we are forced to unite the pyroxenic amygdaloids with the rest of the grunsteins, in one single formation, that which is so common in the transition-mountains of Europe, considered hitherto as not volcanic. In the other solution of the problem, the masses of phonolite, amygdaloid, and grunstein, which are found in the south of the ravine of Piedras Azules, are separated from the grunsteins and serpentine rocks that cover the declivity of the mountains north of the ravine. In the present state of knowledge I find difficulties almost equally great in adopting either of these suppositions; but I have no doubt that, when the real grunsteins (not the hornblende-grunsteins) contained in the gneiss and mica-slates, shall have been more attentively examined in other places; when the basalts (with pyroxene) forming strata in primitive rocks[*] and the diabases and amygdaloids in the transition mountains, shall have been carefully studied; when the texture of the masses shall have been subjected to a kind of mechanical analysis, and the hornblendes better distinguished from the pyroxenes,[†] and the grunsteins from the dolerites; a great number of phenomena which now appear isolated and obscure, will be ranged under general laws. The phonolite and other rocks of igneous origin at Parapara are so much the more interesting, as they indicate ancient eruptions in a granite zone; as they belong to the shore of the basin of the steppes, as the basalts of Harutsh belong to the shore of the desert of Sahara; and lastly, as they are the only rocks of the kind we observed in the mountains of the Capitania-General of Caracas, which are

---

[*] For instance, at Krobsdorf, in Silesia, a stratum of basalt has been recognized in the mica-slate by two celebrated geologists, MM. von Buch and Raumer. (Vom Granite des Riesengebirges, 1813.)

[†] The grunsteins or diabases of the Fichtelgebirge, in Franconia, which belong to the transition-slate, sometimes contain pyroxenes.

also destitute of trachytes or trap-porphyry, basalts, and volcanic productions.[*]

The southern declivity of the western chain is tolerably steep; the steppes, according to my barometrical measurements, being a thousand feet lower than the bottom of the basin of Aragua. From the extensive table-land of the Villa de Cura we descended towards the banks of the Rio Tucutunemo, which has hollowed for itself, in a serpentine rock, a longitudinal valley running from east to west, at nearly the same level as La Victoria. A transverse valley, lying generally north and south, led us into the Llanos, by the villages of Parapara and Ortiz. It grows very narrow in several parts. Basins, the bottoms of which are perfectly horizontal, communicate together by narrow passes with steep declivities. They were, no doubt, formerly small lakes, which, owing to the accumulation of the waters, or some more violent catastrophe, have broken down the dykes by which they were separated. This phenomenon is found in both continents, wherever we examine the longitudinal valleys forming the passages of the Andes, the Alps,[†] or the Pyrenees. It is probable, that the irruption of the waters towards the Llanos have given, by extraordinary rents, the form of ruins to the Morros of San Juan and of San Sebastian. The volcanic tract of Parapara and Ortis is now only 30 or 40 toises above the Llanos. The eruptions consequently took place at the lowest point of the granitic chain.

In the Mesa de Paja, in the ninth degree of latitude, we entered the basin of the Llanos. The sun was almost at its zenith; the earth, wherever it appeared sterile and destitute of vegetation, was at the temperature of 48 or 50°.[‡] Not a breath of air was felt at the height at which we were on our mules; yet, in

---

[*] From the Rio Negro to the coasts of Cumana and Caracas, to the east of the mountains of Merida, which we did not visit.

[†] For example, the road from the valley of Ursern to the Hospice of St. Gothard, and thence to Airolo.

[‡] A thermometer, placed in the sand, rose to 38.4 and 40° Reaumur.

the midst of this apparent calm, whirls of dust incessantly arose, driven on by those small currents of air which glide only over the surface of the ground, and are occasioned by the difference of temperature between the naked sand and the spots covered with grass. These sand-winds augment the suffocating heat of the air. Every grain of quartz, hotter than the surrounding air, radiates heat in every direction; and it is difficult to observe the temperature of the atmosphere, owing to these particles of sand striking against the bulb of the thermometer. All around us the plains seemed to ascend to the sky, and the vast and profound solitude appeared like an ocean covered with sea-weed. According to the unequal mass of vapours diffused through the atmosphere, and the variable decrement in the temperature of the different strata of air, the horizon in some parts was clear and distinct; in other parts it appeared undulating, sinuous, and as if striped. The earth there was confounded with the sky. Through the dry mist and strata of vapour the trunks of palm-trees were seen from afar, stripped of their foliage and their verdant summits, and looking like the masts of a ship descried upon the horizon.

There is something awful, as well as sad and gloomy, in the uniform aspect of these steppes. Everything seems motionless; scarcely does a small cloud, passing across the zenith, and denoting the approach of the rainy season, cast its shadow on the earth. I know not whether the first aspect of the Llanos excite less astonishment than that of the chain of the Andes. Mountainous countries, whatever may be the absolute elevation of the highest summits, have an analogous physiognomy; but we accustom ourselves with difficulty to the view of the Llanos of Venezuela and Casanare, to that of the Pampas of Buenos Ayres and of Chaco, which recal to mind incessantly, and during journeys of twenty or thirty days, the smooth surface of the ocean. I had seen the plains or llanos of La Mancha in Spain, and the heaths (ericeta) that extend from the extremity of Jutland, through Luneburg and Westphalia, to Belgium. These last are really steppes, and, during several ages, only small portions of them have yielded to cultivation; but the plains of the west and north of Europe present only a feeble image of the immense llanos of South America. It is in the south-east of our continent, in Hungary, between the Danube and the Theiss; in Russia, between the Borysthenes, the Don, and the Volga, that we find those vast pastures, which seem to have been levelled by a long abode of the waters, and which meet the horizon on every side. The plains of Hungary, where I traversed them on the frontiers of Germany, between Presburg and Oedenburg, strike the imagination of the traveller by the constant mirage; but their greatest extent is more to the east, between Czegled, Debreczin, and Tittel. There they present the appearance of a vast ocean of verdure, having only two outlets, one near Gran and Waitzen, the other between Belgrade and Widdin.

The different quarters of the world have been supposed to be characterized by the remark, that Europe has its heaths, Asia its steppes, Africa its deserts, and America its savannahs; but by this distinction, contrasts are established that are not founded either on the nature of things, or the genius of languages. The existence of a heath always supposes an association of plants of the family of ericae; the steppes of Asia are not everywhere covered with saline plants; the savannahs of Venezuela furnish not only the gramina, but with them small herbaceous mimosas, legumina, and other dicotyledonous plants. The plains of Songaria, those which extend between the Don and the Volga, and the puszta of Hungary, are real savannahs, pasturages abounding in grasses;[*] while the savannahs to the east and west

---

[*] These vast steppes of Hungary are elevated only thirty or forty toises above the level of the sea, which is more than eighty leagues distant from them. See Wahlenberg's Flora Carpathianica. Baron Podmanitzky, an Hungarian nobleman, highly distinguished for his knowledge of the physical sciences, caused the level of these plains to be taken, to facilitate the formation of a canal then projected between the Danube and the Theiss. He found the line of division, or the convexity of the ground, which slopes on each side towards the beds of the two rivers, to be only thirteen toises above the height of the Danube. The widely extended pastures, which reach in every

of the Rocky Mountains and of New Mexico produce chenopodiums containing carbonate and muriate of soda. Asia has real deserts destitute of vegetation, in Arabia, in Gobi, and in Persia. Since we have become better acquainted with the deserts in the interior of Africa, so long and so vaguely confounded together under the name of desert of Sahara (Zahra); it has been observed, that in this continent, towards the east, savannahs and pastures are found, as in Arabia, situated in the midst of naked and barren tracts. It is these deserts, covered with gravel and destitute of plants, which are almost entirely wanting in the New World. I saw them only in that part of Peru, between Amotape and Coquimbo, on the shores of the Pacific. These are called by the Spaniards, not llanos, but the desiertos of Sechura and Atacamez. This solitary tract is not broad, but it is four hundred and forty leagues long. The rock pierces everywhere through the quicksands. No drop of rain ever falls on it; and, like the desert of Sahara, north of Timbuctoo, the Peruvian desert affords, near Huaura, a rich mine of native salt. Everywhere else, in the New World, there are plains desert because not inhabited, but no real deserts.[*]

The same phenomena are repeated in the most distant regions; and, instead of designating those vast treeless plains in accordance with the nature of the plants they produce, it seems

---

direction to the horizon, are called in the country, Puszta, and, over a distance of many leagues, are without any human habitation. Plains of this kind, intermingled with marshes and sandy tracts, are found on the western side of the Theiss, between Czegled, Csaba, Komloss, and Szarwass; and on the eastern side, between Debreczin, Karczag, and Szoboszlo. The area of these plains of the interior basin of Hungary has been estimated, by a pretty accurate calculation, to be between two thousand five hundred and three thousand square leagues (twenty to a degree). Between Czegled, Szolnok, and Ketskemet, the plain resembles a sea of sand.

[*] We are almost tempted, however, to give the name of desert to that vast and sandy table-land of Brazil, the Campos dos Parecis, which gives birth to the rivers Tapajos, Paraguay, and Madeira, and which reaches the summit of the highest mountains. Almost destitute of vegetation, it reminds us of Gobi, in Mongolia.

natural to class them into deserts, steppes, or savannahs; into bare lands without any appearance of vegetation, and lands covered with gramina or small plants of the dicotyledonous tribe. The savannahs of America, especially those of the temperate zone, have in many works been designated by the French term prairies; but this appears to me little applicable to pastures which are often very dry, though covered with grass of four or five feet in height. The Llanos and the Pampas of South America are really steppes. They are covered with beautiful verdure in the rainy season, but in the time of great drought they assume the aspect of a desert. The grass is then reduced to powder; the earth cracks; the alligators and the great serpents remain buried in the dried mud, till awakened from their long lethargy by the first showers of spring. These phenomena are observed on barren tracts of fifty or sixty leagues in length, wherever the savannahs are not traversed by rivers; for on the borders of rivulets, and around little pools of stagnant water, the traveller finds at certain distances, even during the period of the great droughts, thickets of mauritia, a palm, the leaves of which spread out like a fan, and preserve a brilliant verdure.

The steppes of Asia are all beyond the tropics, and form very elevated table-lands. America also has savannahs of considerable extent on the backs of the mountains of Mexico, Peru, and Quito; but its most extensive steppes, the Llanos of Cumana, Caracas, and Meta, are little raised above the level of the ocean, and all belong to the equinoctial zone. These circumstances give them a peculiar character. They have not, like the steppes of southern Asia, and the deserts of Persia, those lakes without issue, those small systems of rivers which lose themselves either in the sands, or by subterranean filtrations. The Llanos of America incline to the east and south; and their running waters are branches of the Orinoco.

The course of these rivers once led me to believe, that the plains formed table-lands, raised at least from one hundred to one hundred and fifty toises above the level of the ocean. I supposed that the deserts of interior Africa were also at a considerable height; and that they rose one above another as in tiers, from the coast to the interior of the continent. No barometer has yet been carried into the Sahara. With respect to the Llanos of America, I found by barometric heights observed at Calabozo, at the Villa del Pao, and at the mouth of the Meta, that their height is only forty or fifty toises above the level of the sea. The fall of the rivers is extremely gentle, often nearly imperceptible; and therefore the least wind, or the swelling of the Orinoco, causes a reflux in those rivers that flow into it. The Indians believe themselves to be descending during a whole day, when navigating from the mouths of these rivers to their sources. The descending waters are separated from those that flow back by a great body of stagnant water, in which, the equilibrium being disturbed, whirlpools are formed very dangerous for boats.

The chief characteristic of the savannahs or steppes of South America is the absolute want of hills and inequalities—the perfect level of every part of the soil. Accordingly the Spanish conquerors, who first penetrated from Coro to the banks of the Apure, did not call them deserts or savannahs, or meadows, but plains (llanos). Often within a distance of thirty square leagues there is not an eminence of a foot high. This resemblance to the surface of the sea strikes the imagination most powerfully where the plains are altogether destitute of palm-trees; and where the mountains of the shore and of the Orinoco are so distant that they cannot be seen, as in the Mesa de Pavones. A person would be tempted there to take the altitude of the sun with a quadrant, if the horizon of the land were not constantly misty on account of the variable effects of refraction. This equality of surface is still more perfect in the meridian of Calabozo, than towards the east, between Cari, La Villa del Pao, and Nueva Barcelona; but it extends without interruption from the mouths of the Orinoco to La Villa de Araure and to Ospinos, on a parallel of a hundred and eighty leagues in length; and from San Carlos to the savannahs of Caqueat, on a meridian of two hundred leagues. It particularly characterises the New Continent, as it does the low steppes of Asia, between the Borysthenes and the Volga, between the Irtish and the Obi. The deserts of central Africa, of Arabia, Syria, and Persia, Gobi, and Casna, present, on the contrary, many inequalities, ranges of hills, ravines without water, and rocks which pierce the sands.

The Llanos, however, notwithstanding the apparent uniformity of their surface, present two kinds of inequalities, which cannot escape the observation of the traveller. The first is known by the name of banks (bancos); they are in reality shoals in the basin of the steppes, fractured strata of sandstone, or compact limestone, standing four or five feet higher than the rest of the plain. These banks are sometimes three or four leagues in length; they are entirely smooth, with a horizontal surface; their existence is perceived only by examining their margins. The second species of inequality can be recognised only by geodesical or barometric levellings, or by the course of rivers. It is called a mesa or table, and is composed of small flats, or rather convex eminences, that rise insensibly to the height of a few toises. Such are, towards the east, in the province of Cumana, on the north of the Villa de la Merced and Candelaria, the Mesas of Amana, of Guanipa, and of Jonoro, the direction of which is south-west and north-east; and which, in spite of their inconsiderable elevation, divide the waters between the Orinoco and the northern coast of Terra Firma. The convexity of the savannah alone occasions this partition: we there find the dividing of the waters (divortia aquarum[*], as in Poland, where, far from the Carpathian mountains, the plain itself divides the waters between the Baltic and the Black Sea. Geographers, who suppose the existence of a chain of mountains wherever there is a line of division, have not failed to mark one in the maps, at the sources of the Rio Neveri, the Unare, the Guarapiche, and the Pao. Thus the priests of Mongol race, according to ancient and

---

[*] "C. Manlium prope jugis [Tauri] ad divortia aquarum castra posuisse." Livy lib. 38 c. 75.)

superstitious custom, erect oboes, or little mounds of stone, on every point where the rivers flow in an opposite direction.

The uniform landscape of the Llanos; the extremely small number of their inhabitants; the fatigue of travelling beneath a burning sky, and an atmosphere darkened by dust; the view of that horizon, which seems for ever to fly before us; those lonely trunks of palm-trees, which have all the same aspect, and which we despair of reaching, because they are confounded with other trunks that rise by degrees on the visual horizon; all these causes combine to make the steppes appear far more extensive than they are in reality. The planters who inhabit the southern declivity of the chain of the coast see the steppes extend towards the south, as far as the eye can reach, like an ocean of verdure. They know that from the Delta of the Orinoco to the province of Varinas, and thence, by traversing the banks of the Meta, the Guaviare, and the Caguan, they can advance three hundred and eighty leagues* into the plains, first from east to west, and then from north-east to south-east beyond the Equator, to the foot of the Andes of Pasto. They know by the accounts of travellers the Pampas of Buenos Ayres, which are also Llanos covered with fine grass, destitute of trees, and filled with oxen and horses become wild. They suppose that, according to the greater part of our maps of America, this continent has only one chain of mountains, that of the Andes, which stretches from south to north; and they form a vague idea of the contiguity of all the plains from the Orinoco and the Apure to the Rio de la Plata and the Straits of Magellan.

Without stopping here to give a mineralogical description of the transverse chains which divide America from east to west, it will be sufficient to notice the general structure of a continent, the extremities of which, though situated in climates little analogous, nevertheless present several features of resemblance. In order to have an exact idea of the plains, their configuration,

---

* This is the distance from Timbuctoo to the northern coast of Africa.

and their limits, we must know the chains of mountains that form their boundaries. We have already described the Cordillera of the coast, of which the highest summit is the Silla de Caraccas, and which is linked by the Paramo de las Rosas to the Nevada de Merida, and the Andes of New Grenada. We have seen that, in the tenth degree of north latitude, it stretches from Quibor and Barquesimeto as far as the point of Paria. A second chain of mountains, or rather a less elevated but much larger group, extends between the parallels of 3 and 7° from the mouths of the Guaviare and the Meta to the sources of the Orinoco, the Marony, and the Essequibo, towards French and Dutch Guiana. I call this chain the Cordillera of Parime, or of the great cataracts of the Orinoco. It may be followed for a length of two hundred and fifty leagues; but it is less a chain, than a collection of granitic mountains, separated by small plains, without being everywhere disposed in lines. The group of the mountains of Parime narrows considerably between the sources of the Orinoco and the mountains of Demerara, in the Sierras of Quimiropaca and Pacaraimo, which divide the waters between the Carony and the Rio Parime, or Rio de Aguas Blancas. This is the scene of the expeditions which were undertaken in search of El Dorado, and the great city of Manoa, the Timbuctoo of the New Continent. The Cordillera of Parime does not join the Andes of New Grenada, but is separated from them by a space eighty leagues broad. If we suppose it to have been destroyed in this space by some great revolution of the globe (which is scarcely probable) we must admit that it anciently branched off from the Andes between Santa Fe de Bogota and Pamplona. This remark serves to fix more easily in the memory of the reader the geographical position of a Cordillera till now very imperfectly known. A third chain of mountains unites in 16 and 18° south latitude (by Santa Cruz de la Sierra, the Serranias of Aguapehy, and the famous Campos dos Parecis) the Andes of Peru, to the mountains of Brazil. It is the Cordillera of Chiquitos which widens in the Capitania de Minas Geraes, and divides the rivers flowing into the Amazon from those of the Rio de la Plata,[*] not only in the interior of the country, in the meridian of Villa Boa, but also at a few leagues from the coast, between Rio Janeiro and Bahia.[†]

These three transverse chains, or rather these three groups of mountains stretching from west to east, within the limits of the torrid zone, are separated by tracts entirely level, the plains of Caracas, or of the Lower Orinoco; the plains of the Amazon and the Rio Negro; and the plains of Buenos Ayres, or of La Plata. I use the term plains, because the Lower Orinoco and the Amazon, far from flowing in a valley, form but a little furrow in the midst of a vast level. The two basins, placed at the extremities of South America, are savannahs or steppes, pasturage without trees; the intermediate basin, which receives the equatorial rains during the whole year, is almost entirely one vast forest, through which no other roads are known save the rivers. The strong vegetation which conceals the soil, renders also the uniformity of its level less perceptible; and the plains of Caracas and La Plata bear no other name. The three basins we have just described are called, in the language of the colonists, the Llanos of Varinas and of Caracas, the bosques or selvas (forests) of the Amazon, and the Pampas of Buenos Ayres. The trees not only for the most part cover the plains of the Amazon, from the Cordillera de Chiquitos, as far as that of Parime; they also crown these two chains of mountains, which rarely attain the height of the Pyrenees.[‡] On this account, the vast plains of

---

[*] There is only a portage or carrying-place of 5322 bracas between the Guapore (a branch of the Marmore and of the Madeira), and the Rio Aguapehy (a branch of the Jaura and of the Paraguay).

[†] The Cordillera of Chiquitos and of Brazil stretches toward the south-east, in the government of the Rio Grande, beyond the latitude of 30° south.

[‡] We must except the most western part of the Cordillera of Chiquitos, between Cochabamba and Santa Cruz de la Sierra where the summits are covered with snow; but this colossal group almost belongs to the Andes de la Paz, of which it forms a promontory or spur, directed toward the east.

the Amazon, the Madeira, and the Rio Negro, are not so distinctly bounded as the Llanos of Caracas, and the Pampas of Buenos Ayres. As the region of forests comprises at once the plains and the mountains, it extends from 18° south to 7 and 8° north,[*] and occupies an extent of near a hundred and twenty thousand square leagues. This forest of South America, for in fact there is only one, is six times larger than France. It is known to Europeans only on the shores of a few rivers, by which it is traversed; and has its openings, the extent of which is in proportion to that of the forests. We shall soon skirt the marshy savannahs, between the Upper Orinoco, the Conorichite, and the Cassiquiare, in the latitude of 3 and 4°. There are other openings, or as they are called, clear savannahs,[†] in the same parallel, between the sources of the Mao and the Rio de Aguas Blancas, south of the Sierra de Pacaraima. These last savannahs, which are inhabited by Caribs, and nomad Macusis, lie near the frontiers of Dutch and French Guiana.

Having noticed the geological constitution of South America, we shall now mark its principal features. The western coasts are bordered by an enormous wall of mountains, rich in precious metals wherever volcanic fire has not pierced through the eternal snow. This is the Cordillera of the Andes. Summits of trap-porphyry rise beyond three thousand three hundred toises, and the mean height of the chain[‡] is one thousand eight hundred and fifty toises. It stretches in the direction of a meridian, and sends into each hemisphere a lateral branch, in the latitudes of 10° north, and 16 and 18° south. The first of these two branches, that of the coast of Caracas, is of considerable length, and forms

---

[*] To the west, in consequence of the Llanos of Manso, and the Pampas de Huanacos, the forests do not extend generally beyond the parallels of 18 or 19° south latitude; but to the east, in Brazil (in the capitanias of San Pablo and Rio Grande) as well as in Paraguay, on the borders of the Parana, they advance as far as 25° south.

[†] Savannas limpias, that is to say, clear of trees.

[‡] In New Grenada, Quito, and Peru, according to measurements taken by Bouguer, La Condamine, and myself.

in fact a chain. The second branch, the Cordillera of Chiquitos and of the sources of the Guapore, is very rich in gold, and widens toward the east, in Brazil, into vast tablelands, having a mild and temperate climate. Between these two transverse chains, contiguous to the Andes, an isolated group of granitic mountains is situated, from 3 to 7° north latitude; which also runs parallel to the Equator, but, not passing the meridian of 71°, terminates abruptly towards the west, and is not united to the Andes of New Grenada. These three transverse chains have no active volcanoes; we know not whether the most southern, like the two others, be destitute of trachytes or trap-porphyry. None of their summits enter the limit of perpetual snow; and the mean height of the Cordillera of La Parime, and of the littoral chain of Caracas, does not reach six hundred toises, though some of its summits rise fourteen hundred toises above the level of the sea.[*] The three transverse chains are separated by plains entirely closed towards the west, and open towards the east and south-east. When we reflect on their small elevation above the surface of the ocean, we are tempted to consider them as gulfs stretching in the direction of the current of rotation. If, from the effect of some peculiar attraction, the waters of the Atlantic were to rise fifty toises at the mouth of the Orinoco, and two hundred toises at the mouth of the Amazon, the flood would submerge more than the half of South America. The eastern declivity, or the foot of the Andes, now six hundred leagues distant from the coast of Brazil, would become a shore beaten by the waves. This consideration is the result of a barometric measurement, taken in the province of Jaen de Bracamoros, where the river Amazon issues from the Cordilleras. I found the mean height of this immense river only one hundred and ninety-four toises above the present level of the Atlantic. The intermediate plains, however, covered with forests, are still five times higher than the Pampas

---

[*] We do not reckon here, as belonging to the chain of the coast, the Nevados and Paramos of Merida and of Truxillo, which are a prolongation of the Andes of New Grenada.

of Buenos Ayres, and the grass-covered Llanos of Caracas and the Meta.

Those Llanos which form the basin of the Orinoco, and which we crossed twice in one year, in the months of March and July, communicate with the basin of the Amazon and the Rio Negro, bounded on one side by the Cordillera of Chiquitos, and on the other by the mountains of Parime. The opening which is left between the latter and the Andes of New Grenada, occasions this communication. The aspect of the country here reminds us, but on a much larger scale, of the plains of Lombardy, which also are only fifty or sixty toises above the level of the ocean; and are directed first from La Brenta to Turin, east and west; and then from Turin to Coni, north and south. If we were authorized, from other geological facts, to regard the three great plains of the Lower Orinoco, the Amazon, and the Rio de la Plata as basins of ancient lakes,[*] we should imagine we perceived in the plains of the Rio Vichada and the Meta, a channel by which the waters of the upper lake (those of the plains of the Amazon) forced their way towards the lower basin, (that of the Llanos of Caracas,) separating the Cordillera of La Parime from that of the Andes. This channel is a kind of land-strait. The ground, which is perfectly level between the Guaviare, the Meta, and the Apure, displays no vestige of a violent irruption of the waters; but on the edge of the Cordillera of Parime, between the latitudes of 4 and 7°, the Orinoco, flowing in a westerly direction from its source to the mouth of the Guaviare, has forced its way through the rocks, directing its course from south to north. All the great cataracts, as we shall soon see, are within the latitudes just named. When the river has reached the mouth of the Apure in that very low ground where the slope towards the north is met by the counter-slope towards the south-east, that is to say, by the

---

[*] In Siberia, the great steppes between the Irtish and the Obi, especially that of Baraba, full of salt lakes (Tchabakly, Tchany, Karasouk, and Topolony), appear to have been, according to the Chinese traditions, even within historical times, an inland sea.

inclination of the plains which rise imperceptibly towards the mountains of Caracas, the river turns anew and flows eastward. It appeared to me, that it was proper to fix the attention of the reader on these singular inflexions of the Orinoco because, belonging at once to two basins, its course marks, in some sort, even on the most imperfect maps, the direction of that part of the plains intervening between New Grenada and the western border of the mountains of La Parime.

The Llanos or steppes of the Lower Orinoco and of the Meta, like the deserts of Africa, bear different names in different parts. From the mouths of the Dragon the Llanos of Cumana, of Barcelona, and of Caracas or Venezuela,* follow, running from east to west. Where the steppes turn towards the south and south-south-west, from the latitude of 8°, between the meridians of 70 and 73°, we find from north to south, the Llanos of Varinas, Casanare, the Meta, Guaviare, Caguau, and Caqueta.† The plains of Varinas contain some few monuments of the industry of a nation that has disappeared. Between Mijagual and the Cano de la Hacha, we find some real tumuli, called in the country the Serillos de los Indios. They are hillocks in the shape of cones, artificially formed of earth, and probably contain bones, like the tumuli in the steppes of Asia. A fine road is also discovered near Hato de la Calzada, between Varinas and

---

* The following are subdivisions of these three great Llanos, as I marked them down on the spot. The Llanos of Cumana and New Andalusia include those of Maturin and Terecen, of Amana, Guanipa, Jonoro, and Cari. The Llanos of Nueva Barcelona comprise those of Aragua, Pariaguan, and Villa del Pao. We distinguish in the Llanos of Caracas those of Chaguaramas, Uritucu, Calabozo or Guarico, La Portuguesa, San Carlos, and Araure.

† The inhabitants of these plains distinguish as subdivisions, from the Rio Portuguesa to Caqueta, the Llanos of Guanare, Bocono, Nutrius or the Apure, Palmerito near Quintero, Guardalito and Arauca, the Meta, Apiay near the port of Pachaquiaro, Vichada, Guaviare, Arriari, Inirida, the Rio Hacha, and Caguan. The limits between the savannahs and the forests, in the plains that extend from the sources of the Rio Negro to Putumayo, are not sufficiently known.

Canagua, five leagues long, made before the conquest, in the most remote times, by the natives. It is a causeway of earth fifteen feet high, crossing a plain often overflowed. Did nations farther advanced in civilization descend from the mountains of Truxillo and Merido to the plains of the Rio Apure? The Indians whom we now find between this river and the Meta, are in too rude a state to think of making roads or raising tumuli.

I calculated the area of these Llanos from the Caqueta to the Apure, and from the Apure to the Delta of the Orinoco, and found it to be seventeen thousand square leagues twenty to a degree. The part running from north to south is almost double that which stretches from east to west, between the Lower Orinoco and the littoral chain of Caracas. The Pampas on the north and north-west of Buenos Ayres, between this city and Cordova, Jujuy, and the Tucuman, are of nearly the same extent as the Llanos; but the Pampas stretch still farther on to the length of 18° southward; and the land they occupy is so vast, that they produce palm-trees at one of their extremities, while the other, equally low and level, is covered with eternal frost.

The Llanos of America, where they extend in the direction of a parallel of the equator, are three-fourths narrower then the great desert of Africa. This circumstance is very important in a region where the winds constantly blow from east to west. The farther the plains stretch in this direction, the more ardent is their climate. The great ocean of sand in Africa communicates by Yemen[*] with Gedrosia and Beloochistan, as far as the right bank

---

[*] We cannot be surprised that the Arabic should be richer than any other language of the East in words expressing the ideas of desert, uninhabited plains, and plains covered with gramina. I could give a list of thirty-five of these words, which the Arabian authors employ without always distinguishing them by the shades of meaning which each separate word expresses. Makadh and kaah indicate, in preference, plains; bakaak, a table-land; kafr, mikfar, smlis, mahk, and habaucer, a naked desert, covered with sand and gravel; tanufah, a steppe. Zahra means at once a naked desert and a savannah. The word steppe, or step, is Russian, and not Tartarian. In the Turco-Tartar dialect a heath is called tala or tschol. The word gobi, which

of the Indus. It is from the effect of winds that have passed over the deserts situated to the east, that the little basin of the Red Sea, surrounded by plains which send forth from all sides radiant caloric, is one of the hottest regions of the globe. The unfortunate captain Tuckey relates,[*] that the centigrade thermometer keeps there generally in the night at 34°, and by day from 40 to 44°. We shall soon see that, even in the westernmost part of the steppes of Caracas, we seldom found the temperature of the air, in the shade, above 37°.

These physical considerations on the steppes of the New World are linked with others more interesting, inasmuch as they are connected with the history of our species. The great sea of sand in Africa, the deserts without water, are frequented only by caravans, that take fifty days to traverse them.[†] Separating the Negro race from the Moors, and the Berber and Kabyle tribes, the Sahara is inhabited only in the oases. It affords pasturage only in the eastern part, where, from the effect of the trade-winds, the layer of sand being less thick, the springs appear at the surface of the earth. In America, the steppes, less vast, less scorching, fertilized by fine rivers, present fewer obstacles to the intercourse of nations. The Llanos separate the chain of the coast of Caracas and the Andes of New Grenada from the region of forests; from that woody region of the Orinoco which, from the first discovery of America, has been inhabited by nations more rude, and farther removed from civilization, than the inhabitants of the coast, and still more than the mountaineers of the Cordilleras. The steppes, however, were no more heretofore the rampart of civilization than they are now the rampart of the liberty of the hordes that live in the forests. They have not

---

Europeans have corrupted into cobi, signifies in the Mongol tongue a naked desert. It is equivalent to the scha-mo or khan-hai of the Chinese. A steppe, or plain covered with herbs, is in Mongol, kudah; in Chinese, kouana.

[*] Expedition to explore the river Zahir, 1818.

[†] This is the maximum of the time, according to Major Rennell, Travels of Mungo Park volume 2.

hindered the nations of the Lower Orinoco from going up the little rivers and making incursions to the north and the west. If, according to the various distribution of animals on the globe, the pastoral life could have existed in the New World—if, before the arrival of the Spaniards, the Llanos and the Pampas had been filled with those numerous herds of cows and horses that graze there, Columbus would have found the human race in a state quite different. Pastoral nations living on milk and cheese, real nomad races, would have spread themselves over those vast plains which communicate with each other. They would have been seen at the period of great droughts, and even at that of inundations, fighting for the possession of pastures; subjugating one another mutually; and, united by the common tie of manners, language, and worship, they would have risen to that state of demi-civilization which we observe with surprise in the nations of the Mongol and Tartar race. America would then, like the centre of Asia, have had its conquerors, who, ascending from the plains to the tablelands of the Cordilleras, and abandoning a wandering life, would have subdued the civilized nations of Peru and New Grenada, overturned the throne of the Incas and of the Zaque,* and substituted for the despotism which is the fruit of theocracy, that despotism which arises from the patriarchal government of a pastoral people. In the New World the human race has not experienced these great moral and political changes, because the steppes, though more fertile than those of Asia, have remained without herds; because none of the animals that furnish milk in abundance are natives of the plains of South America; and because, in the progressive unfolding of American civilization, the intermediate link is wanting that connects the hunting with the agricultural nations.

We have thought proper to bring together these general notions on the plains of the New Continent, and the contrast they exhibit to the deserts of Africa and the fertile steppes of Asia, in

---

* The Zaque was the secular chief of Cundinamarca. His power was shared with the high priest (lama) of Iraca.

order to give some interest to the narrative of a journey across lands of so monotonous an aspect. Having now accomplished this task, I shall trace the route by which we proceeded from the volcanic mountains of Parapara and the northern side of the Llanos, to the banks of the Apure, in the province of Varinas.

After having passed two nights on horseback, and sought in vain, by day, for some shelter from the heat of the sun beneath the tufts of the moriche palm-trees, we arrived before night at the little Hato del Cayman,[*] called also La Guadaloupe. It was a solitary house in the steppes, surrounded by a few small huts, covered with reeds and skins. The cattle, oxen, horses, and mules are not penned, but wander freely over an extent of several square leagues. There is nowhere any enclosure; men, naked to the waist and armed with a lance, ride over the savannahs to inspect the animals; bringing back those that wander too far from the pastures of the farm, and branding all that do not already bear the mark of their proprietor. These mulattos, who are known by the name of peones llaneros, are partly freed-men and partly slaves. They are constantly exposed to the burning heat of the tropical sun. Their food is meat, dried in the air, and a little salted; and of this even their horses sometimes partake. Being always in the saddle, they fancy they cannot make the slightest excursion on foot. We found an old negro slave, who managed the farm in the absence of his master. He told us of herds composed of several thousand cows, that were grazing in the steppes; yet we asked in vain for a bowl of milk. We were offered, in a calabash, some yellow, muddy, and fetid water, drawn from a neighbouring pool. The indolence of the inhabitants of the Llanos is such that they do not dig wells, though they know that almost everywhere, at ten feet deep, fine springs are found in a stratum of conglomerate, or red sandstone. After suffering during one half of the year from the effect of inundations, they quietly resign themselves, during the other

---

[*] The Farm of the Alligator.

half; to the most distressing deprivation of water. The old negro advised us to cover the cup with a linen cloth, and drink as through a filter, that we might not be incommoded by the smell, and might swallow less of the yellowish mud suspended in the water. We did not then think that we should afterwards be forced, during whole months, to have recourse to this expedient. The waters of the Orinoco are always loaded with earthy particles; they are even putrid, where dead bodies of alligators are found in the creeks, lying on banks of sand, or half-buried in the mud.

No sooner were our instruments unloaded and safely placed, than our mules were set at liberty to go, as they say here, para buscar agua, that is, "to search for water." There are little pools round the farm, which the animals find, guided by their instinct, by the view of some scattered tufts of mauritia, and by the sensation of humid coolness, caused by little currents of air amid an atmosphere which to us appears calm and tranquil. When the pools of water are far distant, and the people of the farm are too lazy to lead the cattle to these natural watering-places, they confine them during five or six hours in a very hot stable before they let them loose. Excess of thirst then augments their sagacity, sharpening as it were their senses and their instinct. No sooner is the stable opened, than the horses and mules, especially the latter (for the penetration of these animals exceeds the intelligence of the horses), rush into the savannahs. With upraised tails and heads thrown back they run against the wind, stopping from time to time as if exploring space; they follow less the impressions of sight than of smell; and at length announce, by prolonged neighings, that there is water in the direction of their course. All these movements are executed more promptly, and with readier success, by horses born in the Llanos, and which have long enjoyed their liberty, than by those that come from the coast, and descend from domestic horses. In animals, for the most part, as in man, the quickness of the senses is diminished by long subjection, and by the habits that arise from a fixed abode and the progress of cultivation.

109

We followed our mules in search of one of those pools, whence the muddy water had been drawn, that so ill quenched our thirst. We were covered with dust, and tanned by the sandy wind, which burns the skin even more than the rays of the sun. We longed impatiently to take a bath, but we found only a great pool of feculent water, surrounded with palm-trees. The water was turbid, though, to our great astonishment, a little cooler than the air. Accustomed during our long journey to bathe whenever we had an opportunity, often several times in one day, we hastened to plunge into the pool. We had scarcely begun to enjoy the coolness of the bath, when a noise which we heard on the opposite bank, made us leave the water precipitately. It was an alligator plunging into the mud.

We were only at the distance of a quarter of a league from the farm, yet we continued walking more than an hour without reaching it. We perceived too late that we had taken a wrong direction. Having left it at the decline of day, before the stars were visible, we had gone forward into the plain at hazard. We were, as usual, provided with a compass, and it might have been easy for us to steer our course from the position of Canopus and the Southern Cross; but unfortunately we were uncertain whether, on leaving the farm, we had gone towards the east or the south. We attempted to return to the spot where we had bathed, and we again walked three quarters of an hour without finding the pool. We sometimes thought we saw fire on the horizon; but it was the light of the rising stars enlarged by the vapours. After having wandered a long time in the savannah, we resolved to seat ourselves beneath the trunk of a palm-tree, in a spot perfectly dry, surrounded by short grass; for the fear of water-snakes is always greater than that of jaguars among Europeans recently disembarked. We could not flatter ourselves that our guides, of whom we knew the insuperable indolence, would come in search of us in the savannah before they had prepared their food and finished their repast. Whilst somewhat perplexed by the uncertainty of our situation, we were agreeably affected by hearing from afar the sound of a horse advancing towards us. The rider was an Indian, armed with a lance, who had just made the rodeo, or round, in order to collect the cattle within a determinate space of ground. The sight of two white men, who said they had lost their way, led him at first to suspect some trick. We found it difficult to inspire him with confidence; he at last consented to guide us to the farm of the Cayman, but without slackening the gentle trot of his horse. Our guides assured us that "they had already begun to be uneasy about us;" and, to justify this inquietude, they gave a long enumeration of persons who, having lost themselves in the Llanos, had been found nearly exhausted. It may be supposed that the danger is imminent only to those who lose themselves far from any habitation, or who, having been stripped by robbers, as has happened of late years, have been fastened by the body and hands to the trunk of a palm-tree.

In order to escape as much as possible from the heat of the day, we set off at two in the morning, with the hope of reaching Calabozo before noon, a small but busy trading-town, situated in the midst of the Llanos. The aspect of the country was still the same. There was no moonlight; but the great masses of nebulae that spot the southern sky enlighten, as they set, a part of the terrestrial horizon. The solemn spectacle of the starry vault, seen in its immense expanse—the cool breeze which blows over the plain during the night—the waving motion of the grass, wherever it has attained any height; everything recalled to our minds the surface of the ocean. The illusion was augmented when the disk of the sun appearing on the horizon, repeated its image by the effects of refraction, and, soon losing its flattened form, ascended rapidly and straight towards the zenith.

Sunrise in the plains is the coolest moment of the day; but this change of temperature does not make a very lively impression on the organs. We did not find the thermometer in general sink below 27.5; while near Acapulco, at Mexico, and in places equally low, the temperature at noon is often 32, and at sunrise only 17 or 18°. The level surface of the ground in the

Llanos, which, during the day, is never in the shade, absorbs so much heat that, notwithstanding the nocturnal radiation toward a sky without clouds, the earth and air have not time to cool very sensibly from midnight to sunrise.

In proportion as the sun rose towards the zenith, and the earth and the strata of superincumbent air took different temperatures, the phenomenon of the mirage displayed itself in its numerous modifications. This phenomenon is so common in every zone, that I mention it only because we stopped to measure with some precision the breadth of the aerial distance between the horizon and the suspended object. There was a constant suspension, without inversion. The little currents of air that swept the surface of the soil had so variable a temperature that, in a drove of wild oxen, one part appeared with the legs raised above the surface of the ground, while the other rested on it. The aerial distance was, according to the distance of the animal, from 3 to 4'. Where tufts of the moriche palm were found growing in long ranges, the extremities of these green rows were suspended like the capes which were, for so long a time, the subject of my observations at Cumana. A well-informed person assured us, that he had seen, between Calabozo and Uritucu, the image of an animal inverted, without there being any direct image. Niebuhr made a similar observation in Arabia. We several times thought we saw on the horizon the figures of tumuli and towers, which disappeared at intervals, without our being able to discern the real shape of the objects. They were perhaps hillocks, or small eminences, situated beyond the ordinary visual horizon. I need not mention those tracts destitute of vegetation, which appear like large lakes with an undulating surface. This phenomenon, observed in very remote times, has occasioned the mirage to receive in Sanscrit the expressive name of desire of the antelope. We admire the frequent allusions in the Indian, Persian, and Arabic poets, to the magical effects of terrestrial refraction. It was scarcely known to the Greeks and Romans. Proud of the riches of their soil, and the mild temperature of the air, they would have felt no envy of this poetry of the desert. It had its birth in Asia; and the oriental poets found its source in the nature of the country they inhabited. They were inspired with the aspect of those vast solitudes, interposed like arms of the sea or gulfs, between lands which nature had adorned with her most luxuriant fertility.

The plain assumes at sunrise a more animated aspect. The cattle, which had reposed during the night along the pools, or beneath clumps of mauritias and rhopalas, were now collected in herds; and these solitudes became peopled with horses, mules, and oxen, that live here free, rather than wild, without settled habitations, and disdaining the care and protection of man. In these hot climates, the oxen, though of Spanish breed, like those of the cold table-lands of Quito, are of a gentle disposition. A traveller runs no risk of being attacked or pursued, as we often were in our excursions on the back of the Cordilleras, where the climate is rude, the aspect of the country more wild, and food less abundant. As we approached Calabozo, we saw herds of roebucks browsing peacefully in the midst of horses and oxen. They are called matacani; their flesh is good; they are a little larger than our roes, and resemble deer with a very sleek skin, of a fawn-colour, spotted with white. Their horns appear to me to have single points. They had little fear of the presence of man: and in herds of thirty or forty we observed several that were entirely white. This variety, common enough among the large stags of the cold climates of the Andes, surprised us in these low and burning plains. I have since learned, that even the jaguar, in the hot regions of Paraguay, sometimes affords albino varieties, the skin of which is of such uniform whiteness that the spots or rings can be distinguished only in the sunshine. The number of matacani, or little deer,[*] is so considerable in the Llanos, that a trade might be carried on with their skins.[†] A skilful hunter

---

[*] They are called in the country Venados de tierras calientes (deer of the warm lands.)

[†] This trade is carried on, but on a very limited scale, at Carora and at Barquesimeto.

could easily kill more than twenty in a day; but such is the indolence of the inhabitants, that often they will not give themselves the trouble of taking the skin. The same indifference is evinced in the chase of the jaguar, a skin of which fetches only one piastre in the steppes of Varinas, while at Cadiz it costs four or five.

The steppes that we traversed are principally covered with grasses of the genera Killingia, Cenchrus, and Paspalum.[*] At this season, near Calabozo and San Jerome del Pirital, these grasses scarcely attain the height of nine or ten inches. Near the banks of the Apure and the Portuguesa they rise to four feet in height, so that the jaguar can conceal himself among them, to spring upon the mules and horses that cross the plain. Mingled with these gramina some plants of the dicotyledonous class are found; as turneras, malvaceae, and, what is very remarkable, little mimosas with irritable leaves,[†] called by the Spaniards dormideras. The same breed of cows, which fatten in Europe on sainfoin and clover, find excellent nourishment in the herbaceous sensitive plants. The pastures where these shrubs particularly abound are sold at a higher price than others. To the east, in the llanos of Cari and Barcelona, the cypura and the craniolaria,[‡] the beautiful white flower of which is from six to eight inches long, rise solitarily amid the gramina. The pastures are richest not only around the rivers subject to inundations, but also wherever the trunks of palm-trees are near each other. The least fertile spots are those destitute of trees; and attempts to cultivate them would be nearly fruitless. We cannot attribute this difference to the shelter afforded by the palm-trees, in preventing the solar rays from drying and burning up the soil. I

---

[*] Killingia monocephala, K. odorata, Cenchrus pilosus, Vilfa tenacissima, Andropogon plumosum, Panicum micranthum, Poa repens, Paspalum leptostachyum, P. conjugatum, Aristida recurvata. (Nova Genera et Species Plantarum, volume 1 pages 84 to 243.)

[†] The sensitive-plant Mimosa dormiens.

[‡] Cypura graminea, Craniolaria annua, the scorzonera of the natives.

have seen, it is true, trees of this family, in the forests of the Orinoco, spreading a tufted foliage; but we cannot say much for the shade of the palm-tree of the llanos, the palma de cobija,[*] which has but a few folded and palmate leaves, like those of the chamaerops, and of which the lower-most are constantly withered. We were surprised to see that almost all these trunks of the corypha were nearly of the same size, namely, from twenty to twenty-four feet high, and from eight to ten inches diameter at the foot. Nature has produced few species of palm-trees in such prodigious numbers. Amidst thousands of trunks loaded with olive-shaped fruits we found about one hundred without fruit. May we suppose that there are some trees with flowers purely monoecious, mingled with others furnished with hermaphrodite flowers?

The Llaneros, or inhabitants of the plains, believe that all these trees, though so low, are many centuries old. Their growth is almost imperceptible, being scarcely to be noticed in the lapse of twenty or thirty years. The wood of the palma de cobija is excellent for building. It is so hard, that it is difficult to drive a nail into it. The leaves, folded like a fan, are employed to cover the roofs of the huts scattered through the Llanos; and these roofs last more than twenty years. The leaves are fixed by bending the extremity of the footstalks, which have been beaten beforehand between two stones, so that they may bend without breaking.

Beside the solitary trunks of this palm-tree, we find dispersed here and there in the steppes a few clumps, real groves (palmares), in which the corypha is intermingled with a tree of the proteaceous family, called chaparro by the natives. It is a new species of rhopala,[†] with hard and resonant leaves. The little groves of rhopala are called chaparales; and it may be supposed

---

[*] The roofing palm-tree Corypha tectorum.
[†] Resembling the Embothrium, of which we found no species in South America. The embothriums are represented in American vegetation by the genera Lomatia and Oreocallis.

that, in a vast plain, where only two or three species of trees are to be found, the chaparro, which affords shade, is considered a highly valuable plant. The corypha spreads through the Llanos of Caracas from Mesa de Peja as far as Guayaval; farther north and north-west, near Guanare and San Carlos, its place is taken by another species of the same genus, with leaves alike palmate but larger. It is called the royal palm of the plains (palma real de los Llanos).[*] Other palm-trees rise south of Guayaval, especially the piritu with pinnate leaves,[†] and the moriche (Mauritia flexuosa), celebrated by Father Gumilla under the name of arbol de la vida, or tree of life. It is the sago-tree of America, furnishing flour, wine, thread for weaving hammocks, baskets, nets, and clothing. Its fruit, of the form of the cones of the pine, and covered with scales, perfectly resembles that of the Calamus rotang. It has somewhat the taste of the apple. When arrived at its maturity it is yellow within and red without. The araguato monkeys eat it with avidity; and the nation of the Guaraounos, whose whole existence, it may be said, is closely linked with that of the moriche palm-tree, produce from it a fermented liquor, slightly acid, and extremely refreshing. This palm-tree, with its large shining leaves, folded like a fan, preserves a beautiful verdure at the period of the greatest drought. The mere sight of it produces an agreeable sensation of coolness, and when loaded with scaly fruit, it contrasts singularly with the mournful aspect of the palma de cobija, the foliage of which is always grey and covered with dust. The Llaneros believe that the former attracts the vapours in the air;[‡] and that for this reason, water is constantly found at its foot, when dug for to a certain depth. The effect is confounded with the cause. The moriche grows best in

[*] This palm-tree of the plains must not be confounded with the palma real of Caracas and of Curiepe, with pinnate leaves.

[†] Perhaps an Aiphanes.

[‡] If the head of the moriche were better furnished with leaves than it generally is, we might perhaps admit that the soil round the tree preserves its humidity through the influence of the shade.

moist places; and it may rather be said that the water attracts the tree. The natives of the Orinoco, by analogous reasoning, admit, that the great serpents contribute to preserve humidity in a province. "You would look in vain for water-serpents," said an old Indian of Javita to us gravely, "where there are no marshes; because the water ceases to collect when you imprudently kill the serpents that attract it."

We suffered greatly from the heat in crossing the Mesa de Calabozo. The temperature of the air augmented sensibly every time that the wind began to blow. The air was loaded with dust; and during these gusts the thermometer rose to 40 or 41°. We went slowly forward, for it would have been dangerous to leave the mules that carried our instruments. Our guides advised us to fill our hats with the leaves of the rhopala, to diminish the action of the solar rays on the hair and the crown of the head. We found relief from this expedient, which was particularly agreeable, when we could procure the thick leaves of the pothos or some other similar plant.

It is impossible to cross these burning plains, without inquiring whether they have always been in the same state; or whether they have been stripped of their vegetation by some revolution of nature. The stratum of mould now found on them is in fact very thin. The natives believe that the palmares and the chaparales (the little groves of palm-trees and rhopala) were more frequent and more extensive before the arrival of the Spaniards. Since the Llanos have been inhabited and peopled with cattle become wild, the savannah is often set on fire, in order to ameliorate the pasturage. Groups of scattered trees are accidentally destroyed with the grasses. The plains were no doubt less bare in the fifteenth century, than they now are; yet the first Conquistadores, who came from Coro, described them then as savannahs, where nothing could be perceived but the sky and the turf, generally destitute of trees, and difficult to traverse on account of the reverberation of heat from the soil. Why does not the great forest of the Orinoco extend to the north, on the left bank of that river? Why does it not fill that vast space that reaches as far as the Cordillera of the coast, and which is fertilized by numerous rivers? These questions are connected with all that relates to the history of our planet. If, indulging in geological reveries, we suppose that the steppes of America, and the desert of Sahara, have been stripped of their vegetation by an irruption of the ocean, or that they formed originally the bottom of an inland sea, we may conceive that thousands of years have not sufficed for the trees and shrubs to advance from the borders of the forests, from the skirts of the plains either naked or covered with turf, toward the centre, and darken so vast a space with their shade. It is more difficult to explain the origin of bare savannahs, encircled by forests, than to recognize the causes that maintain forests and savannahs within their ancient limits, like continents and seas.

We found the most cordial hospitality at Calabozo, in the house of the superintendent of the royal plantations, Don Miguel Cousin. The town, situated between the banks of the Guarico and the Uritucu, contained at this period only five thousand inhabitants; but everything denoted increasing prosperity. The wealth of most of the inhabitants consists in herds, under the management of farmers, who are called hateros, from the word hato, which signifies in Spanish a house or farm placed in the midst of pastures. The scattered population of the Llanos being accumulated on certain points, principally around towns, Calabozo reckons already five villages or missions in its environs. It is computed, that 98,000 head of cattle wander in the pastures nearest to the town. It is very difficult to form an exact idea of the herds contained in the Llanos of Caracas, Barcelona, Cumana, and Spanish Guiana. M. Depons, who lived in the town of Caracas longer than I, and whose statistical statements are generally accurate, reckons in those vast plains, from the mouths of the Orinoco to the lake of Maracaybo, 1,200,000 oxen, 180,000 horses, and 90,000 mules. He estimates the produce of these herds at 5,000,000 francs; adding to the value of the exportation the price of the hides consumed in the country.

There exist, it is believed, in the Pampas of Buenos Ayres, 12,000,000 cows, and 3,000,000 horses, without comprising in this enumeration the cattle that have no acknowledged proprietor.

I shall not hazard any general estimates, which from their nature are too uncertain; but shall only observe that, in the Llanos of Caracas, the proprietors of the great hatos are entirely ignorant of the number of the cattle they possess. They only know that of the young cattle, which are branded every year with a letter or mark peculiar to each herd. The richest proprietors mark as many as 14,000 head every year; and sell to the number of five or six thousand. According to official documents, the exportation of hides from the whole capitania-general of Caracas amounted annually to 174,000 skins of oxen, and 11,500 of goats. When we reflect, that these documents are taken from the books of the custom-houses, where no mention is made of the fraudulent dealings in hides, we are tempted to believe that the estimate of 1,200,000 oxen wandering in the Llanos, from the Rio Carony and the Guarapiche to the lake of Maracaybo, is much underrated. The port of La Guayra alone exported annually from 1789 to 1792, 70,000 or 80,000 hides, entered in the custom-house books, scarcely one-fifth of which was sent to Spain. The exportation from Buenos Ayres, at the end of the eighteenth century, was, according to Don Felix de Azara, 800,000 skins. The hides of Caracas are preferred in the Peninsula to those of Buenos Ayres; because the latter, on account of a longer passage, undergo a loss of twelve per cent in the tanning. The southern part of the savannahs, commonly called the Upper Plains (Llanos de arriba), is very productive in mules and oxen; but the pasturage being in general less good, these animals are obliged to be sent to other plains to be fattened before they are sold. The Llano de Monai, and all the Lower Plains (Llanos de abaxo), abound less in herds, but the pastures are so fertile, that they furnish meat of an excellent quality for the supply of the coast. The mules, which are not fit for labour before the fifth year, are purchased on the spot at the price of fourteen or eighteen piastres. The horses of the Llanos, descending from the fine Spanish breed, are not very large; they are generally of a uniform colour, brown bay, like most of the wild animals. Suffering alternately from drought and floods, tormented by the stings of insects and the bites of the large bats, they lead a sorry life. After having enjoyed for some months the care of man, their good qualities are developed. Here there are no sheep: we saw flocks only on the table-land of Quito.

The hatos of oxen have suffered considerably of late from troops of marauders, who roam over the steppes killing the animals merely to take their hides. This robbery has increased since the trade of the Lower Orinoco has become more flourishing. For half a century, the banks of that great river, from the mouth of the Apure as far as Angostura, were known only to the missionary-monks. The exportation of cattle took place from the ports of the northern coast only, namely from Cumana, Barcelona, Burburata, and Porto Cabello. This dependence on the coast is now much diminished. The southern part of the plains has established an internal communication with the Lower Orinoco; and this trade is the more brisk, as those who devote themselves to it easily escape the trammels of the prohibitory laws.

The greatest herds of cattle in the Llanos of Caracas are those of the hatos of Merecure, La Cruz, Belen, Alta Gracia, and Pavon. The Spanish cattle came from Coro and Tocuyo into the plains. History has preserved the name of the colonist who first conceived the idea of peopling these pasturages, inhabited only by deer, and a large species of cavy.* Christoval Rodriguez sent the first horned cattle into the Llanos, about the year 1548. He was an inhabitant of the town of Tocuyo, and had long resided in New Grenada.

---

* The thick-nosed tapir, or river cavy (Cavia capybara), called chiguire in those countries.

When we hear of the innumerable quantity of oxen, horses, and mules, that are spread over the plains of America, we seem generally to forget that in civilized Europe, on lands of much less extent, there exist, in agricultural countries, quantities no less prodigious. France, according to M. Peuchet, feeds 6,000,000 large horned cattle, of which 3,500,000 are oxen employed in drawing the plough. In the Austrian monarchy, the number of oxen, cows, and calves, has been estimated at 13,400,000 head. Paris alone consumes annually 155,000 horned cattle. Germany receives 150,000 oxen yearly from Hungary. Domestic animals, collected in small herds, are considered by agricultural nations as a secondary object in the riches of the state. Accordingly they strike the imagination much less than those wandering droves of oxen and horses which alone fill the uncultivated tracts of the New World. Civilization and social order favour alike the progress of population, and the multiplication of animals useful to man.

We found at Calabozo, in the midst of the Llanos, an electrical machine with large plates, electrophori, batteries, electrometers; an apparatus nearly as complete as our first scientific men in Europe possess. All these articles had not been purchased in the United States; they were the work of a man who had never seen any instrument, who had no person to consult, and who was acquainted with the phenomena of electricity only by reading the treatise of De Lafond, and Franklin's Memoirs. Senor Carlos del Pozo, the name of this enlightened and ingenious man, had begun to make cylindrical electrical machines, by employing large glass jars, after having cut off the necks. It was only within a few years he had been able to procure, by way of Philadelphia, two plates, to construct a plate machine, and to obtain more considerable effects. It is easy to judge what difficulties Senor Pozo had to encounter, since the first works upon electricity had fallen into his hands, and that he had the courage to resolve to procure himself, by his own industry, all that he had seen described in his books. Till now he had enjoyed only the astonishment and admiration produced by his experiments on persons destitute of all information, and who had never quitted the solitude of the Llanos; our abode at Calabozo gave him a satisfaction altogether new. It may be supposed that he set some value on the opinions of two travellers who could compare his apparatus with those constructed in Europe. I had brought with me electrometers mounted with straw, pith-balls, and gold-leaf; also a small Leyden jar which could be charged by friction according to the method of Ingenhousz, and which served for my physiological experiments. Senor del Pozo could not contain his joy on seeing for the first time instruments which he had not made, yet which appeared to be copied from his own. We also showed him the effect of the contact of heterogeneous metals on the nerves of frogs. The name of Galvani and Volta had not previously been heard in those vast solitudes.

Next to his electrical apparatus, the work of the industry and intelligence of an inhabitant of the Llanos, nothing at Calabozo excited in us so great an interest as the gymnoti, which are animated electrical apparatuses. I was impatient, from the time of my arrival at Cumana, to procure electrical eels. We had been promised them often, but our hopes had always been disappointed. Money loses its value as you withdraw from the coast; and how is the imperturbable apathy of the ignorant people to be vanquished, when they are not excited by the desire of gain?

The Spaniards confound all electric fishes under the name of tembladores.[*] There are some of these in the Caribbean Sea, on the coast of Cumana. The Guayquerie Indians, who are the most skilful and active fishermen in those parts, brought us a fish, which, they said, benumbed their hands. This fish ascends the little river Manzanares. It is a new species of ray, the lateral spots of which are scarcely visible, and which much resembles the torpedo. The torpedos, which are furnished with an electric

---

[*] Literally "tremblers," or "producers of trembling."

organ externally visible, on account of the transparency of the skin, form a genus or subgenus different from the rays properly so called.[*] The torpedo of Cumana was very lively, very energetic in its muscular movements, and yet the electric shocks it gave us were extremely feeble. They became stronger on galvanizing the animal by the contact of zinc and gold. Other tembladores, real gymnoti or electric eels, inhabit the Rio Colorado, the Guarapiche, and several little streams which traverse the Missions of the Chayma Indians. They abound also in the large rivers of America, the Orinoco, the Amazon, and the Meta; but the force of the currents and the depth of the water, prevent them from being caught by the Indians. They see these fish less frequently than they feel shocks from them when swimming or bathing in the river. In the Llanos, particularly in the environs of Calabozo, between the farms of Morichal and the Upper and Lower Missions, the basins of stagnant water and the confluents of the Orinoco (the Rio Guarico and the canos Rastro, Berito, and Paloma) are filled with electric eels. We at first wished to make our experiments in the house we inhabited at Calabozo; but the dread of the shocks caused by the gymnoti is so great, and so exaggerated among the common people, that during three days we could not obtain one, though they are easily caught, and we had promised the Indians two piastres for every strong and vigorous fish. This fear of the Indians is the more extraordinary, as they do not attempt to adopt precautions in which they profess to have great confidence. When interrogated on the effect of the tembladores, they never fail to tell the Whites, that they may be touched with impunity while you are chewing tobacco. This supposed influence of tobacco on

---

[*] Cuvier, Regne Animal volume 2. The Mediterranean contains, according to M. Risso, four species of electrical torpedos, all formerly confounded under the name of Raia torpedo; these are Torpedo narke, T. unimaculata, T. galvanii, and T. marmorata. The torpedo of the Cape of Good Hope, the subject of the recent experiments of Mr. Todd, is, no doubt, a nondescript species.

animal electricity is as general on the continent of South America, as the belief among mariners of the effect of garlic and tallow on the magnetic needle.

Impatient of waiting, and having obtained very uncertain results from an electric eel which had been brought to us alive, but much enfeebled, we repaired to the Cano de Bera, to make our experiments in the open air, and at the edge of the water. We set off on the 19th of March, at a very early hour, for the village of Rastro; thence we were conducted by the Indians to a stream, which, in the time of drought, forms a basin of muddy water, surrounded by fine trees,* the clusia, the amyris, and the mimosa with fragrant flowers. To catch the gymnoti with nets is very difficult, on account of the extreme agility of the fish, which bury themselves in the mud. We would not employ the barbasco, that is to say, the roots of the Piscidea erithyrna, the Jacquinia armillaris, and some species of phyllanthus, which thrown into the pool, intoxicate or benumb the eels. These methods have the effect of enfeebling the gymnoti. The Indians therefore told us that they would "fish with horses," (embarbascar con caballos.†) We found it difficult to form an idea of this extraordinary manner of fishing; but we soon saw our guides return from the savannah, which they had been scouring for wild horses and mules. They brought about thirty with them, which they forced to enter the pool.

The extraordinary noise caused by the horses' hoofs, makes the fish issue from the mud, and excites them to the attack. These yellowish and livid eels, resembling large aquatic serpents, swim on the surface of the water, and crowd under the bellies of the horses and mules. A contest between animals of so different an organization presents a very striking spectacle. The Indians, provided with harpoons and long slender reeds, surround the pool closely; and some climb up the trees, the

---

* Amyris lateriflora, A. coriacea, Laurus pichurin. Myroxylon secundum, Malpighia reticulata.
† Meaning to excite the fish by horses.

branches of which extend horizontally over the surface of the water. By their wild cries, and the length of their reeds, they prevent the horses from running away and reaching the bank of the pool. The eels, stunned by the noise, defend themselves by the repeated discharge of their electric batteries. For a long interval they seem likely to prove victorious. Several horses sink beneath the violence of the invisible strokes which they receive from all sides, in organs the most essential to life; and stunned by the force and frequency of the shocks, they disappear under the water. Others, panting, with mane erect, and haggard eyes expressing anguish and dismay, raise themselves, and endeavour to flee from the storm by which they are overtaken. They are driven back by the Indians into the middle of the water; but a small number succeed in eluding the active vigilance of the fishermen. These regain the shore, stumbling at every step, and stretch themselves on the sand, exhausted with fatigue, and with limbs benumbed by the electric shocks of the gymnoti.

In less than five minutes two of our horses were drowned. The eel being five feet long, and pressing itself against the belly of the horses, makes a discharge along the whole extent of its electric organ. It attacks at once the heart, the intestines, and the caeliac fold of the abdominal nerves. It is natural that the effect felt by the horses should be more powerful than that produced upon man by the touch of the same fish at only one of his extremities. The horses are probably not killed, but only stunned. They are drowned from the impossibility of rising amid the prolonged struggle between the other horses and the eels.

We had little doubt that the fishing would terminate by killing successively all the animals engaged; but by degrees the impetuosity of this unequal combat diminished, and the wearied gymnoti dispersed. They require a long rest, and abundant nourishment, to repair the galvanic force which they have lost.[*]

---

[*] The Indians assured us that when the horses are made to run two days successively into the same pool, none are killed the second day. See, on the

The mules and horses appear less frightened; their manes are no longer bristled, and their eyes express less dread. The gymnoti approach timidly the edge of the marsh, where they are taken by means of small harpoons fastened to long cords. When the cords are very dry the Indians feel no shock in raising the fish into the air. In a few minutes we had five large eels, most of which were but slightly wounded. Some others were taken, by the same means, towards evening.

The temperature of the waters in which the gymnoti habitually live, is from 26 to 27°. Their electric force diminishes it is said, in colder waters; and it is remarkable that, in general, animals endowed with electromotive organs, the effects of which are sensible to man, are not found in the air, but in a fluid that is a conductor of electricity. The gymnotus is the largest of electrical fishes. I measured some that were from five feet to five feet three inches long; and the Indians assert that they have seen them still larger. We found that a fish of three feet ten inches long weighed twelve pounds. The transverse diameter of the body, without reckoning the anal fin, which is elongated in the form of a keel, was three inches and a half. The gymnoti of the Cano de Bera are of a fine olive-green. The under part of the head is yellow mingled with red. Two rows of small yellow spots are placed symmetrically along the back, from the head to the end of the tail. Every spot contains an excretory aperture. In consequence, the skin of the animal is constantly covered with a mucous matter, which, as Volta has proved, conducts electricity twenty or thirty times better than pure water. It is in general somewhat remarkable, that no electric fish yet discovered in the different parts of the world, is covered with scales.[*]

---

fishing for gymnoti Views of Nature Bohn's edition page 18.

[*] We yet know with certainty only seven electric fishes; Torpedo narke, Risso, T. unimaculata, T. marmorata, T. galvanii, Silurus electricus, Tetraodon electricus, Gymnotus electricus. It appears uncertain whether the Trichiurus indicus has electrical properties or not. See Cuvier's Regne Animal volume 2. But the genus Torpedo, very different from that of the rays

The gymnoti, like our eels, are fond of swallowing and breathing air on the surface of the water; but we must not thence conclude that the fish would perish if it could not come up to breathe the air. The European eel will creep during the night upon the grass; but I have seen a very vigorous gymnotus that had sprung out of the water, die on the ground. M. Provencal and myself have proved by our researches on the respiration of fishes, that their humid bronchiae perform the double function of decomposing the atmospheric air, and of appropriating the oxygen contained in water. They do not suspend their respiration in the air; but they absorb the oxygen like a reptile furnished with lungs. It is known that carp may be fattened by being fed, out of the water, if their gills are wet from time to time with humid moss, to prevent them from becoming dry. Fish separate their gill-covers wider in oxygen gas than in water. Their temperature however, does not rise; and they live the same length of time in pure vital air, and in a mixture of ninety parts nitrogen and ten oxygen. We found that tench placed under inverted jars filled with air, absorb half a cubic centimetre of oxygen in an hour. This action takes place in the gills only; for fishes on which a collar of cork has been fastened, and leaving their head out of the jar filled with air, do not act upon the oxygen by the rest of their body.

The swimming-bladder of the gymnotus is two feet five inches long in a fish of three feet ten inches.[*] It is separated by a

---

properly so called, has numerous species in the equatorial seas; and it is probable that there exist several gymnoti specifically different. The Indians mentioned to us a black and very powerful species, inhabiting the marshes of the Apure, which never attains a length of more than two feet, but which we were not able to procure. The raton of the Rio de la Magdalena, which I have described under the name of Gymnotus aequilabiatus (Observations de Zoologie volume 1) forms a particular sub-genus. This is a Carapa, not scaly, and without an electric organ. This organ is also entirely wanting in the Brazilian Carapo, and in all the rays which were carefully examined by Cuvier.

[*] Cuvier has shown that in the Gymnotus electricus there exists, besides the

mass of fat from the external skin; and rests upon the electric organs, which occupy more than two-thirds of the animal's body. The same vessels which penetrate between the plates or leaves of these organs, and which cover them with blood when they are cut transversely, also send out numerous branches to the exterior surface of the air-bladder. I found in a hundred parts of the air of the swimming-bladder four of oxygen and ninety-six of nitrogen. The medullary substance of the brain displays but a feeble analogy with the albuminous and gelatinous matter of the electric organs. But these two substances have in common the great quantity of arterial blood which they receive, and which is deoxidated in them. We may again remark, on this occasion, that an extreme activity in the functions of the brain causes the blood to flow more abundantly towards the head, as the energy of the movement of the muscles accelerates the deoxidation of the arterial blood. What a contrast between the multitude and the diameter of the blood-vessels of the gymnotus, and the small space occupied by its muscular system! This contrast reminds the observer, that three functions of animal life, which appear in other respects sufficiently distinct—the functions of the brain, those of the electrical organ, and those of the muscles, all require the afflux and concourse of arterial or oxygenated blood.

It would be temerity to expose ourselves to the first shocks of a very large and strongly irritated gymnotus. If by chance a stroke be received before the fish is wounded or wearied by long pursuit, the pain and numbness are so violent that it is impossible to describe the nature of the feeling they excite. I do not remember having ever received from the discharge of a large Leyden jar, a more dreadful shock than that which I experienced by imprudently placing both my feet on a gymnotus just taken out of the water. I was affected during the rest of the day with a violent pain in the knees, and in almost every joint. To be aware of the difference that exists between the sensation produced by

---

large swimming-bladder, another situated before it, and much smaller. It looks like the bifurcated swimming-bladder in the Gymnotus aequilabiatus.

the Voltaic battery and an electric fish, the latter should be touched when they are in a state of extreme weakness. The gymnoti and the torpedos then cause a twitching of the muscles, which is propagated from the part that rests on the electric organs, as far as the elbow. We seem to feel, at every stroke, an internal vibration, which lasts two or three seconds, and is followed by a painful numbness. Accordingly, the Tamanac Indians call the gymnotus, in their expressive language, arimna, which means something that deprives of motion.

The sensation caused by the feeble shocks of an electric eel appeared to me analogous to that painful twitching with which I have been seized at each contact of two heterogeneous metals applied to wounds which I had made on my back by means of cantharides. This difference of sensation between the effects of electric fishes and those of a Voltaic battery or a Leyden jar feebly charged has struck every observer; there is, however, nothing in this contrary to the supposition of the identity of electricity and the galvanic action of fishes. The electricity may be the same; but its effects will be variously modified by the disposition of the electrical apparatus, by the intensity of the fluid, by the rapidity of the current, and by the particular mode of action.

In Dutch Guiana, at Demerara for instance, electric eels were formerly employed to cure paralytic affections. At a time when the physicians of Europe had great confidence in the effects of electricity, a surgeon of Essequibo, named Van der Lott, published in Holland a treatise on the medical properties of the gymnotus. These electric remedies are practised among the savages of America, as they were among the Greeks. We are told by Scribonius Largus, Galen, and Dioscorides, that torpedos cure the headache and the gout. I did not hear of this mode of treatment in the Spanish colonies which I visited; and I can assert that, after having made experiments during four hours successively with gymnoti, M. Bonpland and myself felt, till the next day, a debility in the muscles, a pain in the joints, and a general uneasiness, the effect of a strong irritation of the nervous system.

The gymnotus is neither a charged conductor, nor a battery, nor an electromotive apparatus, the shock of which is received every time they are touched with one hand, or when both hands are applied to form a conducting circle between the opposite poles. The electric action of the fish depends entirely on its will; because it does not keep its electric organs always charged, or whether by the secretion of some fluid, or by any other means alike mysterious to us, it be capable of directing the action of its organs to an external object. We often tried, both insulated and otherwise, to touch the fish, without feeling the least shock. When M. Bonpland held it by the head, or by the middle of the body, while I held it by the tail, and, standing on the moist ground, did not take each other's hand, one of us received shocks, which the other did not feel. It depends upon the gymnotus to direct its action towards the point where it finds itself most strongly irritated. The discharge is then made at one point only, and not at the neighbouring points. If two persons touch the belly of the fish with their fingers, at an inch distance, and press it simultaneously, sometimes one, sometimes the other, will receive the shock. In the same manner, when one insulated person holds the tail of a vigorous gymnotus, and another pinches the gills or pectoral fin, it is often the first only by whom the shock is received. It did not appear to us that these differences could be attributed to the dryness or moisture of our hands, or to their unequal conducting power. The gymnotus seemed to direct its strokes sometimes from the whole surface of its body, sometimes from one point only. This effect indicates less a partial discharge of the organ composed of an innumerable quantity of layers, than the faculty which the animal possesses, (perhaps by the instantaneous secretion of a fluid spread through the cellular membrane,) of establishing the communication between its organs and the skin only, in a very limited space.

Nothing proves more strongly the faculty, which the gymnotus possesses, of darting and directing its stroke at will, than the observations made at Philadelphia and Stockholm,[*] on gymnoti rendered extremely tame. When they had been made to fast a long time, they killed small fishes put into the tub. They acted from a distance; that is to say, their electrical shock passed through a very thick stratum of water. We need not be surprised that what was observed in Sweden, on a single gymnotus only, we could not perceive in a great number of individuals in their native country. The electric action of animals being a vital action, and subject to their will, it does not depend solely on their state of health and vigour. A gymnotus that has been kept a long time in captivity, accustoms itself to the imprisonment to which it is reduced; it resumes by degrees the same habits in the tub, which it had in the rivers and marshes. An electrical eel was brought to me at Calabozo: it had been taken in a net, and consequently having no wound. It ate meat, and terribly frightened the little tortoises and frogs which, not aware of their danger, placed themselves on its back. The frogs did not receive the stroke till the moment when they touched the body of the

---

[*] By MM. Williamson and Fahlberg. The following account is given by the latter gentleman. "The gymnotus sent from Surinam to M. Norderling, at Stockholm, lived more than four months in a state of perfect health. It was twenty-seven inches long; and the shocks it gave were so violent, especially in the open air, that I found scarcely any means of protecting myself by non-conductors, in transporting the fish from one place to another. Its stomach being very small, it ate little at a time, but fed often. It approached living fish, first sending them from afar a shock, the energy of which was proportionate to the size of the prey. The gymnotus seldom failed in its aim; one single stroke was almost always sufficient to overcome the resistance which the strata of water, more or less thick according to the distance, opposed to the electrical current. When very much pressed by hunger, it sometimes directed the shocks against the person who daily brought its food of boiled meat. Persons afflicted with rheumatism came to touch it in hopes of being cured. They took it at once by the neck and tail the shocks were in this case stronger than when touched with one hand only. It almost entirely lost its electrical power a short time before its death."

gymnotus. When they recovered, they leaped out of the tub; and when replaced near the fish, they were frightened at the mere sight of it. We then observed nothing that indicated an action at a distance; but our gymnotus, recently taken, was not yet sufficiently tame to attack and devour frogs. On approaching the finger, or the metallic points, very close to the electric organs, no shock was felt. Perhaps the animal did not perceive the proximity of a foreign body; or, if it did, we must suppose that in the commencement of its captivity, timidity prevented it from darting forth its energetic strokes except when strongly irritated by an immediate contact. The gymnotus being immersed in water, I placed my hand, both armed and unarmed with metal, within a very small distance from the electric organs; yet the strata of water transmitted no shock, while M. Bonpland irritated the animal strongly by an immediate contact, and received some very violent shocks. Had we placed a very delicate electroscope in the contiguous strata of water, it might possibly have been influenced at the moment when the gymnotus seemed to direct its stroke elsewhere. Prepared frogs, placed immediately on the body of a torpedo, experience, according to Galvani, a strong contraction at every discharge of the fish.

The electrical organ of the gymnoti acts only under the immediate influence of the brain and the heart. On cutting a very vigorous fish through the middle of the body, the fore part alone gave shocks. These are equally strong in whatever part of the body the fish is touched; it is most disposed, however, to emit them when the pectoral fin, the electrical organ, the lips, the eyes, or the gills, are pinched. Sometimes the animal struggles violently with a person holding it by the tail, without communicating the least shock. Nor did I feel any when I made a slight incision near the pectoral fin of the fish, and galvanized the wound by the contact of two pieces of zinc and silver. The gymnotus bent itself convulsively, and raised its head out of the water, as if terrified by a sensation altogether new; but I felt no vibration in the hands which held the two metals. The most violent muscular movements are not always accompanied by electric discharges.

The action of the fish on the human organs is transmitted and intercepted by the same bodies that transmit and intercept the electrical current of a conductor charged by a Leyden jar, or Voltaic battery. Some anomalies, which we thought we observed, are easily explained, when we recollect that even metals (as is proved from their ignition when exposed to the action of the battery) present a slight obstacle to the passage of electricity; and that a bad conductor annihilates the effect, on our organs, of a feeble electric charge, whilst it transmits to us the effect of a very strong one. The repulsive force which zinc and silver exercise together being far superior to that of gold and silver, I have found that when a frog, prepared and armed with silver, is galvanized under water, the conducting arc of zinc produces contraction as soon as one of its extremities approaches the muscles within three lines distance; while an arc of gold does not excite the organs, when the stratum of water between the gold and the muscles is more than half a line thick. In the same manner, by employing a conducting arc composed of two pieces of zinc and silver soldered together endways; and resting, as before, one of the extremities of the metallic circuit on the femoral nerve, it is necessary, in order to produce contractions, to bring the other extremity of the conductor nearer and nearer to the muscles, in proportion as the irritability of the organs diminishes. Toward the end of the experiment the slightest stratum of water prevents the passage of the electrical current, and it is only by the immediate contact of the arc with the muscles, that the contractions take place. These effects are, however, dependent on three variable circumstances; the energy of the electromotive apparatus, the conductibility of the medium, and the irritability of the organs which receive the impressions: it is because experiments have not been sufficiently multiplied with a view to these three variable elements, that, in the action of electric eels and torpedos, accidental circumstances have been taken for absolute conditions, without which the electric shocks are not felt.

In wounded gymnoti, which give feeble but very equal shocks, these shocks appeared to us constantly stronger on touching the body of the fish with a hand armed with metal, than with the naked hand. They are stronger also, when, instead of touching the fish with one hand, naked, or armed with metal, we press it at once with both hands, either naked or armed. These differences become sensible only when one has gymnoti enough at disposal to be able to choose the weakest; and when the extreme equality of the electric discharges admits of distinguishing between the sensations felt alternately by the hand naked or armed with a metal, by one or both hands naked, and by one or both hands armed with metal. It is also in the case only of small shocks, feeble and uniform, that they are more sensible on touching the gymnotus with one hand (without forming a chain) with zinc, than with copper or iron.

Resinous substances, glass, very dry wood, horn, and even bones, which are generally believed to be good conductors, prevent the action of the gymnoti from being transmitted to man. I was surprised at not feeling the least shock on pressing wet sticks of sealing-wax against the organs of the fish, while the same animal gave me the most violent strokes, when excited by means of a metallic rod. M. Bonpland received shocks, when carrying a gymnotus on two cords of the fibres of the palm-tree, which appeared to us extremely dry. A strong discharge makes its way through very imperfect conductors. Perhaps also the obstacle which the conductor presents renders the discharge more painful. I touched the gymnotus with a wet pot of brown clay, without effect; yet I received violent shocks when I carried the gymnotus in the same pot, because the contact was greater.

When two persons, insulated or otherwise, hold each other's hands, and only one of these persons touches the fish with the hand, either naked or armed with metal, the shock is most commonly felt by both at once. However, it sometimes happens that, in the most severe shocks, the person who comes into immediate contact with the fish alone feels them. When the gymnotus is exhausted, or in a very reduced state of excitability, and will no longer emit strokes on being irritated with one hand, the shocks are felt in a very vivid manner, on forming the chain, and employing both hands. Even then, however, the electric shock takes place only at the will of the animal. Two persons, one of whom holds the tail, and the other the head, cannot, by joining hands and forming a chain, force the gymnotus to dart his stroke.

Though employing the most delicate electrometers in various ways, insulating them on a plate of glass, and receiving very strong shocks which passed through the electrometer, I could never discover any phenomenon of attraction or repulsion. The same observation was made by M. Fahlberg at Stockholm. That philosopher, however, has seen an electric spark, as Walsh and Ingenhousz had before him, in London, by placing the gymnotus in the air, and interrupting the conducting chain by two gold leaves pasted upon glass, and a line distant from each other. No person, on the contrary, has ever perceived a spark issue from the body of the fish itself. We irritated it for a long time during the night, at Calabozo, in perfect darkness, without observing any luminous appearance. Having placed four gymnoti, of unequal strength, in such a manner as to receive the shocks of the most vigorous fish by contact, that is to say, by touching only one of the other fishes, I did not observe that these last were agitated at the moment when the current passed their bodies. Perhaps the current did not penetrate below the humid surface of the skin. We will not, however, conclude from this, that the gymnoti are insensible to electricity; and that they cannot fight with each other at the bottom of the pools. Their nervous system must be subject to the same agents as the nerves of other animals. I have indeed seen, that, on laying open their nerves, they undergo muscular contractions at the mere contact of two opposite metals; and M. Fahlberg, of Stockholm, found that his gymnotus was convulsively agitated when placed in a copper vessel, and feeble discharges from a Leyden jar passed through its skin.

After the experiments I had made on gymnoti, it became highly interesting to me, on my return to Europe, to ascertain with precision the various circumstances in which another electric fish, the torpedo of our seas, gives or does not give shocks. Though this fish had been examined by numerous men of science, I found all that had been published on its electrical effects extremely vague. It has been very arbitrarily supposed, that this fish acts like a Leyden jar, which may be discharged at will, by touching it with both hands; and this supposition appears to have led into error observers who have devoted themselves to researches of this kind. M. Gay-Lussac and myself, during our journey to Italy, made a great number of experiments on torpedos taken in the gulf of Naples. These experiments furnish many results somewhat different from those I collected on the gymnoti. It is probable that the cause of these anomalies is owing rather to the inequality of electric power in the two fishes, than to the different disposition of their organs.

Though the power of the torpedo cannot be compared with that of the gymnotus, it is sufficient to cause very painful sensations. A person accustomed to electric shocks can with difficulty hold in his hands a torpedo of twelve or fourteen inches, and in possession of all its vigour. When the torpedo gives only very feeble strokes under water, they become more sensible if the animal be raised above the surface. I have often observed the same phenomenon in experimenting on frogs.

The torpedo moves the pectoral fins convulsively every time it emits a stroke; and this stroke is more or less painful, according as the immediate contact takes place by a greater or less surface. We observed that the gymnotus gives the strongest shocks without making any movement with the eyes, head, or fins.[*] Is this difference caused by the position of the electric

_____

[*] The anal fin of the gymnoti only has a sensible motion when these fishes

organ, which is not double in the gymnoti? or does the movement of the pectoral fins of the torpedo directly prove that the fish restores the electrical equilibrium by its own skin, discharges itself by its own body, and that we generally feel only the effect of a lateral shock?

We cannot discharge at will either a torpedo or a gymnotus, as we discharge at will a Leyden jar or a Voltaic battery. A shock is not always felt, even on touching the electric fish with both hands. We must irritate it to make it give the shock. This action in the torpedos, as well as in the gymnoti, is a vital action; it depends on the will only of the animal, which perhaps does not always keep its electric organs charged, or does not always employ the action of its nerves to establish the chain between the positive and negative poles. It is certain that the torpedo gives a long series of shocks with astonishing celerity; whether it is that the plates or laminae of its organs are not wholly exhausted, or that the fish recharges them instantaneously.

The electric stroke is felt, when the animal is disposed to give it, whether we touch with a single finger only one of the surfaces of the organs, or apply both hands to the two surfaces, the superior and inferior, at once. In either case it is altogether indifferent whether the person who touches the fish with one finger or both hands be insulated or not. All that has been said on the necessity of a communication with the damp ground to establish a circuit, is founded on inaccurate observations.

M. Gay-Lussac made the important observation that when an insulated person touches the torpedo with one finger, it is indispensible that the contact be direct. The fish may with impunity be touched with a key, or any other metallic instrument; no shock is felt when a conducting or non-conducting body is interposed between the finger and the electrical organ of the torpedo. This circumstance proves a great

---

are excited under the belly, where the electric organ is placed.

difference between the torpedo and the gymnotus, the latter giving his strokes through an iron rod several feet long.

When the torpedo is placed on a metallic plate of very little thickness, so that the plate touches the inferior surface of the organs, the hand that supports the plate never feels any shock, though another insulated person may excite the animal, and the convulsive movement of the pectoral fins may denote the strongest and most reiterated discharges.

If, on the contrary, a person support the torpedo placed upon a metallic plate, with the left hand, as in the foregoing experiment, and the same person touch the superior surface of the electrical organ with the right hand, a strong shock is then felt in both arms. The sensation is the same when the fish is placed between two metallic plates, the edges of which do not touch, and the person applies both hands at once to these plates. The interposition of one metallic plate prevents the communication if that plate be touched with one hand only, while the interposition of two metallic plates does not prevent the shock when both hands are applied. In the latter case it cannot be doubted that the circulation of the fluid is established by the two arms.

If, in this situation of the fish between two plates, there exist any immediate communication between the edges of these two plates, no shock takes place. The chain between the two surfaces of the electric organ is then formed by the plates, and the new communication, established by the contact of the two hands with the two plates, remains without effect. We carried the torpedo with impunity between two plates of metal, and felt the strokes it gave only at the instant when they ceased to touch each other at the edges.

Nothing in the torpedo or in the gymnotus indicates that the animal modifies the electrical state of the bodies by which it is surrounded. The most delicate electrometer is no way affected in whatever manner it is employed, whether bringing it near the organs or insulating the fish, covering it with a metallic plate, and causing the plate to communicate by a conducting wire with the condenser of Volta. We were at great pains to vary the experiments by which we sought to render the electrical tension of the torpedo sensible; but they were constantly without effect, and perfectly confirmed what M. Bonpland and myself had observed respecting the gymnoti, during our abode in South America.

Electrical fishes, when very vigorous, act with equal energy under water and in the air. This observation led us to examine the conducting property of water; and we found that, when several persons form the chain between the superior and inferior surface of the organs of the torpedo, the shock is felt only when these persons join hands. The action is not intercepted if two persons, who support the torpedo with their right hands, instead of taking one another by the left hand, plunge each a metallic point into a drop of water placed on an insulating substance. On substituting flame for the drop of water, the communication is interrupted, and is only re-established, as in the gymnotus, when the two points immediately touch each other in the interior of the flame.

We are, doubtless, very far from having discovered all the secrets of the electrical action of fishes which is modified by the influence of the brain and the nerves; but the experiments we have just described are sufficient to prove that these fishes act by a concealed electricity, and by electromotive organs of a peculiar construction, which are recharged with extreme rapidity. Volta admits that the discharges of the opposite electricities in the torpedos and the gymnoti are made by their own skin, and that when we touch them with one hand only, or by means of a metallic point, we feel the effect of a lateral shock, the electrical current not being directed solely the shortest way. When a Leyden jar is placed on a wet woollen cloth (which is a bad conductor), and the jar is discharged in such a manner that the cloth makes part of the chain, prepared frogs, placed at different distances, indicate by their contractions that the current spreads itself over the whole cloth in a thousand different ways. According to this analogy, the most violent shock given by the gymnotus at a distance would be but a feeble part of the stroke which re-establishes the equilibrium in the interior of the fish.[*] As the gymnotus directs its stroke wherever it pleases, it must also be admitted that the discharge is not made by the whole skin at once, but that the animal, excited perhaps by the motion of a fluid poured into one part of the cellular membrane, establishes at will the communication between its organs and some particular part of the skin. It may be conceived that a lateral stroke, out of the direct current, must become imperceptible under the two conditions of a very weak discharge, or a very great obstacle presented by the nature and length of the conductor. Notwithstanding these considerations, it appears to me very surprising that shocks of the torpedo, strong in appearance, are not propagated to the hand when a very thin plate of metal is interposed between it and the fish.

Schilling declared that the gymnotus approached the magnet involuntarily. We tried in a thousand ways this supposed influence of the magnet on the electrical organs, without having ever observed any sensible effect. The fish no more approached the magnet, than a bar of iron not magnetic. Iron-filings, thrown on its back, remained motionless.

---

[*] The heterogeneous poles of the double electrical organs must exist in each organ. Mr. Todd has recently proved, by experiments made on torpedos at the Cape of Good Hope, that the animal continues to give violent shocks when one of these organs is extirpated. On the contrary, all electrical action is stopped (and this point, as elucidated by Galvani, is of the greatest importance) if injury be inflicted on the brain, or if the nerves which supply the plates of the electrical organs be divided. In the latter case, the nerves being cut, and the brain left untouched, the torpedo continues to live, and perform every muscular movement. A fish, exhausted by too numerous electrical discharges, suffered much more than another fish deprived, by dividing the nerves, of any communication between the brain and the electromotive apparatus. Philosophical Transactions 1816.

The gymnoti, which are objects of curiosity and of the deepest interest to the philosophers of Europe, are at once dreaded and detested by the natives. They furnish, indeed, in their muscular flesh, pretty good aliment; but the electric organ fills the greater part of their body, and this organ is slimy, and disagreeable to the taste; it is accordingly separated with care from the rest of the eel. The presence of gymnoti is also considered as the principal cause of the want of fish in the ponds and pools of the Llanos. They, however, kill many more than they devour: and the Indians told us, that when young alligators and gymnoti are caught at the same time in very strong nets, the latter never show the slightest trace of a wound, because they disable the young alligators before they are attacked by them. All the inhabitants of the waters dread the society of the gymnoti. Lizards, tortoises, and frogs, seek pools where they are secure from the electric action. It became necessary to change the direction of a road near Uritucu, because the electric eels were so numerous in one river, that they every year killed a great number of mules, as they forded the water with their burdens.

Though in the present state of our knowledge we may flatter ourselves with having thrown some light on the extraordinary effects of electric fishes, yet a vast number of physical and physiological researches still remain to be made. The brilliant results which chemistry has obtained by means of the Voltaic battery, have occupied all observers, and turned attention for some time from the examinations of the phenomena of vitality. Let us hope that these phenomena, the most awful and the most mysterious of all, will in their turn occupy the earnest attention of natural philosophers. This hope will be easily realized if they succeed in procuring anew living gymnoti in some one of the great capitals of Europe. The discoveries that will be made on the electromotive apparatus of these fish, much more energetic, and more easy of preservation, than the torpedos,[*] will extend to

_____

[*] In order to investigate the phenomena of the living electromotive apparatus in its greatest simplicity, and not to mistake for general conditions

all the phenomena of muscular motion subject to volition. It will perhaps be found that, in most animals, every contraction of the muscular fibre is preceded by a discharge from the nerve into the muscle; and that the mere simple contact of heterogeneous substances is a source of movement and of life in all organized beings. Did an ingenious and lively people, the Arabians, guess from remote antiquity, that the same force which inflames the vault of Heaven in storms, is the living and invisible weapon of inhabitants of the waters? It is said, that the electric fish of the Nile bears a name in Egypt, that signifies thunder.[*]

We left the town of Calabozo on the 24th of March, highly satisfied with our stay, and the experiments we had made on an object so worthy of the attention of physiologists. I had besides obtained some good observations of the stars; and discovered with surprise, that the errors of maps amounted here also to a quarter of a degree of latitude. No person had taken an observation before me on this spot; and geographers, magnifying as usual the distance from the coast to the islands, have carried back beyond measure all the localities towards the south.

---

circumstances which depend on the degree of energy of the electric organs, it is necessary to perform the experiments on those electrical fishes most easily tamed. If the gymnoti were not known, we might suppose, from the observations made on torpedos, that fishes cannot give their shocks from a distance through very thick strata of water, or through a bar of iron, without forming a circuit. Mr. Williamson has felt strong shocks when he held only one hand in the water, and this hand, without touching the gymnotus, was placed between it and the small fish towards which the stroke was directed from ten or fifteen inches distance. Philosophical Transactions volume 65 pages 99 and 108. When the gymnotus was enfeebled by bad health, the lateral shock was imperceptible; and in order to feel the shock, it was necessary to form a chain, and touch the fish with both hands at once. Cavendish, in his ingenious experiments on an artificial torpedo, had well remarked these differences, depending on the greater or less energy of the charge. Philosophical Transactions 1776 page 212.

[*] It appears, however, that a distinction is to be made between rahd, thunder, and rahadh, the electrical fish; and that this latter word means simply that which causes trembling.

As we advanced into the southern part of the Llanos, we found the ground more dusty, more destitute of herbage, and more cracked by the effect of long drought. The palm-trees disappeared by degrees. The thermometer kept, from eleven in the morning till sunset, at 34 or 35°. The calmer the air appeared at eight or ten feet high, the more we were enveloped in those whirlwinds of dust, caused by the little currents of air that sweep the ground. About four o'clock in the afternoon, we found a young Indian girl stretched upon the savannah. She was almost in a state of nudity, and appeared to be about twelve or thirteen years of age. Exhausted with fatigue and thirst, her eyes, nostrils, and mouth filled with dust, she breathed with a rattling in her throat, and was unable to answer our questions. A pitcher, overturned, and half filled with sand, was lying at her side. Happily one of our mules was laden with water; and we roused the girl from her lethargic state by bathing her face, and forcing her to drink a few drops of wine. She was at first alarmed on seeing herself surrounded by so many persons; but by degrees she took courage, and conversed with our guides. She judged, from the position of the sun, that she must have remained during several hours in that state of lethargy. We could not prevail on her to mount one of our beasts of burden, and she would not return to Uritucu. She had been in service at a neighbouring farm; and she had been discharged, because at the end of a long sickness she was less able to work than before. Our menaces and prayers were alike fruitless; insensible to suffering, like the rest of her race, she persisted in her resolution of going to one of the Indian Missions near the city of Calabozo. We removed the sand from her pitcher, and filled it with water. She resumed her way along the steppe, before we had remounted our horses, and was soon separated from us by a cloud of dust. During the night we forded the Rio Uritucu, which abounds with a breed of crocodiles remarkable for their ferocity. We were advised to prevent our dogs from going to drink in the rivers, for it often happens that the crocodiles of Uritucu come out of the water, and pursue dogs upon the shore. This intrepidity is so much the more striking, as at eight leagues distance, the crocodiles of the Rio Tisnao are extremely timid, and little dangerous. The manners of animals vary in the same species according to local circumstances difficult to be determined. We were shown a hut, or rather a kind of shed, in which our host of Calabozo, Don Miguel Cousin, had witnessed a very extraordinary scene. Sleeping with one of his friends on a bench or couch covered with leather, Don Miguel was awakened early in the morning by a violent shaking and a horrible noise. Clods of earth were thrown into the middle of the hut. Presently a young crocodile two or three feet long issued from under the bed, darted at a dog which lay on the threshold of the door, and, missing him in the impetuosity of his spring, ran towards the beach to gain the river. On examining the spot where the barbacoa, or couch, was placed, the cause of this strange adventure was easily discovered. The ground was disturbed to a considerable depth. It was dried mud, which had covered the crocodile in that state of lethargy, or summer-sleep, in which many of the species lie during the absence of the rains in the Llanos. The noise of men and horses, perhaps the smell of the dog, had aroused the crocodile. The hut being built at the edge of the pool, and inundated during part of the year, the crocodile had no doubt entered, at the time of the inundation of the savannahs, by the same opening at which it was seen to go out. The Indians often find enormous boas, which they call uji, or water-serpents,[*] in the same lethargic state. To reanimate them, they must be irritated, or wetted with water. Boas are killed, and immersed in the streams, to obtain, by means of putrefaction, the tendinous parts of the dorsal muscles, of which excellent guitar-strings are made at Calabozo, preferable to those furnished by the intestines of the alouate monkeys.

The drought and heat of the Llanos act like cold upon animals and plants. Beyond the tropics the trees lose their leaves

---

[*] Culebra de agua, named by the common people traga-venado, the swallower of stags. The word uji belongs to the Tamanac language.

in a very dry air. Reptiles, particularly crocodiles and boas, having very indolent habits, leave with reluctance the basins in which they have found water at the period of great inundations. In proportion as the pools become dry, these animals penetrate into the mud, to seek that degree of humidity which gives flexibility to their skin and integuments. In this state of repose they are seized with stupefaction; but possibly they preserve a communication with the external air; and, however little that communication may be, it possibly suffices to keep up the respiration of an animal of the saurian family, provided with enormous pulmonary sacs, exerting no muscular motion, and in which almost all the vital functions are suspended. It is probable that the mean temperature of the dried mud, exposed to the solar rays, is more than 40°. When the north of Egypt, where the coolest month does not fall below 13.4°, was inhabited by crocodiles, they were often found torpid with cold. They were subject to a winter-sleep, like the European frog, lizard, sand-martin, and marmot. If the hibernal lethargy be observed, both in cold-blooded and in hot-blooded animals, we shall be less surprised to learn, that these two classes furnish alike examples of a summer-sleep. In the same manner as the crocodiles of South America, the tanrecs, or Madagascar hedgehogs, in the midst of the torrid zone, pass three months of the year in lethargy.

On the 25th of March we traversed the smoothest part of the steppes of Caracas, the Mesa de Pavones. It is entirely destitute of the corypha and moriche palm-trees. As far as the eye can reach, not a single object fifteen inches high can be discovered. The air was clear, and the sky of a very deep blue; but the horizon reflected a livid and yellowish light, caused no doubt by the quantity of sand suspended in the atmosphere. We met some large herds of cattle, and with them flocks of birds of a black colour with an olive shade. They are of the genus Crotophaga,[*]

---

[*] The Spanish colonists call the Crotophaga ani, zamurito (little carrion vulture—Vultur aura minuta), or garapatero, the eater of garaparas, insects of

and follow the cattle. We had often seen them perched on the backs of cows, seeking for gadflies and other insects. Like many birds of these desert places, they fear so little the approach of man, that children often catch them in their hands. In the valleys of Aragua, where they are very common, we have seen them perch upon the hammocks on which we were reposing, in open day.

We discover, between Calabozo, Uritucu, and the Mesa de Pavones, wherever there are excavations of some feet deep, the geological constitution of the Llanos. A formation of red sandstone (ancient conglomerate) covers an extent of several thousand square leagues. We shall find it again in the vast plains of the Amazon, on the eastern boundary of the province of Jaen de Bracamoros. This prodigious extension of red sandstone in the low grounds stretching along the east of the Andes, is one of the most striking phenomena I observed during my examination of rocks in the equinoctial regions.

The red sandstone of the Llanos of Caracas lies in a concave position, between the primitive mountains of the shore and of Parime. On the north it is backed by the transition-slates,[*] and on the south it rests immediately on the granites of the Orinoco. We observed in it rounded fragments of quartz (kieselschiefer), and Lydian stone, cemented by an olive-brown ferruginous clay. The cement is sometimes of so bright a red that the people of the country take it for cinnabar. We met a Capuchin monk at Calabozo, who was in vain attempting to extract mercury from this red sandstone. In the Mesa de Paja this rock contains strata of another quartzose sandstone, very fine-grained; more to the south it contains masses of brown iron, and fragments of petrified trees of the monocotyledonous family, but we did not see in it any shells. The red sandstone, called by the Llaneros, the stone of the reefs (piedra de arrecifes), is everywhere

the Acarus family.
[*] At Malpaso and Piedras Azules.

covered with a stratum of clay. This clay, dried and hardened in the sun, splits into separate prismatic pieces with five or six sides. Does it belong to the trap-formation of Parapara? It becomes thicker, and mixed with sand, as we approach the Rio Apure; for near Calabozo it is one toise thick, near the mission of Guayaval five toises, which may lead to the belief that the strata of red sandstone dips towards the south. We gathered in the Mesa de Pavones little nodules of blue iron-ore disseminated in the clay.

A dense whitish-gray limestone, with a smooth fracture, somewhat analogous to that of Caripe, and consequently to that of Jura, lies on the red sandstone between Tisnao and Calabozo.[*] In several other places, for instance in the Mesa de San Diego, and between Ortiz and the Mesa de Paja,[†] we find above the limestone lamellar gypsum alternating with strata of marl. Considerable quantities of this gypsum are sent to the city of Caracas,[‡] which is situated amidst primitive mountains.

This gypsum generally forms only small beds, and is mixed with a great deal of fibrous gypsum. Is it of the same formation as that of Guire, on the coast of Paria, which contains sulphur? or do the masses of this latter substance, found in the valley of Buen Pastor and on the banks of the Orinoco, belong, with the argillaceous gypsum of the Llanos, to a secondary formation much more recent.

These questions are very interesting in the study of the relative antiquity of rocks, which is the principal basis of geology. I know not of any salt-deposits in the Llanos. Horned cattle prosper here without those famous bareros, or

---

[*] Does this formation of secondary limestone of the Llanos contain galena? It has been found in strata of black marl, at Barbacoa, between Truxillo and Barquesimeto, north-west of the Llanos.

[†] Also near Cachipe and San Joacquim, in the Llanos of Barcelona.

[‡] This trade is carried on at Parapara. A load of eight arrobas sells at Caracas for twenty-four piastres.

muriatiferous lands, which abound in the Pampas of Buenos Ayres.[*]

After having wandered for a long time, and without any traces of a road, in the desert savannahs of the Mesa de Pavones, we were agreeably surprised when we came to a solitary farm, the Hato de Alta Gracia, surrounded with gardens and basins of limpid water. Hedges of bead-trees encircled groups of icacoes laden with fruit. Farther on we passed the night near the small village of San Geronymo del Guayaval, founded by Capuchin missionaries. It is situated near the banks of the Rio Guarico, which falls into the Apure. I visited the missionary, who had no other habitation than his church, not having yet built a house. He was a young man, and he received us in the most obliging manner, giving us all the information we desired. His village, or to use the word established among the monks, his Mission, was not easy to govern. The founder, who had not hesitated to establish for his own profit a pulperia, in other words, to sell bananas and guarapo in the church itself, had shown himself to be not very nice in the choice of the new colonists. Many marauders of the Llanos had settled at Guayaval, because the inhabitants of a Mission are exempt from the authority of secular law. Here, as in Australia, it cannot be expected that good colonists will be formed before the second or third generation.

We passed the Guarico, and encamped in the savannahs south of Guayaval. Enormous bats, no doubt of the tribe of Phyllostomas, hovered as usual over our hammocks during a great part of the night. Every moment they seemed to be about to fasten on our faces. Early in the morning we pursued our way over low grounds, often inundated. In the season of rains, a boat may be navigated, as on a lake, between the Guarico and the Apure. We arrived on the 27th of March at the Villa de San Fernando, the capital of the Mission of the Capuchins in the province of Varinas. This was the termination of our journey

---

[*] Known in North America under the name of salt-licks.

over the Llanos; for we passed the three months of April, May, and June on the rivers.

CHAPTER 2.18.

SAN FERNANDO DE APURE. INTERTWININGS AND BIFURCATIONS OF THE RIVERS APURE AND ARAUCA. NAVIGATION ON THE RIO APURE.

Till the second half of the eighteenth century the names of the great rivers Apure, Arauca, and Meta were scarcely known in Europe: certainly less than they had been in the two preceding centuries, when the valiant Felipe de Urre and the conquerors of Tocuyo traversed the Llanos, to seek, beyond the Apure, the great legendary city of El Dorado, and the rich country of the Omeguas, the Timbuctoo of the New Continent. Such daring expeditions could not be carried out without all the apparatus of war; and the weapons, which had been destined for the defence of the new colonists, were employed without intermission against the unhappy natives. When more peaceful times succeeded to those of violence and public calamity, two powerful Indian tribes, the Cabres and the Caribs of the Orinoco, made themselves masters of the country which the Conquistadores had ceased to ravage. None but poor monks were then permitted to advance to the south of the steppes. Beyond the Uritucu an unknown world opened to the Spanish colonists; and the descendants of those intrepid warriors who had extended their conquests from Peru to the coasts of New Grenada and the mouth of the Amazon, knew not the roads that lead from Coro to the Rio Meta. The shore of Venezuela remained a separate country; and the slow conquests of the Jesuit missionaries were successful only by skirting the banks of the Orinoco. These fathers had already penetrated beyond the great cataracts of Atures and Maypures, when the Andalusian Capuchins had scarcely reached the plains of Calabozo, from the coast and the valleys of Aragua. It would be difficult to explain these contrasts by the system according to which the different monastic orders are governed; for the aspect of the country contributes powerfully to the more or less rapid progress of the Missions. They extend but slowly into the interior of the land, over mountains, or in steppes, wherever they do not follow the course of a particular river. It will scarcely be believed, that the Villa de Fernando de Apure, only fifty leagues distant in a direct line from that part of the coast of Caracas which has been longest inhabited, was founded at no earlier a date than 1789. We were shown a parchment, full of fine paintings, containing the privileges of this little town. The parchment was sent from Madrid at the solicitation of the monks, whilst yet only a few huts of reeds were to be seen around a great cross raised in the centre of the hamlet. The missionaries and the secular governments being alike interested in exaggerating in Europe what they have done to augment the culture and population of the provinces beyond the sea, it often happens that names of towns and villages are placed on the list of new conquests, long before their foundation.

The situation of San Fernando, on a large navigable river, near the mouth of another river which traverses the whole province of Varinas, is extremely advantageous for trade. Every production of that province, hides, cacao, cotton, and the indigo of Mijagual, which is of the first quality, passes through this town towards the mouths of the Orinoco. During the season of rains large vessels go from Angostura as far as San Fernando de Apure, and by the Rio Santo Domingo as far as Torunos, the port of the town of Varinas. At that period the inundations of the rivers, which form a labyrinth of branches between the Apure, the Arauca, the Capanaparo, and the Sinaruco, cover a country of nearly four hundred square leagues. At this point, the Orinoco, turned aside from its course, not by neighbouring mountains, but by the rising of counterslopes, runs eastward instead of following its previous direction in the line of the meridian. Considering the surface of the globe as a polyhedron, formed of planes variously inclined, we may conceive by the mere inspection of the maps, that the intersection of these slopes, rising towards the north, the west, and south,[*] between San Fernando de Apure, Caycara, and the mouth of the Meta, must cause a considerable depression. The savannahs in this basin are covered with twelve or fourteen feet of water, and present, at the period of rains, the aspect of a great lake. The farms and villages which seem as if situated on shoals, scarcely rise two or three feet above the surface of the water. Everything here calls to mind the inundations of Lower Egypt, and the lake of Xarayes, heretofore so celebrated among geographers, though it exists only during some months of the year. The swellings of the rivers Apure, Meta, and Orinoco, are also periodical. In the rainy season, the horses that wander in the savannah, and have not time to reach the rising grounds of the Llanos, perish by hundreds. The mares are seen, followed by their colts,[†] swimming during a part of the day to feed upon the grass, the tops of which alone wave above the waters. In this state they are pursued by the crocodiles, and it is by no means uncommon to find the prints of the teeth of these carnivorous reptiles on their thighs. The carcases of horses, mules, and cows, attract an innumerable quantity of vultures. The zamuros are the ibisis of this country, and they render the same service to the inhabitants of the Llanos as the Vultur percnopterus to the inhabitants of Egypt.

---

[*] The risings towards the north and west are connected with two lines of ridges, the mountains of Villa de Cura and of Merida. The third slope, running from north to south, is that of the land-strait between the Andes and the chain of Parime. It determines the general inclination of the Orinoco, from the mouth of the Guaviare to that of the Apure.

[†] The colts are drowned everywhere in large numbers, because they are sooner tired of swimming, and strive to follow the mares in places where the latter alone can touch the ground.

We cannot reflect on the effects of these inundations without admiring the prodigious pliability of the organization of the animals which man has subjected to his sway. In Greenland the dog eats the refuse of the fisheries; and when fish are wanting, feeds on seaweed. The ass and the horse, originally natives of the cold and barren plains of Upper Asia, follow man to the New World, return to the wild state, and lead a restless and weary life in the burning climates of the tropics. Pressed alternately by excess of drought and of humidity, they sometimes seek a pool in the midst of a bare and dusty plain, to quench their thirst; and at other times flee from water, and the overflowing rivers, as menaced by an enemy that threatens them on all sides. Tormented during the day by gadflies and mosquitos, the horses, mules, and cows find themselves attacked at night by enormous bats, which fasten on their backs, and cause wounds that become dangerous, because they are filled with acaridae and other hurtful insects. In the time of great drought the mules gnaw even the thorny cactus* in order to imbibe its cooling juice, and draw it forth as from a vegetable fountain. During the great inundations these same animals lead an amphibious life, surrounded by crocodiles, water-serpents, and manatees. Yet, such are the immutable laws of nature, that their races are preserved in the struggle with the elements, and amid so many sufferings and dangers. When the waters retire, and the rivers return again into their beds, the savannah is overspread with a beautiful scented grass; and the animals of Europe and Upper Asia seem to enjoy, as in their native climes, the renewed vegetation of spring.

During the time of great floods, the inhabitants of these countries, to avoid the force of the currents, and the danger arising from the trunks of trees which these currents bring down, instead of ascending the beds of rivers in their boats, cross the

---

* The asses are particularly adroit in extracting the moisture contained in the Cactus melocatus. They push aside the thorns with their hoofs; but sometimes lame themselves in performing this feat.

savannahs. To go from San Fernando to the villages of San Juan de Payara, San Raphael de Atamaica, or San Francisco de Capanaparo, they direct their course due south, as if they were crossing a single river of twenty leagues broad. The junctions of the Guarico, the Apure, the Cabullare, and the Arauca with the Orinoco, form, at a hundred and sixty leagues from the coast of Guiana, a kind of interior Delta, of which hydrography furnishes few examples in the Old World. According to the height of the mercury in the barometer, the waters of the Apure have only a fall of thirty-four toises from San Fernando to the sea. The fall from the mouths of the Osage and the Missouri to the bar of the Mississippi is not more considerable. The savannahs of Lower Louisiana everywhere remind us of the savannahs of the Lower Orinoco.

During our stay of three days in the little town of San Fernando, we lodged with the Capuchin missionary, who lived much at his ease. We were recommended to him by the bishop of Caracas, and he showed us the most obliging attention. He consulted me on the works that had been undertaken to prevent the flood from undermining the shore on which the town was built. The flowing of the Portuguesa into the Apure gives the latter an impulse towards south-east; and, instead of procuring a freer course for the river, attempts were made to confine it by dykes and piers. It was easy to predict that these would be rapidly destroyed by the swell of the waters, the shore having been weakened by taking away the earth from behind the dyke to employ it in these hydraulic constructions.

San Fernando is celebrated for the excessive heat which prevails there the greater part of the year; and before I begin the recital of our long navigation on the rivers, I shall relate some facts calculated to throw light on the meteorology of the tropics. We went, provided with thermometers, to the flat shores covered with white sand which border the river Apure. At two in the afternoon I found the sand, wherever it was exposed to the sun, at 52.5°. The instrument, raised eighteen inches above the sand, marked 42.8°, and at six feet high 38.7°. The temperature of the air under the shade of a ceiba was 36.2°. These observations were made during a dead calm. As soon as the wind began to blow, the temperature of the air rose 3° higher, yet we were not enveloped by a wind of sand, but the strata of air had been in contact with a soil more strongly heated, or through which whirlwinds of sand had passed. This western part of the Llanos is the hottest, because it receives air that has already crossed the rest of the barren steppe. The same difference has been observed between the eastern and western parts of the deserts of Africa, where the trade-winds blow.

The heat augments sensibly in the Llanos during the rainy season, particularly in the month of July, when the sky is cloudy, and reflects the radiant heat toward the earth. During this season the breeze entirely ceases; and, according to good thermometrical observations made by M. Pozo, the thermometer rises in the shade to 39 and 39.5°, though kept at the distance of more than fifteen feet from the ground. As we approached the banks of the Portuguesa, the Apure, and the Apurito, the air became cooler from the evaporation of so considerable a mass of water. This effect is more especially perceptible at sunset. During the day the shores of the rivers, covered with white sand, reflect the heat in an insupportable degree, even more than the yellowish brown clayey grounds of Calabozo and Tisnao.

On the 28th of March I was on the shore at sunrise to measure the breadth of the Apure, which is two hundred and six toises. The thunder rolled in all directions around. It was the first storm and the first rain of the season. The river was swelled by the easterly wind; but it soon became calm, and then some great cetacea, much resembling the porpoises of our seas, began to play in long files on the surface of the water. The slow and indolent crocodiles seem to dread the neighbourhood of these animals, so noisy and impetuous in their evolutions, for we saw them dive whenever they approached. It is a very extraordinary phenomenon to find cetacea at such a distance from the coast.

The Spaniards of the Missions designate them, as they do the porpoises of the ocean, by the name of toninas. The Tamanacs call them orinucna. They are three or four feet long; and bending their back, and pressing with their tail on the inferior strata of the water, they expose to view a part of the back and of the dorsal fin. I did not succeed in obtaining any, though I often engaged Indians to shoot at them with their arrows. Father Gili asserts that the Gumanos eat their flesh. Are these cetacea peculiar to the great rivers of South America, like the manatee, which, according to Cuvier, is also a fresh water cetaceous animal? or must we admit that they go up from the sea against the current, as the beluga sometimes does in the rivers of Asia? What would lead me to doubt this last supposition is, that we saw toninas above the great cataracts of the Orinoco, in the Rio Atabapo. Did they penetrate into the centre of equinoctial America from the mouth of the Amazon, by the communication of that river with the Rio Negro, the Cassiquiare, and the Orinoco? They are found here at all seasons, and nothing seems to denote that they make periodical migrations like salmon.

While the thunder rolled around us, the sky displayed only scattered clouds, that advanced slowly toward the zenith, and in an opposite direction. The hygrometer of Deluc was at 53°, the centigrade thermometer 23.7°, and Saussure's hygrometer 87.5°. The electrometer gave no sign of electricity. As the storm gathered, the blue of the sky changed at first to deep azure and then to grey. The vesicular vapour became visible, and the thermometer rose three degrees, as is almost always the case, within the tropics, from a cloudy sky which reflects the radiant heat of the soil. A heavy rain fell. Being sufficiently habituated to the climate not to fear the effect of tropical rains, we remained on the shore to observe the electrometer. I held it more than twenty minutes in my hand, six feet above the ground, and observed that in general the pith-balls separated only a few seconds before the lightning was seen. The separation was four lines. The electric charge remained the same during several minutes; and having time to determine the nature of the electricity, by approaching a stick of sealing-wax, I saw here what I had often observed on the ridge of the Andes during a storm, that the electricity of the atmosphere was first positive, then nil, and then negative. These oscillations from positive to negative were often repeated. Yet the electrometer constantly denoted, a little before the lightning, only E., or positive E., and never negative E. Towards the end of the storm the west wind blew very strongly. The clouds dispersed, and the thermometer sunk to 22° on account of the evaporation from the soil, and the freer radiation towards the sky.

I have entered into these details on the electric charge of the atmosphere because travellers in general confine themselves to the description of the impressions produced on a European newly arrived by the solemn spectacle of a tropical storm. In a country where the year is divided into great seasons of drought and wet, or, as the Indians say in their expressive language, of sun[*] and rain[†], it is highly interesting to follow the progress of meteorological phenomena in the transition from one season to another. We had already observed, in the valleys of Aragua from the 18th and 19th of February, clouds forming at the commencement of the night. In the beginning of the month of March the accumulation of the vesicular vapours, visible to the eye, and with them signs of atmospheric electricity, augmented daily. We saw flashes of heat-lightning to the south; and the electrometer of Volta constantly displayed, at sunset, positive electricity. The pith balls, unexcited during the day, separated to the width of three or four lines at the commencement of the night, which is triple what I generally observed in Europe, with the same instrument, in calm weather. Upon the whole, from the 26th of May, the electrical equilibrium of the atmosphere

---

[*] In the Maypure dialect camoti, properly the heat [of the sun]. The Tamanacs call the season of drought uamu, the time of grasshoppers.
[†] In the Tamanac language canepo. The year is designated, among several nations, by the name of one of the two seasons. The Maypures say, so many suns, (or rather so many heats;) the Tamanacs, so many rains.

seemed disturbed. During whole hours the electricity was nil, then it became very strong, and soon after was again imperceptible. The hygrometer of Deluc continued to indicate great dryness (from 33 to 35°), and yet the atmosphere appeared no longer the same. Amidst these perpetual variations of the electric state of the air, the trees, divested of their foliage, already began to unfold new leaves, and seemed to feel the approach of spring.

The variations which we have just described are not peculiar to one year. Everything in the equinoctial zone has a wonderful uniformity of succession, because the active powers of nature limit and balance each other, according to laws that are easily recognized. I shall here note the progress of atmospherical phenomena in the islands to the east of the Cordilleras of Merida and of New Grenada, in the Llanos of Venezuela and the Rio Meta, from four to ten degrees of north latitude, wherever the rains are constant from May to October, and comprehending consequently the periods of the greatest heats, which occur in July and August.[*]

Nothing can equal the clearness of the atmosphere from the month of December to that of February. The sky is then constantly without clouds; and if one should appear, it is a phenomenon that engages the whole attention of the inhabitants. A breeze from the east, and from east-north-east, blows with violence. As it brings with it air always of the same temperature, the vapours cannot become visible by cooling.

About the end of February and the beginning of March, the blue of the sky is less intense, the hygrometer indicates by degrees greater humidity, the stars are sometimes veiled by a slight stratum of vapour, and their light is no longer steady and

---

[*] The maximum of the heat is not felt on the coast, at Cumana, at La Guayra, and in the neighbouring island of Margareta, before the month of September; and the rains, if the name can be given to a few drops that fall at intervals, are observed only in the months of October and November.

planetary; they are seen twinkling from time to time when at 20° above the horizon. The breeze at this period becomes less strong, less regular, and is often interrupted by dead calms. The clouds accumulate towards south-south-east, appearing like distant mountains, with outlines strongly marked. From time to time they detach themselves from the horizon, and traverse the vault of the sky with a rapidity which little corresponds with the feeble wind prevailing in the lower strata of the air. At the end of March, the southern region of the atmosphere is illumined by small electric explosions. They are like phosphorescent gleams, circumscribed by vapour. The breeze then shifts from time to time, and for several hours together, to the west and south-west. This is a certain sign of the approach of the rainy season, which begins at the Orinoco about the end of April. The blue sky disappears, and a grey tint spreads uniformly over it. At the same time the heat of the atmosphere progressively increases; and soon the heavens are no longer obscured by clouds, but by condensed vapours. The plaintive cry of the howling apes begins to be heard before sunrise. The atmospheric electricity, which, during the season of drought, from December to March, had been constantly, in the day-time, from 1.7 to 2 lines, becomes extremely variable from the month of March. It appears nil during whole days; and then for some hours the pith-balls diverge three or four lines. The atmosphere, which is generally, in the torrid as well as in the temperate zone, in a state of positive electricity, passes alternately, for eight or ten minutes, to the negative state. The season of rains is that of storms; and yet a great number of experiments made during three years, prove to me that it is precisely in this season of storms we find the smallest degree of electric tension in the lower regions of the atmosphere. Are storms the effect of this unequal charge of the different superincumbent strata of air? What prevents the electricity from descending towards the earth, in air which becomes more humid after the month of March? The electricity at this period, instead of being diffused throughout the whole atmosphere, appears accumulated on the exterior envelope, at the

159

surface of the clouds. According to M. Gay-Lussac it is the formation of the cloud itself that carries the fluid toward its surface. The storm rises in the plains two hours after the sun has passed the meridian; consequently a short time after the moment of the maximum of diurnal heat within the tropics. It is extremely rare in the islands to hear thunder during the night, or in the morning. Storms at night are peculiar to certain valleys of rivers, having a peculiar climate.

What then are the causes of this rupture of the equilibrium in the electric tension of the air? of this continual condensation of the vapours into water? of this interruption of the breezes? of this commencement and duration of the rainy seasons? I doubt whether electricity has any influence on the formation of vapours. It is rather the formation of these vapours that augments and modifies the electrical tension. North and south of the equator, storms or great explosions take place at the same time in the temperate and in the equinoctial zone. Is there an action propagated through the great aerial ocean from the temperate zone towards the tropics? How can it be conceived, that in that zone where the sun rises constantly to so great a height above the horizon, its passage through the zenith can have so powerful an influence on the meteorological variations? I am of opinion that no local cause determines the commencement of the rains within the tropics; and that a more intimate knowledge of the higher currents of air will elucidate these problems, so complicated in appearance. We can observe only what passes in the lower strata of the atmosphere. The Andes are scarcely inhabited beyond the height of two thousand toises; and at that height the proximity of the soil, and the masses of mountains, which form the shoals of the aerial ocean, have a sensible influence on the ambient air. What we observe on the table-land of Antisana is not what we should find at the same height in a balloon, hovering over the Llanos or the surface of the ocean.

We have just seen that the season of rains and storms in the northern equinoctial zone coincides with the passage of the sun through the zenith of the place,[*] with the cessation of the north-east breezes, and with the frequency of calms and bendavales, which are stormy winds from south-east and south-west, accompanied by a cloudy sky. I believe that, in reflecting on the general laws of the equilibrium of the gaseous masses constituting our atmosphere, we may find, in the interruption of the current that blows from an homonymous pole, in the want of the renewal of air in the torrid zone, and in the continued action of an ascending humid current, a very simple cause of the coincidence of these phenomena. While the north-easterly breeze blows with all its violence north of the equator, it prevents the atmosphere which covers the equinoctial lands and seas from saturating itself with moisture. The hot and moist air of the torrid zone rises aloft, and flows off again towards the poles; while inferior polar currents, bringing drier and colder strata, are every instant taking the place of the columns of ascending air. By this constant action of two opposite currents, the humidity, far from being accumulated in the equatorial region, is carried towards the cold and temperate regions. During this season of breezes, which is that when the sun is in the southern signs, the sky in the northern equinoctial zone is constantly serene. The vesicular vapours are not condensed, because the air, unceasingly renewed, is far from the point of saturation. In proportion as the sun, entering the northern signs, rises towards the zenith, the breeze from the north-east moderates, and by degrees entirely ceases. The difference of temperature between the tropics and the temperate northern zone is then the least possible. It is the summer of the boreal pole; and, if the mean temperature of the winter, between 42 and 52° of north latitude, be from 20 to 26° of the centigrade thermometer less than the equatorial heat, the difference in summer is scarcely from 4 to 6°. The sun being in the zenith, and the breeze having ceased, the causes which

---

[*] These passages take place, in the fifth and tenth degrees of north latitude between the 3rd and the 16th of April, and between the 27th of August and the 8th of September.

produce humidity, and accumulate it in the northern equinoctial zone, become at once more active. The column of air reposing on this zone, is saturated with vapours, because it is no longer renewed by the polar current. Clouds form in this air saturated and cooled by the combined effects of radiation and the dilatation of the ascending air. This air augments its capacity for heat in proportion as it rarefies. With the formation and collection of the vesicular vapours, electricity accumulates in the higher regions of the atmosphere. The precipitation of the vapours is continual during the day; but it generally ceases at night, and frequently even before sunset. The showers are regularly more violent, and accompanied with electric explosions, a short time after the maximum of the diurnal heat. This state of things remains unchanged, till the sun enters into the southern signs. This is the commencement of cold in the northern temperate zone. The current from the north-pole is then re-established, because the difference between the heat of the equinoctial and temperate regions augments daily. The north-east breeze blows with violence, the air of the tropics is renewed, and can no longer attain the degree of saturation. The rains consequently cease, the vesicular vapour is dissolved, and the sky resumes its clearness and its azure tint. Electrical explosions are no longer heard, doubtless because electricity no longer comes in contact with the groups of vesicular vapours in the high regions of the air, I had almost said the coating of clouds, on which the fluid can accumulate.

We have here considered the cessation of the breezes as the principal cause of the equatorial rains. These rains in each hemisphere last only as long as the sun has its declination in that hemisphere. It is necessary to observe, that the absence of the breeze is not always succeeded by a dead calm; but that the calm is often interrupted, particularly along the western coast of America, by bendavales, or south-west and south-east winds. This phenomenon seems to demonstrate that the columns of humid air which rise in the northern equatorial zone, sometimes flow off toward the south pole. In fact, the countries situated in the torrid zone, both north and south of the equator, furnish, during their summer, while the sun is passing through their zenith, the maximum of difference of temperature with the air of the opposite pole. The southern temperate zone has its winter, while it rains on the north of the equator; and while a mean heat prevails from 5 to 6° greater than in the time of drought, when the sun is lower.[*] The continuation of the rains, while the bendavales blow, proves that the currents from the remoter pole do not act in the northern equinoctial zone like the currents of the nearer pole, on account of the greater humidity of the southern polar current. The air, wafted by this current, comes from a hemisphere consisting almost entirely of water. It traverses all the southern equatorial zone to reach the parallel of 8° north latitude; and is consequently less dry, less cold, less adapted to act as a counter-current to renew the equinoctial air and prevent its saturation, than the northern polar current, or the breeze from the north-east.[†] We may suppose that the bendavales are impetuous winds which, on some coasts, for instance on that of Guatimala, (because they are not the effect of a regular and progressive descent of the air of the tropics towards the south pole, but they alternate with calms), are accompanied by electrical explosions, and are in fact squalls, that indicate a reflux, an abrupt and instantaneous rupture, of equilibrium in the aerial ocean.

We have here discussed one of the most important phenomena of the meteorology of the tropics, considered in its most general view. In the same manner as the limits of the trade-winds do not form circles parallel with the equator, the action of the polar currents is variously felt in different meridians. The

[*] From the equator to 10° of north latitude the mean temperatures of the summer and winter months scarcely differ 2 or 3°; but at the limits of the torrid zone, toward the tropic of Cancer, the difference amounts to 8 or 9°.

[†] In the two temperate zones the air loses its transparency every time that the wind blows from the opposite pole, that is to say, from the pole that has not the same denomination as the hemisphere in which the wind blows.

chains of mountains and the coasts in the same hemisphere have often opposite seasons. There are several examples of these anomalies; but, in order to discover the laws of nature, we must know, before we examine into the causes of local perturbations, the average state of the atmosphere, and the constant type of its variations.

The aspect of the sky, the progress of the electricity, and the shower of the 28th of March, announced the commencement of the rainy season; we were still advised, however, to go from San Fernando de Apure by San Francisco de Capanaparo, the Rio Sinaruco, and the Hato de San Antonio, to the village of the Ottomacs, recently founded near the banks of the Meta, and to embark on the Orinoco a little above Carichana. This way by land lies across an unhealthy and feverish country. An old farmer named Francisco Sanchez obligingly offered to conduct us. His dress denoted the great simplicity of manners prevailing in those distant countries. He had acquired a fortune of more than 100,000 piastres, and yet he mounted on horseback with his feet bare, and wearing large silver spurs. We knew by the experience of several weeks the dull uniformity of the vegetation of the Llanos, and preferred the longer road, which leads by the Rio Apure to the Orinoco. We chose one of those very large canoes called lanchas by the Spaniards. A pilot and four Indians were sufficient to manage it. They constructed, near the stern, in the space of a few hours, a cabin covered with palm-leaves, sufficiently spacious to contain a table and benches. These were made of ox-hides, strained tight, and nailed to frames of brazil-wood. I mention these minute circumstances, to prove that our accommodations on the Rio Apure were far different from those to which we were afterwards reduced in the narrow boats of the Orinoco. We loaded the canoe with provision for a month. Fowls, eggs, plantains, cassava, and cacao, are found in abundance at San Fernando. The good Capuchin, Fray Jose Maria de Malaga, gave us sherry wine, oranges, and tamarinds, to make cooling beverages. We could easily foresee that a roof constructed of palm-tree leaves would become excessively hot on a large river, where we were almost always exposed to the perpendicular rays of the sun. The Indians relied less on the provision we had purchased, than on their hooks and nets. We took also some fire-arms, which we found in general use as far as the cataracts; but farther south the great humidity of the air prevents the missionaries from using them. The Rio Apure abounds in fish, manatees, and turtles, the eggs of which afford an aliment more nutritious than agreeable to the taste. Its banks are inhabited by an innumerable quantity of birds, among which the pauxi and the guacharaca, which may be called the turkeys and pheasants of those countries, are found to be the most useful. Their flesh appeared to be harder and less white than that of the gallinaceous tribe in Europe, because they use much more muscular exercise. We did not forget to add to our provision, fishing-tackle, fire-arms, and a few casks of brandy, to serve as a medium of barter with the Indians of the Orinoco.

We departed from San Fernando on the 30th of March, at four in the afternoon. The weather was extremely hot; the thermometer rising in the shade to 34°, though the breeze blew very strongly from the south-east. Owing to this contrary wind we could not set our sails. We were accompanied, in the whole of this voyage on the Apure, the Orinoco, and the Rio Negro, by the brother-in-law of the governor of the province of Varinas, Don Nicolas Soto, who had recently arrived from Cadiz. Desirous of visiting countries so calculated to excite the curiosity of a European, he did not hesitate to confine himself with us during seventy-four days in a narrow boat infested with mosquitos. His amiable disposition and gay temper often helped to make us forget the sufferings of a voyage which was not wholly exempt from danger. We passed the mouth of the Apurito, and coasted the island of the same name, formed by the Apure and the Guarico. This island is in fact only a very low spot of ground, bordered by two great rivers, both of which, at a little distance from each other, fall into the Orinoco, after having formed a junction below San Fernando by the first bifurcation of the Apure. The Isla del Apurito is twenty-two leagues in length, and two or three leagues in breadth. It is divided by the Cano de la Tigrera and the Cano del Manati into three parts, the two extremes of which bear the names of Isla de Blanco and Isla de los Garzitas. The right bank of the Apure, below the Apurito, is somewhat better cultivated than the left bank, where the Yaruros, or Japuin Indians, have constructed a few huts with reeds and stalks of palm-leaves. These people, who live by hunting and fishing, are very skilful in killing jaguars. It is they who principally carry the skins, known in Europe by the name of tiger-skins, to the Spanish villages. Some of these Indians have been baptized, but they never visit the Christian churches. They are considered as savages because they choose to remain independent. Other tribes of Yaruros live under the rule of the missionaries, in the village of Achaguas, situated south of the Rio Payara. The individuals of this nation, whom I had an opportunity of seeing at the Orinoco, have a stern expression of countenance; and some features in their physiognomy, erroneously called Tartarian, belong to branches of the Mongol race, the eye very long, the cheekbones high, but the nose prominent throughout its whole length. They are taller, browner, and less thick-set than the Chayma Indians. The missionaries praise the intellectual character of the Yaruros, who were formerly a powerful and numerous nation on the banks of the Orinoco, especially in the environs of Cuycara, below the mouth of the Guarico. We passed the night at Diamante, a small sugar-plantation formed opposite the island of the same name.

During the whole of my voyage from San Fernando to San Carlos del Rio Negro, and thence to the town of Angostura, I noted down day by day, either in the boat or where we disembarked at night, all that appeared to me worthy of observation. Violent rains, and the prodigious quantity of mosquitos with which the air is filled on the banks of the Orinoco and the Cassiquiare, necessarily occasioned some interruptions; but I supplied the omission by notes taken a few days after. I here subjoin some extracts from my journal. Whatever is written while the objects we describe are before our eyes bears a character of truth and individuality which gives attraction to things the least important.

On the 31st March a contrary wind obliged us to remain on shore till noon. We saw a part of some cane-fields laid waste by the effect of a conflagration which had spread from a neighbouring forest. The wandering Indians everywhere set fire to the forest where they have encamped at night; and during the season of drought, vast provinces would be the prey of these conflagrations if the extreme hardness of the wood did not prevent the trees from being entirely consumed. We found trunks of desmanthus and mahogany which were scarcely charred two inches deep.

Having passed the Diamante we entered a land inhabited only by tigers, crocodiles, and chiguires; the latter are a large species of the genus Cavia of Linnaeus. We saw flocks of birds, crowded so closely together as to appear against the sky like a dark cloud which every instant changed its form. The river widens by degrees. One of its banks is generally barren and sandy from the effect of inundations; the other is higher, and covered with lofty trees. In some parts the river is bordered by forests on each side, and forms a straight canal a hundred and fifty toises broad. The manner in which the trees are disposed is very remarkable. We first find bushes of sauso,[*] forming a kind of hedge four feet high, and appearing as if they had been clipped by the hand of man. A copse of cedar, brazilletto, and lignum-vitae, rises behind this hedge. Palm-trees are rare; we saw only a few scattered trunks of the thorny piritu and corozo. The large quadrupeds of those regions, the jaguars, tapirs, and peccaries, have made openings in the hedge of sauso which we have just described. Through these the wild animals pass when they come to drink at the river. As they fear but little the approach of a boat, we had the pleasure of viewing them as they

---

[*] Hermesia castaneifolia. This is a new genus, approaching the alchornea of Swartz.

paced slowly along the shore till they disappeared in the forest, which they entered by one of the narrow passes left at intervals between the bushes. These scenes, which were often repeated, had ever for me a peculiar attraction. The pleasure they excite is not owing solely to the interest which the naturalist takes in the objects of his study, it is connected with a feeling common to all men who have been brought up in the habits of civilization. You find yourself in a new world, in the midst of untamed and savage nature. Now the jaguar—the beautiful panther of America—appears upon the shore; and now the hocco,[*] with its black plumage and tufted head, moves slowly along the sausos. Animals of the most different classes succeed each other. "Esse como en el Paradiso," "It is just as it was in Paradise," said our pilot, an old Indian of the Missions. Everything, indeed, in these regions recalls to mind the state of the primitive world with its innocence and felicity. But in carefully observing the manners of animals among themselves, we see that they mutually avoid and fear each other. The golden age has ceased; and in this Paradise of the American forests, as well as everywhere else, sad and long experience has taught all beings that benignity is seldom found in alliance with strength.

When the shore is of considerable breadth, the hedge of sauso remains at a distance from the river. In the intermediate space we see crocodiles, sometimes to the number of eight or ten, stretched on the sand. Motionless, with their jaws wide open, they repose by each other, without displaying any of those marks of affection observed in other animals living in society. The troop separates as soon as they quit the shore. It is, however, probably composed of one male only, and many females; for as M. Descourtils, who has so much studied the crocodiles of St. Domingo, observed to me, the males are rare, because they kill one another in fighting during the season of their loves. These monstrous creatures are so numerous, that throughout the whole

---

[*] Ceyx alector, the peacock-pheasant; C. pauxi, the cashew-bird.

course of the river we had almost at every instant five or six in view. Yet at this period the swelling of the Rio Apure was scarcely perceived; and consequently hundreds of crocodiles were still buried in the mud of the savannahs. About four in the afternoon we stopped to measure a dead crocodile which had been cast ashore. It was only sixteen feet eight inches long; some days after M. Bonpland found another, a male, twenty-two feet three inches long. In every zone, in America as in Egypt, this animal attains the same size. The species so abundant in the Apure, the Orinoco,[*] and the Rio de la Magdalena, is not a cayman, but a real crocodile, analogous to that of the Nile, having feet dentated at the external edges. When it is recollected that the male enters the age of puberty only at ten years, and that its length is then eight feet, we may presume that the crocodile measured by M. Bonpland was at least twenty-eight years old. The Indians told us, that at San Fernando scarcely a year passes, without two or three grown-up persons, particularly women who fetch water from the river, being drowned by these carnivorous reptiles. They related to us the history of a young girl of Uritucu, who by singular intrepidity and presence of mind, saved herself from the jaws of a crocodile. When she felt herself seized, she sought the eyes of the animal, and plunged her fingers into them with such violence, that the pain forced the crocodile to let her go, after having bitten off the lower part of her left arm. The girl, notwithstanding the enormous quantity of blood she lost, reached the shore, swimming with the hand that still remained to her. In those desert countries, where man is ever wrestling with nature, discourse daily turns on the best means that may be employed to escape from a tiger, a boa, or a crocodile; every one prepares himself in some sort for the dangers that may await him. "I knew," said the young girl of Uritucu coolly, "that the cayman lets go his hold, if you push your fingers into his eyes." Long after my return to Europe, I learned that in the interior of

---

[*] It is the arua of the Tamanac Indians, the amana of the Maypure Indians, the Crocodilus acutus of Cuvier.

169

Africa the negroes know and practise the same means of defence. Who does not recollect, with lively interest, Isaac, the guide of the unfortunate Mungo Park, who was seized twice by a crocodile, and twice escaped from the jaws of the monster, having succeeded in thrusting his fingers into the creature's eyes while under water. The African Isaac, and the young American girl, owed their safety to the same presence of mind, and the same combination of ideas.

The movements of the crocodile of the Apure are sudden and rapid when it attacks any object; but it moves with the slowness of a salamander, when not excited by rage or hunger. The animal in running makes a rustling noise, which seems to proceed from the rubbing of the scales of its skin one against another. In this movement it bends its back, and appears higher on its legs than when at rest. We often heard this rattling of the scales very near us on the shore; but it is not true, as the Indians pretend, that, like the armadillo, the old crocodiles "can erect their scales, and every part of their armour." The motion of these animals is no doubt generally in a straight line, or rather like that of an arrow, supposing it to change its direction at certain distances. However, notwithstanding the little apparatus of false ribs, which connects the vertebrae of the neck, and seems to impede the lateral movement, crocodiles can turn easily when they please. I often saw young ones biting their tails; and other observers have seen the same action in crocodiles at their full growth. If their movements almost always appear to be straight forward, it is because, like our small lizards, they move by starts. Crocodiles are excellent swimmers; they go with facility against the most rapid current. It appeared to me, however, that in descending the river, they had some difficulty in turning quickly about. A large dog, which had accompanied us in our journey from Caracas to the Rio Negro, was one day pursued in swimming by an enormous crocodile. The latter had nearly reached its prey, when the dog escaped by turning round suddenly and swimming against the current. The crocodile performed the same movement, but much more slowly than the dog, which succeeded in gaining the shore.

The crocodiles of the Apure find abundant food in the chiguires (thick-nosed tapirs),* which live fifty or sixty together in troops on the banks of the river. These animals, as large as our pigs, have no weapons of defence; they swim somewhat better than they run: yet they become the prey of the crocodiles in the water, and of the tigers on land. It is difficult to conceive, how, being thus persecuted by two powerful enemies, they become so numerous; but they breed with the same rapidity as the little cavies or guinea-pigs, which come to us from Brazil.

We stopped below the mouth of the Cano de la Tigrera, in a sinuosity called la Vuelta del Joval, to measure the velocity of the water at its surface. It was not more than 3.2 feet† in a second, which gives 2.56 feet for the mean velocity. The height of the barometer indicated barely a slope of seventeen inches in a mile of nine hundred and fifty toises. The velocity is the simultaneous effect of the slope of the ground, and the accumulation of the waters by the swelling of the upper parts of the river. We were again surrounded by chiguires, which swim like dogs, raising their heads and necks above the water. We saw with surprise a large crocodile on the opposite shore, motionless, and sleeping in the midst of these nibbling animals. It awoke at the approach of our canoe, and went into the water slowly, without frightening the chiguires. Our Indians accounted for this

---

* Cavia capybara, Linn. The word chiguire belongs to the language of the Palenkas and the Cumanagotos. The Spaniards call this animal guardatinaja; the Caribs, capigua; the Tamanacs, cappiva; and the Maypures, chiato. According to Azara, it is known at Buenos Ayres by the Indian names of capiygua and capiguara. These various denominations show a striking analogy between the languages of the Orinoco and those of the Rio de la Plata.

† In order to measure the velocity of the surface of a river, I generally measured on the beach a base of 250 feet, and observed with the chronometer the time that a floating body, abandoned to the current, required to reach this distance.

indifference by the stupidity of the animals, but it is more probable that the chiguires know by long experience, that the crocodile of the Apure and the Orinoco does not attack upon land, unless he finds the object he would seize immediately in his way, at the instant when he throws himself into the water.

Near the Joval nature assumes an awful and extremely wild aspect. We there saw the largest jaguar we had ever met with. The natives themselves were astonished at its prodigious length, which surpassed that of any Bengal tiger I had ever seen in the museums of Europe. The animal lay stretched beneath the shade of a large zamang.* It had just killed a chiguire, but had not yet touched its prey, on which it kept one of its paws. The zamuro vultures were assembled in great numbers to devour the remains of the jaguar's repast. They presented the most curious spectacle, by a singular mixture of boldness and timidity. They advanced within the distance of two feet from the animal, but at the least movement he made they drew back. In order to observe more nearly the manners of these creatures, we went into the little skiff that accompanied our canoe. Tigers very rarely attack boats by swimming to them; and never but when their ferocity is heightened by a long privation of food. The noise of our oars led the animal to rise slowly, and hide itself behind the sauso bushes that bordered the shore. The vultures tried to profit by this moment of absence to devour the chiguire; but the tiger, notwithstanding the proximity of our boat, leaped into the midst of them, and in a fit of rage, expressed by his gait and the movement of his tail, carried off his prey to the forest. The Indians regretted that they were not provided with their lances, in order to go on shore and attack the tiger. They are accustomed to this weapon, and were right in not trusting to our fire-arms. In so excessively damp an atmosphere muskets often miss fire.

Continuing to descend the river, we met with the great herd of chiguires which the tiger had put to flight, and from which he

---

* A species of mimosa.

had selected his prey. These animals saw us land very unconcernedly; some of them were seated, and gazed upon us, moving the upper lip like rabbits. They seemed not to be afraid of man, but the sight of our dog put them to flight. Their hind legs being longer than their fore legs, their pace is a slight gallop, but with so little swiftness that we succeeded in catching two of them. The chiguire, which swims with the greatest agility, utters a short moan in running, as if its respiration were impeded. It is the largest of the family of rodentia or gnawing animals. It defends itself only at the last extremity, when it is surrounded and wounded. Having great strength in its grinding teeth,[*] particularly the hinder ones, which are pretty long, it can tear the paw of a tiger, or the leg of a horse, with its bite. Its flesh has a musky smell somewhat disagreeable; yet hams are made of it in this country, a circumstance which almost justifies the name of water-hog, given to the chiguire by some of the older naturalists. The missionary monks do not hesitate to eat these hams during Lent. According to their zoological classification they place the armadillo, the thick-nosed tapir, and the manatee, near the tortoises; the first, because it is covered with a hard armour like a sort of shell; and the others because they are amphibious. The chiguires are found in such numbers on the banks of the rivers Santo Domingo, Apure, and Arauca, in the marshes and in the inundated savannahs[†] of the Llanos, that the pasturages suffer from them. They browze the grass which fattens the horses best, and which bears the name of chiguirero, or chiguire-grass. They feed also upon fish; and we saw with surprise, that, when scared by the approach of a boat, the animal in diving remains eight or ten minutes under water.

---

[*] We counted eighteen on each side. On the hind feet, at the upper end of the metatarsus, there is a callosity three inches long and three quarters of an inch broad, destitute of hair. The animal, when seated, rests upon this part. No tail is visible externally; but on putting aside the hair we discover a tubercle, a mass of naked and wrinkled flesh, of a conical figure, and half an inch long.

[†] Near Uritucu, in the Cano del Ravanal, we saw a flock of eighty or one hundred of these animals.

We passed the night as usual, in the open air, though in a plantation, the proprietor of which employed himself in hunting tigers. He wore scarcely any clothing, and was of a dark brown complexion like a Zambo. This did not prevent his classing himself amongst the Whites. He called his wife and his daughter, who were as naked as himself, Dona Isabella and Dona Manuela. Without having ever quitted the banks of the Apure, he took a lively interest in the news of Madrid—enquiring eagerly respecting those never-ending wars, and everything down yonder (todas las cosas de alla). He knew, he said, that the king was soon to come and visit the grandees of the country of Caracas, but he added with some pleasantry, as the people of the court can eat only wheaten bread, they will never pass beyond the town of Victoria, and we shall not see them here. I had brought with me a chiguire, which I had intended to have roasted; but our host assured us, that such Indian game was not food fit for nos otros caballeros blancos, (white gentlemen like ourselves and him). Accordingly he offered us some venison, which he had killed the day before with an arrow, for he had neither powder nor fire-arms.

We supposed that a small wood of plantain-trees concealed from us the hut of the farm; but this man, so proud of his nobility and the colour of his skin, had not taken the trouble of constructing even an ajoupa, or hut of palm-leaves. He invited us to have our hammocks hung near his own, between two trees; and he assured us, with an air of complacency, that, if we came up the river in the rainy season, we should find him beneath a roof (baxo techo). We soon had reason to complain of a system of philosophy which is indulgent to indolence, and renders a man indifferent to the conveniences of life. A furious wind arose after midnight, lightnings flashed over the horizon, thunder rolled, and we were wet to the skin. During this storm a whimsical incident served to amuse us for a moment. Dona Isabella's cat had perched upon the tamarind-tree, at the foot of which we lay. It fell into the hammock of one of our companions, who, being hurt by the claws of the cat, and suddenly aroused from a profound sleep, imagined he was attacked by some wild beast of the forest. We ran to him on hearing his cries, and had some trouble to convince him of his error. While it rained in torrents on our hammocks and on our instruments which we had brought ashore, Don Ignacio congratulated us on our good fortune in not sleeping on the strand, but finding ourselves in his domain, among whites and persons of respectability (entre gente blanca y de trato). Wet as we were, we could not easily persuade ourselves of the advantages of our situation, and we listened with some impatience to the long narrative our host gave us of his pretended expedition to the Rio Meta, of the valour he had displayed in a sanguinary combat with the Guahibo Indians, and "the services that he had rendered to God and his king, in carrying away Indian children (los Indiecitos) from their parents, to distribute them in the Missions." We were struck with the singularity of finding in that vast solitude a man believing himself to be of European race and knowing no other shelter than the shade of a tree, and yet having all the vain pretensions, hereditary prejudices, and errors of long-standing civilization!

On the 1st of April, at sunrise, we quitted Senor Don Ignacio and Senora Dona Isabella his wife. The weather was cooler, for the thermometer (which generally kept up in the daytime to 30 or 35°) had sunk to 24°. The temperature of the river was little changed: it continued constantly at 26 or 27°. The current carried with it an enormous number of trunks of trees. It might be imagined that on ground entirely smooth, and where the eye cannot distinguish the least hill, the river would have formed by the force of its current a channel in a straight line; but a glance at the map, which I traced by the compass, will prove the contrary. The two banks, worn by the waters, do not furnish an equal resistance; and almost imperceptible inequalities of the level suffice to produce great sinuosities. Yet below the Joval, where the bed of the river enlarges a little, it forms a channel that appears perfectly straight, and is shaded on each side by very tall trees. This part of the river is called Cano Rico. I found it to be one hundred and thirty-six toises broad. We passed a low island, inhabited by thousands of flamingos, rose-coloured spoonbills, herons, and moorhens, which displayed plumage of the most various colours. These birds were so close together that they seemed to be unable to stir. The island they frequent is called Isla de Aves, or Bird Island. Lower down we passed the point where the Rio Arichuna, an arm of the Apure, branches off to the Cabulare, carrying away a considerable body of its waters. We stopped, on the right bank, at a little Indian mission, inhabited by the tribe of the Guamos, called the village of Santa Barbara de Arichuna.

The Guamos[*] are a race of Indians very difficult to fix on a settled spot. They have great similarity of manners with the Achaguas, the Guajibos,[†] and the Ottomacs, partaking their disregard of cleanliness, their spirit of vengeance, and their taste for wandering; but their language differs essentially. The greater part of these four tribes live by fishing and hunting, in plains often inundated, situated between the Apure, the Meta, and the Guaviare. The nature of these regions seems to invite the natives to a wandering life. On entering the mountains of the Cataracts of the Orinoco, we shall soon find, among the Piraoas, the Macos, and the Maquiritaras, milder manners, a love of agriculture, and great cleanliness in the interior of their huts. On mountain ridges, in the midst of impenetrable forests, man is compelled to fix himself; and cultivate a small spot of land. This cultivation requires little care; while, in a country where there are no other roads than rivers, the life of the hunter is laborious and difficult. The Guamos of the mission of Santa Barbara could not furnish us with the provision we wanted. They cultivate only a little cassava. They appeared hospitable; and when we entered their huts, they offered us dried fish, and water cooled in porous vessels.

---

[*] Father Gili observes that their Indian name is Uamu and Pau, and that they originally dwelt on the Upper Apure.

[†] Their Indian name is Guahiva.

Beyond the Vuelta del Cochino Roto, in a spot where the river has scooped itself a new bed, we passed the night on a bare and very extensive strand. The forest being impenetrable, we had the greatest difficulty to find dry wood to light fires, near which the Indians believe themselves in safety from the nocturnal attacks of the tiger. Our own experience seems to bear testimony in favour of this opinion; but Azara asserts that, in his time, a tiger in Paraguay carried off a man who was seated near a fire lighted in the savannah.

The night was calm and serene, and there was a beautiful moonlight. The crocodiles, stretched along the shore, placed themselves in such a manner as to be able to see the fire. We thought we observed that its blaze attracted them, as it attracts fishes, crayfish, and other inhabitants of the water. The Indians showed us the tracks of three tigers in the sand, two of which were very young. A female had no doubt conducted her little ones to drink at the river. Finding no tree on the strand, we stuck our oars in the ground, and to these we fastened our hammocks. Everything passed tranquilly till eleven at night; and then a noise so terrific arose in the neighbouring forest, that it was almost impossible to close our eyes. Amid the cries of so many wild beasts howling at once, the Indians discriminated such only as were at intervals heard separately. These were the little soft cries of the sapajous, the moans of the alouate apes, the howlings of the jaguar and couguar, the peccary, and the sloth, and the cries of the curassao, the parraka, and other gallinaceous birds. When the jaguars approached the skirt of the forest, our dog, which till then had never ceased barking, began to howl and seek for shelter beneath our hammocks. Sometimes, after a long silence, the cry of the tiger came from the tops of the trees; and then it was followed by the sharp and long whistling of the monkeys, which appeared to flee from the danger that threatened them. We heard the same noises repeated, during the course of whole months, whenever the forest approached the bed of the river. The security evinced by the Indians inspires confidence in the minds of travellers, who readily persuade themselves that the tigers are afraid of fire, and that they do not attack a man lying in his hammock. These attacks are in fact extremely rare; and, during a long abode in South America, I remember only one example, of a llanero, who was found mutilated in his hammock opposite the island of Achaguas.

When the natives are interrogated on the causes of the tremendous noise made by the beasts of the forest at certain hours of the night, the answer is, "They are keeping the feast of the full moon."

I believe this agitation is most frequently the effect of some conflict that has arisen in the depths of the forest. The jaguars, for instance, pursue the peccaries and the tapirs, which, having no defence but in their numbers, flee in close troops, and break down the bushes they find in their way. Terrified at this struggle, the timid and mistrustful monkeys answer, from the tops of the trees, the cries of the large animals. They awaken the birds that live in society, and by degrees the whole assembly is in commotion. It is not always in a fine moonlight, but more particularly at the time of a storm and violent showers, that this tumult takes place among the wild beasts. "May Heaven grant them a quiet night and repose, and us also!" said the monk who accompanied us to the Rio Negro, when, sinking with fatigue, he assisted in arranging our accommodations for the night. It was indeed strange, to find no silence in the solitude of woods. In the inns of Spain we dread the sound of guitars from the next apartment; on the Orinoco, where the traveller's resting-place is the open beach, or beneath the shelter of a solitary tree, his slumbers are disturbed by a serenade from the forest.

We set sail before sunrise, on the 2nd of April. The morning was beautiful and cool, according to the feelings of those who are accustomed to the heat of these climates. The thermometer rose only to 28° in the air, but the dry and white sand of the beach, notwithstanding its radiation towards a cloudless sky, retained a temperature of 36°. The porpoises (toninas) ploughed the river in long files. The shore was covered with fishing-birds.

Some of these perched on the floating wood as it passed down the river, and surprised the fish that preferred the middle of the stream. Our canoe was aground several times during the morning. These shocks are sufficiently violent to split a light bark. We struck on the points of several large trees, which remain for years in an oblique position, sunk in the mud. These trees descend from Sarare, at the period of great inundations, and they so fill the bed of the river, that canoes in going up find it difficult sometimes to make their way over the shoals, or wherever there are eddies. We reached a spot near the island of Carizales, where we saw trunks of the locust-tree, of an enormous size, above the surface of the water. They were covered with a species of plotus, nearly resembling the anhinga, or white bellied darter. These birds perch in files, like pheasants and parrakas, and they remain for hours entirely motionless, with their beaks raised toward the sky.

Below the island of Carizales we observed a diminution of the waters of the river, at which we were the more surprised, as, after the bifurcation at la Boca de Arichuna, there is no branch, no natural drain, which takes away water from the Apure. The loss is solely the effect of evaporation, and of filtration on a sandy and wet shore. Some idea of the magnitude of these effects may be formed, from the fact that we found the heat of the dry sands, at different hours of the day, from 36 to 52°, and that of sands covered with three or four inches of water 32°. The beds of rivers are heated as far as the depth to which the solar rays can penetrate without undergoing too great an extinction in their passage through the superincumbent strata of water. Besides, filtration extends in a lateral direction far beyond the bed of the river. The shore, which appears dry to us, imbibes water as far up as to the level of the surface of the river. We saw water gush out at the distance of fifty toises from the shore, every time that the Indians struck their oars into the ground. Now these sands, wet below, but dry above, and exposed to the solar rays, act like sponges, and lose the infiltrated water every instant by evaporation. The vapour that is emitted, traverses the upper stratum of sand strongly heated, and becomes sensible to the eye when the air cools towards evening. As the beach dries, it draws from the river new portions of water; and it may be easily conceived that this continual alternation of vaporization and lateral absorption must cause an immense loss, difficult to submit to exact calculation. The increase of these losses would be in proportion to the length of the course of the rivers, if from their source to their mouth they were equally surrounded by a flat shore; but these shores being formed by deposits from the water, and the water having less velocity in proportion as it is more remote from its source, throwing down more sediment in the lower than in the upper part of its course, many rivers in hot climates undergo a diminution in the quantity of their water, as they approach their outlets. Mr. Barrow observed these curious effects of sands in the southern part of Africa, on the banks of the Orange River. They have also become the subject of a very important discussion, in the various hypotheses that have been formed respecting the course of the Niger.[*]

Near the Vuelta de Basilio, where we landed to collect plants, we saw on the top of a tree two beautiful little monkeys, black as jet, of the size of the sai, with prehensile tails. Their physiognomy and their movements sufficiently showed that they were neither the quato (Simia beelzebub) nor the chamek, nor any of the Ateles. Our Indians themselves had never seen any that resembled them. Monkeys, especially those living in troops, make long emigrations at certain periods, and consequently it happens that at the beginning of the rainy seasons the natives discover round their huts different kinds which they have not

---

[*] Geographers supposed, for a long period, that the Niger was entirely absorbed by the sands, and evaporated by the heat of the tropical sun, as no embouchure could be found on the western coast of Africa to meet the requirements of so enormous a river. It was discovered, however, by the Landers, in 1830, that it does really flow into the Atlantic; yet the cause mentioned above is so powerful, that of all the numerous branches into which it separates at its mouth, only one (the Nun River) is navigable even for light ships, and for half the year even those are unable to enter.

before observed. On this same bank our guides showed us a nest of young iguanas only four inches long. It was difficult to distinguish them from common lizards. There was no distinguishing mark yet formed but the dewlap below the throat. The dorsal spines, the large erect scales, all those appendages that render the iguana so remarkable when it attains its full growth, were scarcely traceable.

The flesh of this animal of the saurian family appeared to us to have an agreeable taste in every country where the climate is very dry; we even found it so at periods when we were not in want of other food. It is extremely white, and next to the flesh of the armadillo, one of the best kinds of food to be found in the huts of the natives.

It rained toward evening, and before the rain fell, swallows, exactly resembling our own, skimmed over the surface of the water. We saw also a flock of paroquets pursued by little goshawks without crests. The piercing cries of these paroquets contrasted singularly with the whistling of the birds of prey. We passed the night in the open air, upon the beach, near the island of Carizales. There were several Indian huts in the neighbourhood, surrounded with plantations. Our pilot assured us beforehand that we should not hear the cries of the jaguar, which, when not extremely pressed by hunger, withdraws from places where he does not reign unmolested. "Men put him out of humour" (los hombres lo enfadan), say the people in the Missions. A pleasant and simple expression, that marks a well-observed fact.

Since our departure from San Fernando we had not met a single boat on this fine river. Everything denoted the most profound solitude. On the morning of the 3rd of April our Indians caught with a hook the fish known in the country by the name of caribe,[*] or caribito, because no other fish has such a thirst for blood. It attacks bathers and swimmers, from whom it

---

[*] Caribe in the Spanish language signifies cannibal.

181

often bites away considerable pieces of flesh. The Indians dread extremely these caribes; and several of them showed us the scars of deep wounds in the calf of the leg and in the thigh, made by these little animals. They swim at the bottom of rivers; but if a few drops of blood be shed on the water, they rise by thousands to the surface, so that if a person be only slightly bitten, it is difficult for him to get out of the water without receiving a severer wound. When we reflect on the numbers of these fish, the largest and most voracious of which are only four or five inches long, on the triangular form of their sharp and cutting teeth, and on the amplitude of their retractile mouths, we need not be surprised at the fear which the caribe excites in the inhabitants of the banks of the Apure and the Orinoco. In places where the river was very limpid, where not a fish appeared, we threw into the water little morsels of raw flesh, and in a few minutes a perfect cloud of caribes had come to dispute their prey. The belly of this fish has a cutting edge, indented like a saw, a characteristic which may be also traced in the serra-salmes, the myletes, and the pristigastres. The presence of a second adipous dorsal fin, and the form of the teeth, covered by lips distant from each other, and largest in the lower jaw, place the caribe among the serra-salmes. Its mouth is much wider than that of the myletes of Cuvier. Its body, toward the back, is ash-coloured with a tint of green, but the belly, the gill-covers, and the pectoral, anal, and ventral fins, are of a fine orange hue. Three species are known in the Orinoco, and are distinguished by their size. The intermediate appears to be identical with the medium species of the piraya, or piranha, of Marcgrav.[*] The caribito has a very agreeable flavour. As no one dares to bathe where it is found, it may be considered as one of the greatest scourges of those climates, in which the sting of the mosquitos and the general irritation of the skin render the use of baths so necessary.

---

[*] Salmo rhombeus, Linn.

We stopped at noon in a desert spot called Algodonal. I left my companions while they drew the boat ashore and were occupied in preparing our dinner. I went along the beach to get a near view of a group of crocodiles sleeping in the sun, and lying in such a manner as to have their tails, which were furnished with broad plates, resting on one another. Some little herons,[*] white as snow, walked along their backs, and even upon their heads, as if passing over trunks of trees. The crocodiles were of a greenish grey, half covered with dried mud; from their colour and immobility they might have been taken for statues of bronze. This excursion had nearly proved fatal to me. I had kept my eyes constantly turned towards the river; but, whilst picking up some spangles of mica agglomerated together in the sand, I discovered the recent footsteps of a tiger, easily distinguishable from their form and size. The animal had gone towards the forest, and turning my eyes on that side, I found myself within eighty paces of a jaguar that was lying under the thick foliage of a ceiba. No tiger had ever appeared to me so large.

There are accidents in life against which we may seek in vain to fortify our reason. I was extremely alarmed, yet sufficiently master of myself and of my motions to enable me to follow the advice which the Indians had so often given us as to how we ought to act in such cases. I continued to walk on without running, avoided moving my arms, and I thought I observed that the jaguar's attention was fixed on a herd of capybaras which was crossing the river. I then began to return, making a large circuit toward the edge of the water. As the distance increased, I thought I might accelerate my pace. How often was I tempted to look back in order to assure myself that I was not pursued! Happily I yielded very tardily to this desire. The jaguar had

---

[*] Garzon chico. It is believed, in Upper Egypt, that herons have an affection for crocodiles, because they take advantage in fishing of the terror that monstrous animal causes among the fishes, which he drives from the bottom to the surface of the water; but on the banks of the Nile, the heron keeps prudently at some distance from the crocodile.

remained motionless. These enormous cats with spotted robes are so well fed in countries abounding in capybaras, pecaries, and deer, that they rarely attack men. I arrived at the boat out of breath, and related my adventure to the Indians. They appeared very little interested by my story; yet, after having loaded our guns, they accompanied us to the ceiba beneath which the jaguar had lain. He was there no longer, and it would have been imprudent to have pursued him into the forest, where we must have dispersed, or advanced in single file, amidst the intertwining lianas.

In the evening we passed the mouth of the Cano del Manati, thus named on account of the immense quantity of manatees caught there every year. This herbivorous animal of the cetaceous family, is called by the Indians apcia and avia,[*] and it attains here generally ten or twelve feet in length. It usually weighs from five hundred to eight hundred pounds, but it is asserted that one has been taken of eight thousand pounds weight. The manatee abounds in the Orinoco below the cataracts, in the Rio Meta, and in the Apure, between the two islands of Carizales and Conserva. We found no vestiges of nails on the external surface or the edges of the fins, which are quite smooth; but little rudiments of nails appear at the third phalanx, when the skin of the fins is taken off. We dissected one of these animals, which was nine feet long, at Carichana, a Mission of the Orinoco. The upper lip was four inches longer than the lower one. It was covered with a very fine skin, and served as a proboscis. The inside of the mouth, which has a sensible warmth in an animal newly killed, presented a very singular conformation. The tongue was almost motionless; but in front of

---

[*] The first of these words belongs to the Tamanac language, and the second to the Ottomac. Father Gili proves, in opposition to Oviedo, that manati (fish with hands) is not Spanish, but belongs to the languages of Hayti (St. Domingo) and the Maypures. I believe also that, according to the genius of the Spanish tongue, the animal would have been called manudo or manon, but not manati.

the tongue there was a fleshy excrescence in each jaw, and a cavity lined with a very hard skin, into which the excrescence fitted. The manatee eats such quantities of grass, that we have found its stomach, which is divided into several cavities, and its intestines, (one hundred and eight feet long,) filled with it. On opening the animal at the back, we were struck with the magnitude, form, and situation of its lungs. They have very large cells, and resemble immense swimming-bladders. They are three feet long. Filled with air, they have a bulk of more than a thousand cubic inches. I was surprised to see that, possessing such considerable receptacles for air, the manatee comes so often to the surface of the water to breathe. Its flesh is very savoury, though, from what prejudice I know not, it is considered unwholesome and apt to produce fever. It appeared to me to resemble pork rather than beef. It is most esteemed by the Guamos and the Ottomacs; and these two nations are particularly expert in catching the manatee. Its flesh, when salted and dried in the sun, can be preserved a whole year; and, as the clergy regard this mammiferous animal as a fish, it is much sought during Lent. The vital principal is singularly strong in the manatee; it is tied after being harpooned, but is not killed till it has been taken into the canoe. This is effected, when the animal is very large, in the middle of the river, by filling the canoe two-thirds with water, sliding it under the animal, and then baling out the water by means of a calabash. This fishery is most easy after great inundations, when the manatee has passed from the great rivers into the lakes and surrounding marshes, and the waters diminish rapidly. At the period when the Jesuits governed the Missions of the Lower Orinoco, they assembled every year at Cabruta, below the mouth of the Apure, to have a grand fishing for manatees, with the Indians of their Missions, at the foot of the mountain now called El Capuchino. The fat of the animal, known by the name of manatee-butter (manteca de manati,) is used for lamps in the churches; and is also employed in preparing food. It has not the fetid smell of whale-oil, or that of the other cetaceous animals which spout water. The hide of the manati, which is more than an inch and a half thick, is cut into slips, and serves, like thongs of ox-leather, to supply the place of cordage in the Llanos. When immersed in water, it has the defect of undergoing a slight degree of putrefaction. Whips are made of it in the Spanish colonies. Hence the words latigo and manati are synonymous. These whips of manatee-leather are a cruel instrument of punishment for the unhappy slaves, and even for the Indians of the Missions, though, according to the laws, the latter ought to be treated like freemen.

We passed the night opposite the island of Conserva. In skirting the forest we were surprised by the sight of an enormous trunk of a tree seventy feet high, and thickly set with branching thorns. It is called by the natives barba de tigre. It was perhaps a tree of the berberideous family.[*] The Indians had kindled fires at the edge of the water. We again perceived that their light attracted the crocodiles, and even the porpoises (toninas), the noise of which interrupted our sleep, till the fire was extinguished. A female jaguar approached our station whilst taking her young one to drink at the river. The Indians succeeded in chasing her away, but we heard for a long time the cries of the little jaguar, which mewed like a young cat. Soon after, our great dog was bitten, or, as the Indians say, stung, at the point of the nose, by some enormous bats that hovered around our hammocks. These bats had long tails, like the Molosses: I believe, however, that they were Phyllostomes, the tongue of which, furnished with papillae, is an organ of suction, and is capable of being considerably elongated. The dog's wound was very small and round; and though he uttered a plaintive cry when he felt himself bitten, it was not from pain, but because he was

---

[*] We found, on the banks of the Apure, Ammania apurensis, Cordia cordifolia, C. grandiflora, Mollugo sperguloides, Myosotis lithospermoides, Spermacocce diffusa, Coronilla occidentalis, Bignonia apurensis, Pisonia pubescens, Ruellia viscosa, some new species of Jussieua, and a new genus of the composite family, approximating to Rolandra, the Trichospira menthoides of M. Kunth.

frightened at the sight of the bats, which came out from beneath our hammocks. These accidents are much more rare than is believed even in the country itself. In the course of several years, notwithstanding we slept so often in the open air, in climates where vampire-bats,[*] and other analogous species are so common, we were never wounded. Besides, the puncture is no-way dangerous, and in general causes so little pain, that it often does not awaken the person till after the bat has withdrawn.

The 4th of April was the last day we passed on the Rio Apure. The vegetation of its banks became more and more uniform. During several days, and particularly since we had left the Mission of Arichuna, we had suffered cruelly from the stings of insects, which covered our faces and hands. They were not mosquitos, which have the appearance of little flies, or of the genus Simulium, but zancudos, which are really gnats, though very different from our European species.[†] These insects appear only after sunset. Their proboscis is so long that, when they fix on the lower surface of a hammock, they pierce through it and the thickest garments with their sting.

We had intended to pass the night at the Vuelta del Palmito, but the number of jaguars at that part of the Apure is so great, that our Indians found two hidden behind the trunk of a locust-tree, at the moment when they were going to sling our hammocks. We were advised to re-embark, and take our station in the island of Apurito, near its junction with the Orinoco. That portion of the island belongs to the province of Caracas, while the right banks of the Apure and the Orinoco form a part, the one of the province of Varinas, the other of Spanish Guiana. We found no trees to which we could suspend our hammocks, and were obliged to sleep on ox-hides spread on the ground. The

---

[*] Verspertilio spectrum.
[†] M. Latreille has discovered that the mosquitos of South Carolina are of the genus Simulium (Atractocera meigen).

boats were too narrow and too full of zancudos to permit us to pass the night in them.

In the place where we had landed our instruments, the banks being steep, we saw new proofs of the indolence of the gallinaceous birds of the tropics. The curassaos and cashew-birds[*] have the habit of going down several times a day to the river to allay their thirst. They drink a great deal, and at short intervals. A vast number of these birds had joined, near our station, a flock of parraka pheasants. They had great difficulty in climbing up the steep banks; they attempted it several times without using their wings. We drove them before us, as if we had been driving sheep. The zamuro vultures raise themselves from the ground with great reluctance.

We were singularly struck at the small quantity of water which the Rio Apure furnishes at this season to the Orinoco. The Apure, which, according to my measurements, was still one hundred and thirty-six toises broad at the Cano Rico, was only sixty or eighty at its mouth.[†] Its depth here was only three or four toises. It loses, no doubt, a part of its waters by the Rio Arichuna and the Cano del Manati, two branches of the Apure that flow into the Payara and the Guarico; but its greatest loss appears to be caused by filtrations on the beach, of which we have before spoken. The velocity of the Apure near its mouth was only 3.2 feet per second; so that I could easily have calculated the whole quantity of the water if I had taken, by a series of proximate soundings, the whole dimensions of the transverse section.

We touched several times on shoals before we entered the Orinoco. The ground gained from the water is immense towards the confluence of the two rivers. We were obliged to be towed

---

[*] The latter (Crax pauxi) is less common than the former.

[†] Not quite so broad as the Seine at the Pont Royal, opposite the palace of the Tuileries, and a little more than half the width of the Thames at Westminster Bridge.

along by the bank. What a contrast between this state of the river immediately before the entrance of the rainy season, when all the effects of dryness of the air and of evaporation have attained their maximum, and that autumnal state when the Apure, like an arm of the sea, covers the savannahs as far as the eye can reach! We discerned towards the south the lonely hills of Coruato; while to the east the granite rocks of Curiquima, the Sugar Loaf of Caycara, and the mountains of the Tyrant[*] (Cerros del Tirano) began to rise on the horizon. It was not without emotion that we beheld for the first time, after long expectation, the waters of the Orinoco, at a point so distant from the coast.

---

[*] This name alludes, no doubt, to the expedition of Antonio Sedeno. The port of Caycara, opposite Cabruta, still bears the name of that Conquistador.

CHAPTER 2.19.
JUNCTION OF THE APURE AND THE ORINOCO. MOUNTAINS OF ENCARAMADA. URUANA. BARAGUAN. CARICHANA. MOUTH OF THE META. ISLAND OF PANUMANA.

On leaving the Rio Apure we found ourselves in a country presenting a totally different aspect. An immense plain of water stretched before us like a lake, as far as we could see. White-topped waves rose to the height of several feet, from the conflict of the breeze and the current. The air resounded no longer with the piercing cries of herons, flamingos, and spoonbills, crossing in long files from one shore to the other. Our eyes sought in vain those waterfowls, the habits of which vary in each tribe. All nature appeared less animated. Scarcely could we discover in the hollows of the waves a few large crocodiles, cutting obliquely, by the help of their long tails, the surface of the agitated waters. The horizon was bounded by a zone of forests, which nowhere reached so far as the bed of the river. A vast beach, constantly parched by the heat of the sun, desert and bare as the shores of the sea, resembled at a distance, from the effect of the mirage, pools of stagnant water. These sandy shores, far from fixing the limits of the river, render them uncertain, by enlarging or contracting them alternately, according to the variable action of the solar rays.

In these scattered features of the landscape, in this character of solitude and of greatness, we recognize the course of the Orinoco, one of the most majestic rivers of the New World. The water, like the land, displays everywhere a characteristic and peculiar aspect. The bed of the Orinoco resembles not the bed of the Meta, the Guaviare, the Rio Negro, or the Amazon. These differences do not depend altogether on the breadth or the velocity of the current; they are connected with a multitude of impressions which it is easier to perceive upon the spot than to define with precision. Thus, the mere form of the waves, the tint of the waters, the aspect of the sky and the clouds, would lead an experienced navigator to guess whether he were in the Atlantic, in the Mediterranean, or in the equinoctial part of the Pacific.

The wind blew fresh from east-north-east. Its direction was favourable for sailing up the Orinoco, towards the Mission of Encaramada; but our canoes were so ill calculated to resist the shocks of the waves, that, from the violence of the motion, those who suffered habitually at sea were equally incommoded on the river. The short, broken waves are caused by the conflict of the waters at the junction of the two rivers. This conflict is very violent, but far from being so dangerous as Father Gumilla describes. We passed the Punta Curiquima, which is an isolated mass of quartzose granite, a small promontory composed of rounded blocks. There, on the right bank of the Orinoco, Father Rotella founded, in the time of the Jesuits, a Mission of the Palenka and Viriviri or Guire Indians. But during inundations, the rock Curiquima and the village at its foot were entirely surrounded by water; and this serious inconvenience, together with the sufferings of the missionaries and Indians from the innumerable quantity of mosquitos and niguas,[*] led them to forsake this humid spot. It is now entirely deserted, while opposite to it, on the right bank of the river, the little mountains of Coruato are the retreat of wandering Indians, expelled either from the Missions, or from tribes that are not subject to the government of the monks.

---

[*] The chego (Pulex penetrans) which penetrates under the nails of the toe in men and monkeys, and there deposits its eggs.

Struck with the extreme breadth of the Orinoco, between the mouth of the Apure and the rock Curiquima, I ascertained it by means of a base measured twice on the western beach. The bed of the Orinoco, at low water, was 1906 toises broad; but this breadth increases to 5517 toises, when, in the rainy season, the rock Curiquima, and the farm of Capuchino near the hill of Pocopocori, become islands. The swelling of the Orinoco is augmented by the impulse of the waters of the Apure, which, far from forming, like other rivers, an acute angle with the upper part of that into which it flows, meets it at right angles.

We first proceeded south-west, as far as the shore inhabited by the Guaricoto Indians on the left bank of the Orinoco, and then we advanced straight toward the south. The river is so broad that the mountains of Encaramada appear to rise from the water, as if seen above the horizon of the sea. They form a continued chain from east to west. These mountains are composed of enormous blocks of granite, cleft and piled one upon another. Their division into blocks is the effect of decomposition. What contributes above all to embellish the scene at Encaramada is the luxuriance of vegetation that covers the sides of the rocks, leaving bare only their rounded summits. They look like ancient ruins rising in the midst of a forest. The mountain immediately at the back of the Mission, the Tepupano[*] of the Tamanac Indians is terminated by three enormous granitic cylinders, two of which are inclined, while the third, though worn at its base, and more than eighty feet high, has preserved a vertical position. This rock, which calls to mind the form of the Schnarcher in the Hartz mountains, or that of the Organs of Actopan in Mexico,[†] composed formerly a part of the rounded

---

[*] Tepu-pano, place of stones, in which we recognize tepu stone, rock, as in tepu-iri, mountain. We here perceive that Lesgian Oigour-Tartar root tep, stone (found in America among the Americans, in teptl; among the Caribs, in tebou; among the Tamanacs, in tepuiri); a striking analogy between the languages of Caucasus and Upper Asia and those of the banks of the Orinoco.

[†] In Captain Tuckey's Voyage on the river Congo, we find represented a

summit of the mountain. In every climate, unstratified granite separates by decomposition into blocks of prismatic, cylindric, or columnar figures.

Opposite the shore of the Guaricotos, we drew near another heap of rocks, which is very low, and three or four toises long. It rises in the midst of the plain, and has less resemblance to a tumulus than to those masses of granitic stone, which in North Holland and Germany bear the name of hunenbette, beds (or tombs) of heroes. The shore, at this part of the Orinoco, is no longer of pure and quartzose sand; but is composed of clay and spangles of mica, deposited in very thin strata, and generally at an inclination of forty or fifty degrees. It looks like decomposed mica-slate. This change in the geological configuration of the shore extends far beyond the mouth of the Apure. We had begun to observe it in this latter river as far off as Algodonal and the Cano del Manati. The spangles of mica come, no doubt, from the granite mountains of Curiquima and Encaramada; since further north-east we find only quartzose sand, sandstone, compact limestone, and gypsum. Alluvial earth carried successively from south to north need not surprise us in the Orinoco; but to what shall we attribute the same phenomenon in the bed of the Apure, seven leagues west of its mouth? In the present state of things, notwithstanding the swellings of the Orinoco, the waters of the Apure never retrograde so far; and, to explain this phenomenon, we are forced to admit that the micaceous strata were deposited at a time when the whole of the very low country lying between Caycara, Algodonal, and the mountains of Encaramada, formed the basin of an inland lake.

We stopped some time at the port of Encaramada, which is a sort of embarcadero, a place where boats assemble. A rock of forty or fifty feet high forms the shore. It is composed of blocks of granite, heaped one upon another, as at the Schneeberg in

---

granitic rock, Taddi Enzazi, which bears a striking resemblance to the mountain of Encaramada.

Franconia, and in almost all the granitic mountains of Europe. Some of these detached masses have a spheroidal form; they are not balls with concentric layers, but merely rounded blocks, nuclei separated from their envelopes by the effect of decomposition. This granite is of a greyish lead-colour, often black, as if covered with oxide of manganese; but this colour does not penetrate one fifth of a line into the rock, which is of a reddish white colour within, coarse-grained, and destitute of hornblende.

The Indian names of the Mission of San Luis del Encaramada, are Guaja and Caramana.[*] This small village was founded in 1749 by Father Gili, the Jesuit, author of the Storia dell' Orinoco, published at Rome. This missionary, learned in the Indian tongues, lived in these solitudes during eighteen years, till the expulsion of the Jesuits. To form a precise idea of the savage state of these countries it must be recollected that Father Gili speaks of Carichana,[†] which is forty leagues from Encaramada, as of a spot far distant; and that he never advanced so far as the first cataract in the river of which he ventured to undertake the description.

---

[*] All the Missions of South America have names composed of two words, the first of which is necessarily the name of a saint, the patron of the church, and the second an Indian name, that of the nation, or the spot where the establishment is placed. Thus we say, San Jose de Maypures, Santa Cruz de Cachipo, San Juan Nepomuceno de los Atures, etc. These compound names appear only in official documents; the Inhabitants adopt but one of the two names, and generally, provided it be sonorous, the Indian. As the names of saints are several times repeated in neighbouring places, great confusion in geography arises from these repetitions. The names of San Juan, San Diego, and San Pedro, are scattered in our maps as if by chance. It is pretended that the Mission of Guaja affords a very rare example of the composition of two Spanish words. The word Encaramada means things raised one upon another, from encaramar, to raise up. It is derived from the figure of Tepupano and the neighbouring rocks: perhaps it is only an Indian word caramana, in which, as in manati, a Spanish signification was believed to be discovered.

[†] Saggio di Storia Americana volume 1 page 122.

In the port of Encaramada we met with some Caribs of Panapana. A cacique was going up the Orinoco in his canoe, to join in the famous fishing of turtles' eggs. His canoe was rounded toward the bottom like a bongo, and followed by a smaller boat called a curiara. He was seated beneath a sort of tent, constructed, like the sail, of palm-leaves. His cold and silent gravity, the respect with which he was treated by his attendants, everything denoted him to be a person of importance. He was equipped, however, in the same manner as his Indians. They were all equally naked, armed with bows and arrows, and painted with onoto, which is the colouring fecula of the Bixa orellana. The chief, the domestics, the furniture, the boat, and the sail, were all painted red. These Caribs are men of an almost athletic stature; they appeared to us much taller than any Indians we had hitherto seen. Their smooth and thick hair, cut short on the forehead like that of choristers, their eyebrows painted black, their look at once gloomy and animated, gave a singular expression to their countenances. Having till then seen only the skulls of some Caribs of the West India Islands preserved in the collections of Europe, we were surprised to find that these Indians, who were of pure race, had foreheads much more rounded than they are described. The women, who were very tall, and disgusting from their want of cleanliness, carried their infants on their backs. The thighs and legs of the infants were bound at certain distances by broad strips of cotton cloth, and the flesh, strongly compressed beneath the ligatures, was swelled in the interstices. It is generally to be observed, that the Caribs are as attentive to their exterior and their ornaments, as it is possible for men to be, who are naked and painted red. They attach great importance to certain configurations of the body; and a mother would be accused of culpable indifference toward her children, if she did not employ artificial means to shape the calf of the leg after the fashion of the country. As none of our Indians of Apure understood the Caribbee language, we could obtain no information from the cacique of Panama respecting the encampments that are made at this season in several islands of the Orinoco for collecting turtles' eggs.

Near Encaramada a very long island divides the river into two branches. We passed the night in a rocky creek, opposite the mouth of the Rio Cabullare, which is formed by the Payara and the Atamaica, and is sometimes considered as one of the branches of the Apure, because it communicates with that river by the Rio Arichuna. The evening was beautiful. The moon illumined the tops of the granite rocks. The heat was so uniformly distributed, that, notwithstanding the humidity of the air, no twinkling of the stars was observable, even at four or five degrees above the horizon. The light of the planets was singularly dimmed; and if, on account of the smallness of the apparent diameter of Jupiter, I had not suspected some error in the observation, I should say, that here, for the first time, we thought we distinguished the disk of Jupiter with the naked eye. Towards midnight, the north-east wind became extremely violent. It brought no clouds, but the vault of the sky was covered more and more with vapours. Strong gusts were felt, and made us fear for the safety of our canoe. During this whole day we had seen very few crocodiles, but all of an extraordinary size, from twenty to twenty-four feet. The Indians assured us that the young crocodiles prefer the marshes, and the rivers that are less broad, and less deep. They crowd together particularly in the Canos, and we may say of them, what Abdallatif says of the crocodiles of the Nile,[*] "that they swarm like worms in the shallow waters of the river, and in the shelter of uninhabited islands."

On the 6th of April, whilst continuing to ascend the Orinoco, first southward and then to south-west, we perceived the southern side of the Serrania, or chain of the mountains of Encaramada. The part nearest the river is only one hundred and forty or one hundred and sixty toises high; but from its abrupt

---

[*] Description de l'Egypte translated by De Sacy.

declivities, its situation in the midst of a savannah, and its rocky summits, cut into shapeless prisms, the Serrania appears singularly elevated. Its greatest breadth is only three leagues. According to information given me by the Indians of the Pareka nation, it is considerably wider toward the east. The summits of Encaramada form the northernmost link of a group of mountains which border the right bank of the Orinoco, between the latitudes of 5° and 7° 30′ from the mouth of the Rio Zama to that of the Cabullare. The different links into which this group is divided are separated by little grassy plains. They do not preserve a direction perfectly parallel to each other; for the most northern stretch from west to east, and the most southern from north-west to south-east. This change of direction sufficiently explains the increase of breadth observed in the Cordillera of Parime towards the east, between the sources of the Orinoco and of the Rio Paruspa. On penetrating beyond the great cataracts of Atures and of Maypures, we shall see seven principal links, those of Encaramada or Sacuina, of Chaviripa, of Baraguan, of Carichana, of Uniama, of Calitamini, and of Sipapo, successively appear. This sketch may serve to give a general idea of the geological configuration of the ground. We recognize everywhere on the globe a tendency toward regular forms, in those mountains that appear the most irregularly grouped. Every link appears, in a transverse section, like a distinct summit, to those who navigate the Orinoco; but this division is merely in appearance. The regularity in the direction and separation of the links seems to diminish in proportion as we advance towards the east. The mountains of Encaramada join those of Mato, which give birth to the Rio Asiveru or Cuchivero; those of Chaviripe are prolonged by the granite chain of the Corosal, of Amoco, and of Murcielago, towards the sources of the Erevato and the Ventuari.

It was across these mountains, which are inhabited by Indians of gentle character, employed in agriculture,[*] that, at the time of the expedition for settling boundaries, General Iturriaga took some horned cattle for the supply of the new town of San Fernando de Atabapo. The inhabitants of Encaramada then showed the Spanish soldiers the way by the Rio Manapiari,[†] which falls into the Ventuari. By descending these two rivers, the Orinoco and the Atabapo may be reached without passing the great cataracts, which present almost insurmountable obstacles to the conveyance of cattle. The spirit of enterprise which had so eminently distinguished the Castilians at the period of the discovery of America, was again roused for a time in the middle of the eighteenth century, when Ferdinand VI was desirous of knowing the true limits of his vast possessions; and in the forests of Guiana, that land of fiction and fabulous tradition, the wily Indians revived the chimerical idea of the wealth of El Dorado, which had so much occupied the imagination of the first conquerors.

Amidst the mountains of Encaramada, which, like most coarse-grained granite rocks, are destitute of metallic veins, we cannot help inquiring whence came those grains of gold which Juan Martinez[‡] and Raleigh profess to have seen in such abundance in the hands of the Indians of the Orinoco. From what I observed in that part of America, I am led to think that gold, like tin,[§] is sometimes disseminated in an almost imperceptible

---

[*] The Mapoyes, Parecas, Javaranas, and Curacicanas, who possess fine plantations (conucos) in the savannahs by which these forests are bounded.

[†] Between Encaramada and the Rio Manapiare, Don Miguel Sanchez, chief of this little expedition, crossed the Rio Guainaima, which flows into the Cuchivero. Sanchez died, from the fatigue of this journey, on the borders of the Ventuari.

[‡] The companion of Diego Ordaz.

[§] Thus tin is found in granite of recent formation, at Geyer; in hyalomicte or graisen, at Zinnwald; and in syenitic porphyry, at Altenberg, in Saxony, as well as near Naila, in the Fichtelgebirge. I have also seen, in the Upper Palatinate, micaceous iron, and black earthy cobalt, far from any kind of vein,

manner in the very mass of granite rocks, without our being able to perceive that there is a ramification and an intertwining of small veins. Not long ago the Indians of Encaramada found in the Quebrada del Tigre[*] a piece of native gold two lines in diameter. It was rounded, and appeared to have been washed along by the waters. This discovery excited the attention of the missionaries much more than of the natives; it was followed by no other of the same kind.

I cannot quit this first link of the mountains of Encaramada without recalling to mind a fact that was not unknown to Father Gili, and which was often mentioned to me during our abode in the Missions of the Orinoco. The natives of those countries have retained the belief that, "at the time of the great waters, when their fathers were forced to have recourse to boats, to escape the general inundation, the waves of the sea beat against the rocks of Encaramada." This belief is not confined to one nation singly, the Tamanacs; it makes part of a system of historical tradition, of which we find scattered notions among the Maypures of the great cataracts; among the Indians of the Rio Erevato, which runs into the Caura; and among almost all the tribes of the Upper Orinoco. When the Tamanacs are asked how the human race survived this great deluge, the age of water, of the Mexicans, they say, a man and a woman saved themselves on a high mountain, called Tamanacu, situated on the banks of the Asiveru; and casting behind them, over their heads, the fruits of the mauritia palm-tree, they saw the seeds contained in those fruits produce men and women, who repeopled the earth. Thus we find in all its simplicity, among nations now in a savage state, a tradition which the Greeks embellished with all the charms of imagination! A few leagues from Encaramada, a rock, called Tepu-mereme, or the painted rock, rises in the midst of the savannah. Upon it are traced representations of animals, and

---

disseminated in a granite destitute of mica, as magnetic iron-sand is in volcanic rocks.

[*] The Tiger-ravine.

symbolic figures resembling those we saw in going down the Orinoco, at a small distance below Encaramada, near the town Caycara. Similar rocks in Africa are called by travellers fetish stones. I shall not make use of this term, because fetishism does not prevail among the natives of the Orinoco; and the figures of stars, of the sun, of tigers, and of crocodiles, which we found traced upon the rocks in spots now uninhabited, appeared to me in no way to denote the objects of worship of those nations. Between the banks of the Cassiquiare and the Orinoco, between Encaramada, the Capuchino, and Caycara, these hieroglyphic figures are often seen at great heights, on rocky cliffs which could be accessible only by constructing very lofty scaffolds. When the natives are asked how those figures could have been sculptured, they answer with a smile, as if relating a fact of which only a white man could be ignorant, that "at the period of the great waters, their fathers went to that height in boats."

These ancient traditions of the human race, which we find dispersed over the whole surface of the globe, like the relics of a vast shipwreck, are highly interesting in the philosophical study of our own species. Like certain families of the vegetable kingdom, which, notwithstanding the diversity of climates and the influence of heights, retain the impression of a common type, the traditions of nations respecting the origin of the world, display everywhere the same physiognomy, and preserve features of resemblance that fill us with astonishment. How many different tongues, belonging to branches that appear totally distinct, transmit to us the same facts! The traditions concerning races that have been destroyed, and the renewal of nature, scarcely vary in reality, though every nation gives them a local colouring. In the great continents, as in the smallest islands of the Pacific Ocean, it is always on the loftiest and nearest mountain that the remains of the human race have been saved; and this event appears the more recent, in proportion as the nations are uncultivated, and as the knowledge they have of their own existence has no very remote date. After having studied with attention the Mexican monuments anterior to the discovery of the New World; after having penetrated into the forests of the Orinoco, and observed the diminutive size of the European establishments, their solitude, and the state of the tribes that have remained independent; we cannot allow ourselves to attribute the analogies just cited to the influence exercised by the missionaries, and by Christianity, on the national traditions. Nor is it more probable, that the discovery of sea-shells on the summit of mountains gave birth, among the nations of the Orinoco, to the tradition of some great inundation which extinguished for a time the germs of organic life on our globe. The country that extends from the right bank of the Orinoco to the Cassiquiare and the Rio Negro, is a country of primitive rocks. I saw there one small formation of sandstone or conglomerate; but no secondary limestone, and no trace of petrifactions.

A fresh north-east breeze carried us full-sail towards the Boca de la Tortuga. We landed, at eleven in the morning, on an island which the Indians of the Missions of Uruana considered as their property, and which lies in the middle of the river. This island is celebrated for the turtle fishery, or, as they say here, the cosecha, the harvest [of eggs,] that takes place annually. We here found an assemblage of Indians, encamped under huts made of palm-leaves. This encampment contained more than three hundred persons. Accustomed, since we had left San Fernando de Apure, to see only desert shores, we were singularly struck by the bustle that prevailed here. We found, besides the Guamos and the Ottomacs of Uruana, who are both considered as savage races, Caribs and other Indians of the Lower Orinoco. Every tribe was separately encamped, and was distinguished by the pigments with which their skins were painted. Some white men were seen amidst this tumultuous assemblage, chiefly pulperos, or little traders of Angostura, who had come up the river to purchase turtle oil from the natives. The missionary of Uruana, a native of Alcala, came to meet us, and he was extremely astonished at seeing us. After having admired our instruments, he gave us an exaggerated picture of the sufferings to which we should be necessarily exposed in ascending the Orinoco beyond the cataracts. The object of our journey appeared to him very mysterious. "How is it possible to believe," said he, "that you have left your country, to come and be devoured by mosquitos on this river, and to measure lands that are not your own?" We were happily furnished with recommendations from the Superior of the Franciscan Missions, and the brother-in-law of the governor of Varinas, who accompanied us, soon dissipated the doubts to which our dress, our accent, and our arrival in this sandy island, had given rise among the Whites. The missionary invited us to partake a frugal repast of fish and plantains. He told us that he had come to encamp with the Indians during the time of the harvest of eggs, "to celebrate mass every morning in the open air, to procure the oil necessary for the church-lamps, and especially to govern this mixed republic (republica de Indios y Castellanos) in which every one wished to profit singly by what God had granted to all."

We made the tour of the island, accompanied by the missionary and by a pulpero, who boasted of having, for ten successive years, visited the camp of the Indians, and attended the turtle-fishery. We were on a plain of sand perfectly smooth; and were told that, as far as we could see along the beach, turtles' eggs were concealed under a layer of earth. The missionary carried a long pole in his hand. He showed us, that by means of this pole, the extent of the stratum of eggs could be determined as accurately as the miner determines the limits of a bed of marl, of bog iron-ore, or of coal. On thrusting the rod perpendicularly into the ground, the sudden want of resistance shows that the cavity or layer of loose earth containing the eggs, has been reached. We saw that the stratum is generally spread with so much uniformity, that the pole finds it everywhere in a radius of ten toises around any given spot. Here they talk continually of square perches of eggs; it is like a mining-country, divided into lots, and worked with the greatest regularity. The stratum of eggs, however, is far from covering the whole island: they are not found wherever the ground rises abruptly, because the turtle cannot mount heights. I related to my guides the emphatic description of Father Gumilla, who asserts, that the shores of the Orinoco contain fewer grains of sand than the river contains turtles; and that these animals would prevent vessels from advancing, if men and tigers did not annually destroy so great a number.[*] "Son cuentos de frailes," "they are monkish legends," said the pulpero of Angostura, in a low voice; for the only travellers in this country being the missionaries, they here call monks' stories, what we call travellers' tales, in Europe.

The Indians assured us that, in going up the Orinoco from its mouth to its junction with the Apure, not one island or one beach is to be found, where eggs can be collected in abundance. The great turtle (arrau[†]) dreads places inhabited by men, or much frequented by boats. It is a timid and mistrustful animal, raising only its head above the water, and hiding itself at the least noise. The shores where almost all the turtles of the Orinoco appear to assemble annually, are situated between the junction of the Orinoco with the Apure, and the great cataracts; that is to say, between Cabruta and the Mission of Atures. There are found the three famous fisheries; those of Encaramada, or Boca del Cabullare; of Cucuruparu, or Boca de la Tortuga; and of Pararuma, a little below Carichana. It seems that the arrau does not pass beyond the cataracts; and we were assured, that only the turtles called terekay, (in Spanish terecayas,) are found above Atures and Maypures.

---

[*] "It would be as difficult to count the grains of sand on the shores of the Orinoco, as to count the immense number of tortoises which inhabit its margins and waters. Were it not for the vast consumption of tortoises and their eggs, the river Orinoco, despite its great magnitude, would be unnavigable, for vessels would be impeded by the enormous multitude of the tortoises." Gumilla, Orinoco Illustrata volume 1 pages 331 to 336.

[†] This word belongs to the Maypure language, and must not be confounded with arua, which means a crocodile, among the Tamanacs, neighbours of the Maypures. The Ottomacs call the turtle of Uruana, achea; the Tamanacs, peje.

The arrau, called by the Spaniards of the Missions simply tortuga, is an animal whose existence is of great importance to the nations on the Lower Orinoco. It is a large freshwater tortoise, with palmate and membraneous feet; the head very flat, with two fleshy and acutely-pointed appendages under the chin; five claws to the fore feet, and four to the hind feet, which are furrowed underneath. The upper shell has five central, eight lateral, and twenty-four marginal plates. The colour is darkish grey above, and orange beneath. The feet are yellow, and very long. There is a deep furrow between the eyes. The claws are very strong and crooked. The anus is placed at the distance of one-fifth from the extremity of the tail. The full-grown animal weighs from forty to fifty pounds. Its eggs are much larger than those of pigeons, and less elongated than the eggs of the terekay. They are covered with a calcareous crust, and, it is said, they have sufficient firmness for the children of the Ottomac Indians, who are great players at ball, to throw them into the air from one to another. If the arrau inhabited the bed of the river above the cataracts, the Indians of the Upper Orinoco would not travel so far to procure the flesh and the eggs of this tortoise. Yet, formerly, whole tribes from the Atabapo and the Cassiquiare have been known to pass the cataracts, in order to take part in the fishery at Uruana.

The terekay is less than the arrau. It is in general only fourteen inches in diameter. The number of plates in the upper shell is the same, but they are somewhat differently arranged. I counted three in the centre of the disk, and five hexagonal on each side. The margins contain twenty-four, all quadrangular, and much curved. The upper shell is of a black colour inclining to green; the feet and claws are like those of the arrau. The whole animal is of an olive-green, but it has two spots of red mixed with yellow on the top of the head. The throat is also yellow, and furnished with a prickly appendage. The terekays do not assemble in numerous societies like the arraus, to lay their eggs in common, and deposit them upon the same shore. The eggs of the terekay have an agreeable taste, and are much sought after by the inhabitants of Spanish Guiana. They are found in the Upper Orinoco, as well as below the cataracts, and even in the Apure, the Uritucu, the Guarico, and the small rivers that traverse the Llanos of Caracas. The form of the feet and head, the appendages of the chin and throat, and the position of the anus, seem to indicate that the arrau, and probably the terekay also, belong to a new subdivision of the tortoises, that may be separated from the emydes. The period at which the large arrau tortoise lays its eggs coincides with the period of the lowest waters. The Orinoco beginning to increase from the vernal equinox, the lowest flats are found uncovered from the end of January till the 20th or 25th of March. The arrau tortoises collect in troops in the month of January, then issue from the water, and warm themselves in the sun, reposing on the sands. The Indians believe that great heat is indispensable to the health of the animal, and that its exposure to the sun favours the laying of the eggs. The arraus are found on the beach a great part of the day during the whole month of February. At the beginning of March the straggling troops assemble, and swim towards the small number of islands on which they habitually deposit their eggs. It is probable that the same tortoise returns every year to the same locality. At this period, a few days before they lay their eggs, thousands of these animals may be seen ranged in long files, on the borders of the islands of Cucuruparu, Uruana, and Pararuma, stretching out their necks and holding their heads above water, to see whether they have anything to dread. The Indians, who are anxious that the bands when assembled should not separate, that the tortoises should not disperse, and that the laying of the eggs should be performed tranquilly, place sentinels at certain distances along the shore. The people who pass in boats are told to keep in the middle of the river, and not frighten the tortoises by cries. The laying of the eggs takes place always during the night, and it begins soon after sunset. With its hind feet, which are very long, and furnished with crooked claws, the animal digs a hole of three feet in diameter and two in depth. These tortoises feel so pressing a desire to lay their eggs, that some of them descend into holes that have been dug by others, but which are not yet covered with earth. There they deposit a new layer of eggs on that which has been recently laid. In this tumultuous movement an immense number of eggs are broken. The missionary showed us, by removing the sand in several places, that this loss probably amounts to a fifth of the whole quantity. The yolk of the broken eggs contributes, in drying, to cement the sand; and we found very large concretions of grains of quartz and broken shells. The number of animals working on the beach during the night is so considerable, that day surprises many of them before the laying of their eggs is terminated. They are then urged on by the double necessity of depositing their eggs, and closing the holes they have dug, that they may not be perceived by the jaguars. The tortoises that thus remain too late are insensible to their own danger. They work in the presence of the Indians, who visit the beach at a very early hour, and who call them mad tortoises. Notwithstanding the rapidity of their movements, they are then easily caught with the hand.

The three encampments formed by the Indians, in the places indicated above, begin about the end of March or commencement of April. The gathering of the eggs is conducted in a uniform manner, and with that regularity which characterises all monastic institutions. Before the arrival of the missionaries on the banks of the river, the Indians profited much less from a production which nature has supplied in such abundance. Every tribe searched the beach in its own way; and an immense number of eggs were uselessly broken, because they were not dug up with precaution, and more eggs were uncovered than could be carried away. It was like a mine worked by unskilful hands. The Jesuits have the merit of having reduced this operation to regularity; and though the Franciscan monks, who succeeded the Jesuits in the Missions of the Orinoco, boast of having followed the example of their predecessors, they unhappily do not effect all that prudence requires. The Jesuits did not suffer the whole beach to be searched; they left a part untouched, from the fear of seeing the breed of tortoises, if not destroyed, at least considerably diminished. The whole beach is now dug up without reserve; and accordingly it seems to be perceived that the gathering is less productive from year to year.

When the camp is formed, the missionary of Uruana names his lieutenant, or commissary, who divides the ground where the eggs are found into different portions, according to the number of the Indian tribes who take part in the gathering. They are all Indians of Missions, as naked and rude as the Indians of the woods; though they are called reducidos and neofitos, because they go to church at the sound of the bell, and have learned to kneel down during the consecration of the host.

The lieutenant (commissionado del Padre) begins his operations by sounding. He examines by means of a long wooden pole or a cane of bamboo, how far the stratum of eggs extends. This stratum, according to our measurements, extended to the distance of one hundred and twenty feet from the shore. Its average depth is three feet. The commissionado places marks to indicate the point where each tribe should stop in its labours. We were surprised to hear this harvest of eggs estimated like the produce of a well-cultivated field. An area accurately measured of one hundred and twenty feet long, and thirty feet wide, has been known to yield one hundred jars of oil, valued at about forty pounds sterling. The Indians remove the earth with their hands; they place the eggs they have collected in small baskets, carry them to their encampment, and throw them into long troughs of wood filled with water. In these troughs the eggs, broken and stirred with shovels, remain exposed to the sun till the oily part, which swims on the surface, has time to inspissate. As fast as this collects on the surface of the water, it is taken off and boiled over a quick fire. This animal oil, called tortoise butter (manteca de tortugas[*]) keeps the better, it is said, in proportion as it has undergone a strong ebullition. When well

---

[*] The Tamanac Indians give it the name of carapa; the Maypures call it timi.

prepared, it is limpid, inodorous, and scarcely yellow. The missionaries compare it to the best olive oil, and it is used not merely for burning in lamps, but for cooking. It is not easy, however, to procure oil of turtles' eggs quite pure. It has generally a putrid smell, owing to the mixture of eggs in which the young are already formed.

I acquired some general statistical notions on the spot, by consulting the missionary of Uruana, his lieutenant, and the traders of Angostura. The shore of Uruana furnishes one thousand botijas, or jars of oil, annually. The price of each jar at Angostura varies from two piastres to two and a half. We may admit that the total produce of the three shores, where the cosecha, or gathering of eggs, is annually made, is five thousand botijas. Now as two hundred eggs yield oil enough to fill a bottle (limeta), it requires five thousand eggs for a jar or botija of oil. Estimating at one hundred, or one hundred and sixteen, the number of eggs that one tortoise produces, and reckoning that one third of these is broken at the time of laying, particularly by the mad tortoises, we may presume that, to obtain annually five thousand jars of oil, three hundred and thirty thousand arrau tortoises, the weight of which amounts to one hundred and sixty-five thousand quintals, must lay thirty-three millions of eggs on the three shores where this harvest is gathered. The results of these calculations are much below the truth. Many tortoises lay only sixty or seventy eggs; and a great number of these animals are devoured by jaguars at the moment they emerge from the water. The Indians bring away a great number of eggs to eat them dried in the sun; and they break a considerable number through carelessness during the gathering. The number of eggs that are hatched before the people can dig them up is so prodigious, that near the encampment of Uruana I saw the whole shore of the Orinoco swarming with little tortoises an inch in diameter, escaping with difficulty from the pursuit of the Indian children. If to these considerations be added, that all the arraus do not assemble on the three shores of the encampments; and that there are many which lay their eggs in solitude, and some weeks later,[*] between the mouth of the Orinoco and the confluence of the Apure; we must admit that the number of turtles which annually deposit their eggs on the banks of the Lower Orinoco, is near a million. This number is very great for so large an animal. In general large animals multiply less considerably than the smaller ones.

The labour of collecting the eggs, and preparing the oil, occupies three weeks. It is at this period only that the missionaries have any communication with the coast and the civilized neighbouring countries. The Franciscan monks who live south of the cataracts, come to the harvest of eggs less to procure oil, than to see, as they say, white faces; and to learn whether the king inhabits the Escurial or San Ildefonso, whether convents are still suppressed in France, and above all, whether the Turks continue to keep quiet. On these subjects, (the only ones interesting to a monk of the Orinoco), the small traders of Angostura, who visit the encampments, can give, unfortunately, no very exact information. But in these distant countries no doubt is ever entertained of the news brought by a white man from the capital. The profit of the traders in oil amounts to seventy or eighty per cent; for the Indians sell it them at the price of a piastre a jar or botija, and the expense of carriage is not more than two-fifths of a piastre per jar. The Indians bring away also a considerable quantity of eggs dried in the sun, or slightly boiled. Our rowers had baskets or little bags of cotton-cloth filled with these eggs. Their taste is not disagreeable, when well preserved. We were shown large shells of turtles, which had been destroyed by the jaguars. These animals follow the arraus

---

[*] The arraus, which lay their eggs before the beginning of March, (for in the same species the more or less frequent basking in the sun, the food, and the peculiar organization of each individual, occasion differences,) come out of the water with the terekays, which lay in January and February. Father Gumilla believes them to be arraus that were not able to lay their eggs the preceding year. It is difficult to find the eggs of the terekays, because these animals, far from collecting in thousands on the same beach, deposit their eggs as they are scattered about.

209

towards those places on the beach where the eggs are laid. They surprise the arraus on the sand; and, in order to devour them at their ease, turn them in such a manner that the under shell is uppermost. In this situation the turtles cannot rise; and as the jaguar turns many more than he can eat in one night, the Indians often avail themselves of his cunning and avidity.

When we reflect on the difficulty experienced by the naturalist in getting out the body of the turtle without separating the upper and under shells, we cannot sufficiently wonder at the suppleness of the tiger's paw, which is able to remove the double armour of the arrau, as if the adhering parts of the muscles had been cut by a surgical instrument. The jaguar pursues the turtle into the water when it is not very deep. It even digs up the eggs; and together with the crocodile, the heron, and the galinazo vulture, is the most cruel enemy of the little turtles recently hatched. The island of Pararuma had been so much infested with crocodiles the preceding year, during the egg-harvest, that the Indians in one night caught eighteen, of twelve or fifteen feet long, by means of curved pieces of iron, baited with the flesh of the manatee. Besides the beasts of the forests we have just named, the wild Indians also very much diminish the quantity of the oil. Warned by the first slight rains, which they call turtle-rains (peje canepori[*]), they hasten to the banks of the Orinoco, and kill the turtles with poisoned arrows, whilst, with upraised heads and paws extended, the animals are warming themselves in the sun.

Though the little turtles (tortuguillos) may have burst the shells of their eggs during the day, they are never seen to come out of the ground but at night. The Indians assert that the young animal fears the heat of the sun. They tried also to show us, that when the tortuguillo is carried in a bag to a distance from the shore, and placed in such a manner that its tail is turned to the river, it takes without hesitation the shortest way to the water. I

---

[*] In the Tamanac language, from peje, a tortoise, and canepo, rain.

confess, that this experiment, of which Father Gumilla speaks, does not always succeed equally well: yet in general it does appear that at great distances from the shore, and even in an island, these little animals feel with extreme delicacy in what direction the most humid air prevails.

Reflecting on the almost uninterrupted layer of eggs that extends along the beach, and on the thousands of little turtles that seek the water as soon as they are hatched, it is difficult to admit that the many turtles which have made their nests in the same spot, can distinguish their own young, and lead them, like the crocodiles, to the lakes in the vicinity of the Orinoco. It is certain, however, that the animal passes the first years of its life in pools where the water is shallow, and does not return to the bed of the great river till it is full-grown. How then do the tortuguillos find these pools? Are they led thither by female turtles, which adopt the young as by chance? The crocodiles, less numerous, deposit their eggs in separate holes; and, in this family of saurians, the female returns about the time when the incubation is terminated, calls her young, which answer to her voice, and often assists them to get out of the ground. The arrau tortoise, no doubt, like the crocodile, knows the spot where she has made her nest; but, not daring to return to the beach on which the Indians have formed their encampment, how can she distinguish her own young from those which do not belong to her? On the other hand, the Ottomac Indians declare that, at the period of inundation, they have met with female turtles followed by a great number of young ones. These were perhaps arraus whose eggs had been deposited on a desert beach to which they could return. Males are extremely rare among these animals. Scarcely is one male found among several hundred females. The cause of this disparity cannot be the same as with the crocodiles, which fight in the coupling season.

Our pilot had anchored at the Playa de huevos, to purchase some provisions, our store having begun to run short. We found there fresh meat, Angostura rice, and even biscuit made of wheat-flour. Our Indians filled the boat with little live turtles, and eggs dried in the sun, for their own use. Having taken leave of the missionary of Uruana, who had treated us with great kindness, we set sail about four in the afternoon. The wind was fresh, and blew in squalls. Since we had entered the mountainous part of the country, we had discovered that our canoe carried sail very badly; but the master was desirous of showing the Indians who were assembled on the beach, that, by going close to the wind, he could reach, at one single tack, the middle of the river. At the very moment when he was boasting of his dexterity, and the boldness of his manoeuvre, the force of the wind upon the sail became so great that we were on the point of going down. One side of the boat was under water, which rushed in with such violence that it was soon up to our knees. It washed over a little table at which I was writing at the stern of the boat. I had some difficulty to save my journal, and in an instant we saw our books, papers, and dried plants, all afloat. M. Bonpland was lying asleep in the middle of the canoe. Awakened by the entrance of the water and the cries of the Indians, he understood the danger of our situation, whilst he maintained that coolness which he always displayed in the most difficult circumstances. The lee-side righting itself from time to time during the squall, he did not consider the boat as lost. He thought that, were we even forced to abandon it, we might save ourselves by swimming, since there was no crocodile in sight. Amidst this uncertainty the cordage of the sail suddenly gave way. The same gust of wind, that had thrown us on our beam, served also to right us. We laboured to bale the water out of the boat with calabashes, the sail was again set, and in less than half an hour we were in a state to proceed. The wind now abated a little. Squalls alternating with dead calms are common in that part of the Orinoco which is bordered by mountains. They are very dangerous for boats deeply laden, and without decks. We had escaped as if by miracle. To the reproaches that were heaped on our pilot for having kept too near the wind, he replied with the phlegmatic coolness peculiar to the Indians, observing "that the whites would find sun enough on those banks to dry their papers." We lost only one book—the first volume of the Genera Plantarum of Schreber—which had fallen overboard. At nightfall we landed on a barren island in the middle of the river, near the Mission of Uruana. We supped in a clear moonlight, seating ourselves on some large turtle-shells that were found scattered about the beach. What satisfaction we felt on finding ourselves thus comfortably landed! We figured to ourselves the situation of a man who had been saved alone from shipwreck, wandering on these desert shores, meeting at every step with other rivers which fall into the Orinoco, and which it is dangerous to pass by swimming, on account of the multitude of crocodiles and caribe fishes. We pictured to ourselves such a man, alive to the most tender affections of the soul, ignorant of the fate of his companions, and thinking more of them than of himself. If we love to indulge such melancholy meditations, it is because, when just escaped from danger, we seem to feel as it were the necessity of strong emotions. Our minds were full of what we had just witnessed. There are periods in life when, without being discouraged, the future appears more uncertain. It was only three days since we had entered the Orinoco, and there yet remained three months for us to navigate rivers encumbered with rocks, and in boats smaller than that in which we had so nearly perished.

The night was intensely hot. We lay upon skins spread on the ground, there being no trees to which we could fasten our hammocks. The torments of the mosquitos increased every day; and we were surprised to find that on this spot our fires did not prevent the approach of the jaguars. They swam across the arm of the river that separated us from the mainland. Towards morning we heard their cries very near. They had come to the island where we passed the night. The Indians told us that, during the collecting of the turtles' eggs, tigers are always more frequent in those regions, and display at that period the greatest intrepidity.

On the following day, the 7th, we passed, on our right, the mouth of the great Rio Arauca, celebrated for the immense number of birds that frequent it; and, on our left, the Mission of Uruana, commonly called La Concepcion de Urbana. This small village, which contains five hundred souls, was founded by the Jesuits, about the year 1748, by the union of the Ottomac and Cavere Indians. It lies at the foot of a mountain composed of detached blocks of granite, which, I believe, bears the name of Saraguaca. Masses of rock, separated one from the other by the effect of decomposition, form caverns, in which we find indubitable proofs of the ancient civilization of the natives. Hieroglyphic figures, and even characters in regular lines, are seen sculptured on their sides; though I doubt whether they bear any analogy to alphabetic writing. We visited the Mission of Uruana on our return from the Rio Negro, and saw with our own eyes those heaps of earth which the Ottomacs eat, and which have become the subject of such lively discussion in Europe.[*]

On measuring the breadth of the Orinoco between the islands called Isla de Uruana and Isla de la Manteca, we found it, during the high waters, 2674 toises, which make nearly four nautical miles. This is eight times the breadth of the Nile at Manfalout and Syout, yet we were at the distance of a hundred and ninety-four leagues from the mouth of the Orinoco.

The temperature of the water at its surface was 27.8° of the centigrade thermometer, near Uruana. That of the river Zaire, or Congo, in Africa, at an equal distance from the equator, was

---

[*] This earth is a greasy kind of clay, which, in seasons of scarcity, the natives use to assuage the cravings of hunger; it having been proved by their experience as well as by physiological researches, that want of food can be more easily borne by filling the cavity of the stomach with some substance, even although it may be in itself very nearly or totally innutritious. The Indian hunters of North America, for the same purpose, tie boards tightly across the abdomen; and most savage races are found to have recourse to expedients that answer the same end.

found by Captain Tuckey, in the months of July and August, to be only from 23.9 to 25.6°.

The western bank of the Orinoco remains low farther than the mouth of the Meta; while from the Mission of Uruana the mountains approach the eastern bank more and more. As the strength of the current increases in proportion as the river grows narrower, the progress of our boat became much slower. We continued to ascend the Orinoco under sail, but the high and woody grounds deprived us of the wind. At other times the narrow passes between the mountains by which we sailed, sent us violent gusts, but of short duration. The number of crocodiles increased below the junction of the Rio Arauca, particularly opposite the great lake of Capanaparo, which communicates with the Orinoco, as the Laguna de Cabullarito communicates at the same time with the Orinoco and the Rio Arauca. The Indians told us that the crocodiles came from the inlands, where they had been buried in the dried mud of the savannahs. As soon as the first showers arouse them from their lethargy, they crowd together in troops, and hasten toward the river, there to disperse again. Here, in the equinoctial zone, it is the increase of humidity that recalls them to life; while in Georgia and Florida, in the temperate zone, it is the augmentation of heat that rouses these animals from a state of nervous and muscular debility, during which the active powers of respiration are suspended or singularly diminished. The season of great drought, improperly called the summer of the torrid zone, corresponds with the winter of the temperate zone; and it is a curious physiological phenomenon to observe the alligators of North America plunged into a winter-sleep by excess of cold, at the same period when the crocodiles of the Llanos begin their siesta or summer-sleep. If it were probable that these animals of the same family had heretofore inhabited the same northern country, we might suppose that, in advancing towards the equator, they feel the want of repose after having exercised their muscles for seven or eight months, and that they retain under a new sky the habits which appear to be essentially linked with their organization.

Having passed the mouths of the channels communicating with the lake of Capanaparo, we entered a part of the Orinoco, where the bed of the river is narrowed by the mountains of Baraguan. It is a kind of strait, reaching nearly to the confluence of the Rio Suapure. From these granite mountains the natives heretofore gave the name of Baraguan to that part of the Orinoco comprised between the mouths of the Arauca and the Atabapo. Among savage nations great rivers bear different denominations in the different portions of their course. The Passage of Baraguan presents a picturesque scene. The granite rocks are perpendicular. They form a range of mountains lying north-west and south-east; and the river cutting this dyke nearly at a right angle, the summits of the mountains appear like separate peaks. Their elevation in general does not surpass one hundred and twenty toises; but their situation in the midst of a small plain, their steep declivities, and their flanks destitute of vegetation, give them a majestic character. They are composed of enormous masses of granite of a parallelopipedal figure, but rounded at the edges, and heaped one upon another. The blocks are often eighty feet long, and twenty or thirty broad. They would seem to have been piled up by some external force, if the proximity of a rock identical in its composition, not separated into blocks but filled with veins, did not prove that the parallelopipedal form is owing solely to the action of the atmosphere. These veins, two or three inches thick, are distinguished by a fine-grained quartz-granite crossing a coarse-grained granite almost porphyritic, and abounding in fine crystals of red feldspar. I sought in vain, in the Cordillera of Baraguan, for hornblende, and those steatitic masses that characterise several granites of the Higher Alps in Switzerland.

We landed in the middle of the strait of Baraguan to measure its breadth. The rocks project so much towards the river that I measured with difficulty a base of eighty toises. I found the river eight hundred and eighty-nine toises broad. In order to conceive how this passage bears the name of a strait, we must recollect that the breadth of the river from Uruana to the junction of the

Meta is in general from 1500 to 2500 toises. In this place, which is extremely hot and barren, I measured two granite summits, much rounded: one was only a hundred and ten, and the other eighty-five, toises. There are higher summits in the interior of the group, but in general these mountains, of so wild an aspect, have not the elevation that is assigned to them by the missionaries.

We looked in vain for plants in the clefts of the rocks, which are as steep as walls, and furnish some traces of stratification. We found only an old trunk of aubletia[*], with large apple-shaped fruit, and a new species of the family of the apocyneae.[†] All the stones were covered with an innumerable quantity of iguanas and geckos with spreading and membranous fingers. These lizards, motionless, with heads raised, and mouths open, seemed to suck in the heated air. The thermometer placed against the rock rose to 50.2°. The soil appeared to undulate, from the effect of mirage, without a breath of wind being felt. The sun was near the zenith, and its dazzling light, reflected from the surface of the river, contrasted with the reddish vapours that enveloped every surrounding object. How vivid is the impression produced by the calm of nature, at noon, in these burning climates! The beasts of the forests retire to the thickets; the birds hide themselves beneath the foliage of the trees, or in the crevices of the rocks. Yet, amidst this apparent silence, when we lend an attentive ear to the most feeble sounds transmitted through the air, we hear a dull vibration, a continual murmur, a hum of insects, filling, if we may use the expression, all the lower strata of the air. Nothing is better fitted to make man feel the extent and power of organic life. Myriads of insects creep upon the soil, and flutter round the plants parched by the heat of the sun. A confused noise issues from every bush, from the decayed trunks of trees, from the clefts of the rocks, and from the ground undermined by lizards, millepedes, and cecilias. These are so

---

[*] Aubletia tiburba.
[†] Allamanda salicifolia.

many voices proclaiming to us that all nature breathes; and that, under a thousand different forms, life is diffused throughout the cracked and dusty soil, as well as in the bosom of the waters, and in the air that circulates around us.

The sensations which I here recall to mind are not unknown to those who, without having advanced to the equator, have visited Italy, Spain, or Egypt. That contrast of motion and silence, that aspect of nature at once calm and animated, strikes the imagination of the traveller when he enters the basin of the Mediterranean, within the zone of olives, dwarf palms, and date-trees.

We passed the night on the eastern bank of the Orinoco, at the foot of a granitic hill. Near this desert spot was formerly seated the Mission of San Regis. We could have wished to find a spring in the Baraguan, for the water of the river had a smell of musk, and a sweetish taste extremely disagreeable. In the Orinoco, as well as in the Apure, we are struck with the difference observable in the various parts of the river near the most barren shore. The water is sometimes very drinkable, and sometimes seems to be loaded with a slimy matter. "It is the bark (meaning the coriaceous covering) of the putrefied cayman that is the cause," say the natives. "The more aged the cayman, the more bitter is his bark." I have no doubt that the carcasses of these large reptiles, those of the manatees, which weigh five hundred pounds, and the presence of the porpoises (toninas) with their mucilaginous skin, may contaminate the water, especially in the creeks, where the river has little velocity. Yet the spots where we found the most fetid water, were not always those where dead animals were accumulated on the beach. When, in such burning climates, where we are constantly tormented by thirst, we are reduced to drink the water of a river at the temperature of 27 or 28°, we cannot help wishing at least that water so hot, and so loaded with sand, should be free from smell.

On the 8th of April we passed the mouths of the Suapure or Sivapuri, and the Caripo, on the east, and the outlet of the

Sinaruco on the west. This last river is, next to the Rio Arauca, the most considerable between the Apure and the Meta. The Suapure, full of little cascades, is celebrated among the Indians for the quantity of wild honey obtained from the forests in its neighbourhood. The melipones there suspend their enormous hives to the branches of trees. Father Gili, in 1766, made an excursion on the Suapure, and on the Turiva, which falls into it. He there found tribes of the nation of Areverians. We passed the night a little below the island Macapina.

Early on the following morning we arrived at the beach of Pararuma, where we found an encampment of Indians similar to that we had seen at the Boca de la Tortuga. They had assembled to search the sands, for collecting the turtles' eggs, and extracting the oil; but they had unfortunately made a mistake of several days. The young turtles had come out of their shells before the Indians had formed their camp; and consequently the crocodiles and the garzes, a species of large white herons, availed themselves of the delay. These animals, alike fond of the flesh of the young turtles, devour an innumerable quantity. They fish during the night, for the tortuguillos do not come out of the earth to gain the neighbouring river till after the evening twilight. The zamuro vultures are too indolent to hunt after sunset. They stalk along the shores in the daytime, and alight in the midst of the Indian encampment to steal provisions; but they often find no other means of satisfying their voracity than by attacking young crocodiles of seven or eight inches long, either on land or in water of little depth. It is curious to see the address with which these little animals defend themselves for a time against the vultures. As soon as they perceive the enemy, they raise themselves on their fore paws, bend their backs, and lift up their heads, opening their wide jaws. They turn continually, though slowly, toward their assailant to show him their teeth, which, even when the animal has but recently issued from the egg, are very long and sharp. Often while the attention of a young crocodile is wholly engaged by one of the zamuros, another seizes the favourable opportunity for an unforeseen attack. He pounces on the crocodile, grasps him by the neck, and bears him off to the higher regions of the air. We had an opportunity of observing this manoeuvre during several mornings, at Mompex, on the banks of the Magdalena, where we had collected more than forty very young crocodiles, in a spacious court surrounded by a wall.

We found among the Indians assembled at Pararuma some white men, who had come from Angostura to purchase the tortoise-butter. After having wearied us for a long time with their complaints of the bad harvest, and the mischief done by the tigers among the turtles, at the time of laying their eggs, they conducted us beneath an ajoupa, that rose in the centre of the Indian camp. We here found the missionary-monks of Carichana and the Cataracts seated on the ground, playing at cards, and smoking tobacco in long pipes. Their ample blue garments, their shaven heads, and their long beards, might have led us to mistake them for natives of the East. These poor priests received us in the kindest manner, giving us every information necessary for the continuation of our voyage. They had suffered from tertian fever for some months; and their pale and emaciated aspect easily convinced us that the countries we were about to visit were not without danger to the health of travellers.

The Indian pilot, who had brought us from San Fernando de Apure as far as the shore of Pararuma, was unacquainted with the passage of the rapids[*] of the Orinoco, and would not undertake to conduct our bark any farther. We were obliged to conform to his will. Happily for us, the missionary of Carichana consented to sell us a fine canoe at a very moderate price: and Father Bernardo Zea, missionary of the Atures and Maypures near the great cataracts, offered, though still unwell, to accompany us as far as the frontiers of Brazil. The number of natives who can assist in guiding boats through the Raudales is so inconsiderable that, but for the presence of the monk, we

---

[*] Little cascades, chorros raudalitos.

should have risked spending whole weeks in these humid and unhealthy regions. On the banks of the Orinoco, the forests of the Rio Negro are considered as delicious spots. The air is indeed cooler and more healthful. The river is free from crocodiles; one may bathe without apprehension, and by night as well as by day there is less torment from the sting of insects than on the Orinoco. Father Zea hoped to reestablish his health by visiting the Missions of Rio Negro. He talked of those places with that enthusiasm which is felt in all the colonies of South America for everything far off.

The assemblage of Indians at Pararuma again excited in us that interest, which everywhere attaches man in a cultivated state to the study of man in a savage condition, and the successive development of his intellectual faculties. How difficult to recognize in this infancy of society, in this assemblage of dull, silent, inanimate Indians, the primitive character of our species! Human nature does not here manifest those features of artless simplicity, of which poets in every language have drawn such enchanting pictures. The savage of the Orinoco appeared to us to be as hideous as the savage of the Mississippi, described by that philosophical traveller Volney, who so well knew how to paint man in different climates. We are eager to persuade ourselves that these natives, crouching before the fire, or seated on large turtle-shells, their bodies covered with earth and grease, their eyes stupidly fixed for whole hours on the beverage they are preparing, far from being the primitive type of our species, are a degenerate race, the feeble remains of nations who, after having been long dispersed in the forests, are replunged into barbarism.

Red paint being in some sort the only clothing of the Indians, two kinds may be distinguished among them, according as they are more or less affluent. The common decoration of the Caribs, the Ottomacs, and the Jaruros, is onoto,[*] called by the Spaniards

---

[*] Properly anoto. This word belongs to the Tamanac Indians. The Maypures call it majepa. The Spanish missionaries say onotarse, to rub the skin with

achote, and by the planters of Cayenne, rocou. It is the colouring matter extracted from the pulp of the Bixa orellana.[*] The Indian women prepare the anato by throwing the seeds of the plant into a tub filled with water. They beat this water for an hour, and then leave it to deposit the colouring fecula, which is of an intense brick-red. After having separated the water, they take out the fecula, dry it between their hands, knead it with oil of turtles' eggs, and form it into round cakes of three or four ounces weight. When turtle oil is wanting, some tribes mix with the anato the fat of the crocodile.

Another pigment, much more valuable, is extracted from a plant of the family of the bignoniae, which M. Bonpland has made known by the name of Bignonia chica. It climbs up and clings to the tallest trees by the aid of tendrils. Its bilabiate flowers are an inch long, of a fine violet colour, and disposed by twos or threes. The bipinnate leaves become reddish in drying. The fruit is a pod, filled with winged seeds, and is two feet long. This plant grows spontaneously, and in great abundance, near Maypures; and in going up the Orinoco, beyond the mouth of the Guaviare, from Santa Barbara to the lofty mountain of Duida, particularly near Esmeralda. We also found it on the banks of the Cassiquiare. The red pigment of chica is not obtained from the fruit, like the onoto, but from the leaves macerated in water. The colouring matter separates in the form of a light powder. It is collected, without being mixed with turtle-oil, into little lumps eight or nine inches long, and from two to three high, rounded at the edges. These lumps, when heated, emit an agreeable smell of benzoin. When the chica is subjected to distillation, it yields no sensible traces of ammonia. It is not, like indigo, a substance combined with azote. It dissolves slightly in sulphuric and muriatic acids, and even in alkalis. Ground with oil, the chica

---

anato.
[*] The word bixa, adopted by botanists, is derived from the ancient language of Haiti (the island of St. Domingo). Rocou, the term commonly used by the French, is derived from the Brazilian word, urucu.

furnishes a red colour that has a tint of lake. Applied to wool, it might be confounded with madder-red. There is no doubt but that the chica, unknown in Europe before our travels, may be employed usefully in the arts. The nations on the Orinoco, by whom this pigment is best prepared, are the Salivas, the Guipunaves,* the Caveres, and the Piraoas. The processes of infusion and maceration are in general very common among all the nations on the Orinoco. Thus the Maypures carry on a trade of barter with the little loaves of puruma, which is a vegetable fecula, dried in the manner of indigo, and yielding a very permanent yellow colour. The chemistry of the savage is reduced to the preparation of pigments, that of poisons, and the dulcification of the amylaceous roots, which the aroides and the euphorbiaceous plants afford.

Most of the missionaries of the Upper and Lower Orinoco permit the Indians of their Missions to paint their skins. It is painful to add, that some of them speculate on this barbarous practice of the natives. In their huts, pompously called conventos,† I have often seen stores of chica, which they sold as high as four francs the cake. To form a just idea of the extravagance of the decoration of these naked Indians, I must observe, that a man of large stature gains with difficulty enough by the labour of a fortnight, to procure in exchange the chica necessary to paint himself red. Thus as we say, in temperate climates, of a poor man, "he has not enough to clothe himself," you hear the Indians of the Orinoco say, "that man is so poor, that he has not enough to paint half his body." The little trade in chica is carried on chiefly with the tribes of the Lower Orinoco, whose country does not produce the plant which furnishes this much-valued substance. The Caribs and the Ottomacs paint only the head and the hair with chica, but the Salives possess this pigment in sufficient abundance to cover their whole bodies. When the missionaries send on their own account small cargoes

---

* Or Guaypunaves; they call themselves Uipunavi.
† In the Missions, the priest's house bears the name of the convent.

of cacao, tobacco, and chiquichiqui[*] from the Rio Negro to Angostura, they always add some cakes of chica, as being articles of merchandise in great request.

The custom of painting is not equally ancient among all the tribes of the Orinoco. It has increased since the time when the powerful nation of the Caribs made frequent incursions into those countries. The victors and the vanquished were alike naked; and to please the conqueror it was necessary to paint like him, and to assume his colour. The influence of the Caribs has now ceased, and they remain circumscribed between the rivers Carony, Cuyuni, and Paraguamuzi; but the Caribbean fashion of painting the whole body is still preserved. The custom has survived the conquest.

Does the use of the anato and chica derive its origin from the desire of pleasing, and the taste for ornament, so common among the most savage nations? or must we suppose it to be founded on the observation, that these colouring and oily matters with which the skin is plastered, preserve it from the sting of the mosquitos? I have often heard this question discussed in Europe; but in the Missions of the Orinoco, and wherever, within the tropics, the air is filled with venomous insects, the inquiry would appear absurd. The Carib and the Salive, who are painted red, are not less cruelly tormented by the mosquitos and the zancudos, than the Indians whose bodies are plastered with no colour. The sting of the insect causes no swelling in either; and scarcely ever produces those little pustules which occasion such smarting and itching to Europeans recently arrived. But the native and the White suffer equally from the sting, till the insect has withdrawn its sucker from the skin. After a thousand useless essays, M. Bonpland and myself tried the expedient of rubbing our hands and arms with the fat of the crocodile, and the oil of turtle-eggs, but we never felt the least relief, and were stung as before. I know that the Laplanders boast of oil and fat as the

[*] Ropes made with the petioles of a palm-tree with pinnate leaves.

most useful preservatives; but the insects of Scandinavia are not of the same species as those of the Orinoco. The smoke of tobacco drives away our gnats, while it is employed in vain against the zancudos. If the application of fat and astringent[*] substances preserved the inhabitants of these countries from the torment of insects, as Father Gumilla alleges, why has not the custom of painting the skin become general on these shores? Why do so many naked natives paint only the face, though living in the neighbourhood of those who paint the whole body?[†]

We are struck with the observation, that the Indians of the Orinoco, like the natives of North America, prefer the substances that yield a red colour to every other. Is this predilection founded on the facility with which the savage procures ochreous earths, or the colouring fecula of anato and of chica? I doubt this much. Indigo grows wild in a great part of equinoctial America. This plant, like so many other leguminous plants, would have furnished the natives abundantly with pigments to colour themselves blue like the ancient Britons.[‡] Yet we see no American tribe painted with indigo. It appears to me probable, as I have already hinted above, that the preference given by the Americans to the red colour is generally founded on the tendency which nations feel to attribute the idea of beauty to whatever characterises their national physiognomy. Men whose skin is naturally of a brownish red, love a red colour. If they be born with a forehead little raised, and the head flat, they endeavour to depress the foreheads of their children. If they be distinguished from other nations by a thin beard, they try to eradicate the few hairs that nature has given them. They think themselves embellished in proportion as they heighten the

---

[*] The pulp of the anato, and even the chica, are astringent and slightly purgative.

[†] The Caribs, the Salives, the Tamanacs, and the Maypures.

[‡] The half-clad nations of the temperate zone often paint their skin of the same colour as that with which their clothes are dyed.

characteristic marks of their race, or of their national conformation.

We were surprised to see, that, in the camp of Pararuma, the women far advanced in years were more occupied with their ornaments than the youngest women. We saw an Indian female of the nation of the Ottomacs employing two of her daughters in the operation of rubbing her hair with the oil of turtles' eggs, and painting her back with anato and caruto. The ornament consisted of a sort of lattice-work formed of black lines crossing each other on a red ground. Each little square had a black dot in the centre. It was a work of incredible patience. We returned from a very long herborization, and the painting was not half finished. This research of ornament seems the more singular when we reflect that the figures and marks are not produced by the process of tattooing, but that paintings executed with so much care are effaced,[*] if the Indian exposes himself imprudently to a heavy shower. There are some nations who paint only to celebrate festivals; others are covered with colour during the whole year: and the latter consider the use of anato as so indispensable, that both men and women would perhaps be less ashamed to present themselves without a guayaco[†] than destitute of paint. These guayucos of the Orinoco are partly bark of trees, and partly cotton-cloth. Those of the men are broader than those worn by the women, who, the missionaries say, have in general a less lively feeling of modesty. A similar observation was made by Christopher Columbus. May we not attribute this in difference, this want of delicacy in women belonging to nations of which the manners are not much depraved, to that rude state

---

[*] The black and caustic pigment of the caruto (Genipa americana) however, resists a long time the action of water, as we found with regret, having one day, in sport with the Indians, caused our faces to be marked with spots and strokes of caruto. When we returned to Angostura, in the midst of Europeans, these marks were still visible.

[†] A word of the Caribbean language. The perizoma of the Indians of the Orinoco is rather a band than an apron.

of slavery to which the sex is reduced in South America by male injustice and tyranny?

When we speak in Europe of a native of Guiana, we figure to ourselves a man whose head and waist are decorated with the fine feathers of the macaw, the toucan, and the humming-bird. Our painters and sculptors have long since regarded these ornaments as the characteristic marks of an American. We were surprised at not finding in the Chayma Missions, in the encampments of Uruana and of Pararuma (I might almost say on all the shores of the Orinoco and the Cassiquiare) those fine plumes, those feathered aprons, which are so often brought by travellers from Cayenne and Demerara. These tribes for the most part, even those whose intellectual faculties are most expanded, who cultivate alimentary plants, and know how to weave cotton, are altogether as naked,[*] as poor, and as destitute of ornaments as the natives of New Holland. The excessive heat of the air, the profuse perspiration in which the body is bathed at every hour of the day and a great part of the night, render the use of clothes insupportable. Their objects of ornament, and particularly their plumes of feathers, are reserved for dances and solemn festivals. The plumes worn by the Guipunaves[†] are the most celebrated; being composed of the fine feathers of manakins and parrots.

The Indians are not always satisfied with one colour uniformly spread; they sometimes imitate, in the most whimsical manner, in painting their skin, the form of European garments. We saw some at Pararuma, who were painted with blue jackets and black buttons. The missionaries related to us that the Guaynaves of the Rio Caura are accustomed to stain themselves red with anato, and to make broad transverse stripes on the body, on which they stick spangles of silvery mica. Seen at a distance, these naked men appear to be dressed in laced clothes. If painted

---

[*] For instance, the Macos and the Piraoas. The Caribs must be excepted, whose perizoma is a cotton cloth, so broad that it might cover the shoulders.

[†] These came originally from the banks of the Inirida, one of the rivers that fall into the Guaviare.

nations had been examined with the same attention as those who are clothed, it would have been perceived that the most fertile imagination, and the most mutable caprice, have created the fashions of painting, as well as those of garments.

Painting and tattooing are not restrained, in either the New or the Old World, to one race or one zone only. These ornaments are most common among the Malays and American races; but in the time of the Romans they were also employed by the white race in the north of Europe. As the most picturesque garments and modes of dress are found in the Grecian Archipelago and western Asia, so the type of beauty in painting and tattooing is displayed by the islanders of the Pacific. Some clothed nations still paint their hands, their nails, and their faces. It would seem that painting is then confined to those parts of the body that remain uncovered; and while rouge, which recalls to mind the savage state of man, is disappearing by degrees in Europe, in some towns of the province of Peru the ladies think they embellish their delicate skins by covering them with colouring vegetable matter, starch, white-of-egg, and flour. After having lived a long time among men painted with anato and chica, we are singularly struck with these remains of ancient barbarism retained amidst all the usages of civilization.

The encampment at Pararuma afforded us an opportunity of examining several animals in their natural state, which, till then, we had seen only in the collections of Europe. These little animals form a branch of commerce for the missionaries. They exchange tobacco, the resin called mani, the pigment of chica, gallitos (rock-manakins), orange monkeys, capuchin monkeys, and other species of monkeys in great request on the coast, for cloth, nails, hatchets, fishhooks, and pins. The productions of the Orinoco are bought at a low price from the Indians, who live in dependence on the monks; and these same Indians purchase fishing and gardening implements from the monks at a very high price, with the money they have gained at the egg-harvest. We ourselves bought several animals, which we kept with us throughout the rest of our passage on the river, and studied their manners.

The gallitos, or rock-manakins, are sold at Pararuma in pretty little cages made of the footstalks of palm-leaves. These birds are infinitely more rare on the banks of the Orinoco, and in the north and west of equinoctial America, than in French Guiana. They have hitherto been found only near the Mission of Encaramada, and in the Raudales or cataracts of Maypures. I say expressly IN the cataracts, because the gallitos choose for their habitual dwelling the hollows of the little granitic rocks that cross the Orinoco and form such numerous cascades. We sometimes saw them appear in the morning in the midst of the foam of the river, calling their females, and fighting in the manner of our cocks, folding the double moveable crest that decorates the crown of the head. As the Indians very rarely take the full-grown gallitos, and those males only are valued in Europe, which from the third year have beautiful saffron-coloured plumage, purchasers should be on their guard not to confound young females with young males. Both the male and female gallitos are of an olive-brown; but the pollo, or young male, is distinguishable at the earliest age, by its size and its yellow feet. After the third year the plumage of the males assumes a beautiful saffron tint; but the female remains always of a dull dusky brown colour, with yellow only on the wing-coverts and tips of the wings.[*] To preserve in our collections the fine tint of the plumage of a male and full-grown rock-manakin, it must not be exposed to the light. This tint grows pale more easy than in the other genera of the passerine order. The young males, as in most other birds, have the plumage or livery of their mother. I am surprised to see that so skilful a naturalist as Le Vaillant can doubt whether the females always remain of a

---

[*] Especially the part which ornithologists call the carpus.

dusky olive tint.* The Indians of the Raudales all assured me that they had never seen a saffron-coloured female.

Among the monkeys, brought by the Indians to the fair of Pararuma, we distinguished several varieties of the sai,† belonging to the little groups of creeping monkeys called matchi in the Spanish colonies; marimondes‡, or ateles with a red belly; titis, and viuditas. The last two species particularly attracted our attention, and we purchased them to send to Europe.

The titi of the Orinoco (Simia sciurea), well-known in our collections, is called bititeni by the Maypure Indians. It is very common on the south of the cataracts. Its face is white; and a little spot of bluish-black covers the mouth and the point of the nose. The titis of the most elegant form, and the most beautiful colour (with hair of a golden yellow), come from the banks of the Cassiquiare. Those that are taken on the shores of the Guaviare are large and difficult to tame. No other monkey has so much the physiognomy of a child as the titi; there is the same expression of innocence, the same playful smile, the same rapidity in the transition from joy to sorrow. Its large eyes are instantly filled with tears, when it is seized with fear. It is extremely fond of insects, particularly of spiders. The sagacity of this little animal is so great, that one of those we brought in our boat to Angostura distinguished perfectly the different plates annexed to Cuvier's Tableau elementaire d'Histoire naturelle. The engravings of this work are not coloured; yet the titi advanced rapidly its little hand in the hope of catching a grasshopper or a wasp, every time that we showed it the eleventh plate, on which these insects are represented. It remained perfectly indifferent when it was shown engravings of skeletons or heads of mammiferous animals.§ When several of these little

---

* Oiseaux de Paradis volume 2 page 61.
† Simia capucina the capuchin monkey.
‡ Simia belzebuth.
§ I may observe, that I have never heard of an instance in which a picture, representing, in the greatest perfection, hares or deer of their natural size, has

monkeys, shut up in the same cage, are exposed to the rain, and the habitual temperature of the air sinks suddenly two or three degrees, they twist their tail (which, however, is not prehensile) round their neck, and intertwine their arms and legs to warm one another. The Indian hunters told us, that in the forests they often met groups of ten or twelve of these animals, whilst others sent forth lamentable cries, because they wished to enter amid the group to find warmth and shelter. By shooting arrows dipped in weak poison at one of these groups, a great number of young monkeys are taken alive at once. The titi in falling remains clinging to its mother, and if it be not wounded by the fall, it does not quit the shoulder or the neck of the dead animal. Most of those that are found alive in the huts of the Indians have been thus taken from the dead bodies of their mothers. Those that are full grown, when cured of a slight wound, commonly die before they can accustom themselves to a domestic state. The titis are in general delicate and timid little animals. It is very difficult to convey them from the Missions of the Orinoco to the coast of Caracas, or of Cumana. They become melancholy and dejected in proportion as they quit the region of the forests, and enter the Llanos. This change cannot be attributed to the slight elevation of the temperature; it seems rather to depend on a greater intensity of light, a less degree of humidity, and some chemical property of the air of the coast.

The saimiri, or titi of the Orinoco, the atele, the sajou, and other quadrumanous animals long known in Europe, form a striking contrast, both in their gait and habits, with the macavahu, called by the missionaries viudita, or widow in mourning. The hair of this little animal is soft, glossy, and of a fine black. Its face is covered with a mask of a square form and a whitish colour tinged with blue. This mask contains the eyes,

---

made the least impression even on sporting dogs, the intelligence of which appears the most improved. Is there any authenticated instance of a dog having recognized a full length picture of his master? In all these cases, the sight is not assisted by the smell.

nose, and mouth. The ears have a rim: they are small, very pretty, and almost bare. The neck of the widow presents in front a white band, an inch broad, and forming a semicircle. The feet, or rather the hinder hands, are black like the rest of the body; but the fore paws are white without, and of a glossy black within. In these marks, or white spots, the missionaries think they recognize the veil, the neckerchief, and the gloves of a widow in mourning. The character of this little monkey, which sits up on its hinder extremities only when eating, is but little indicated in its appearance. It has a wild and timid air; it often refuses the food offered to it, even when tormented by a ravenous appetite. It has little inclination for the society of other monkeys. The sight of the smallest saimiri puts it to flight. Its eye denotes great vivacity. We have seen it remain whole hours motionless without sleeping, and attentive to everything that was passing around. But this wildness and timidity are merely apparent. The viudita, when alone, and left to itself, becomes furious at the sight of a bird. It then climbs and runs with astonishing rapidity; darts upon its prey like a cat; and kills whatever it can seize. This rare and delicate monkey is found on the right bank of the Orinoco, in the granite mountains which rise behind the Mission of Santa Barbara. It inhabits also the banks of the Guaviare, near San Fernando de Atabapo.

The viudita accompanied us on our whole voyage on the Cassiquiare and the Rio Negro, passing the cataracts twice. In studying the manners of animals, it is a great advantage to observe them during several months in the open air, and not in houses, where they lose all their natural vivacity.

The new canoe intended for us was, like all Indian boats, a trunk of a tree hollowed out partly by the hatchet and partly by fire. It was forty feet long, and three broad. Three persons could not sit in it side by side. These canoes are so crank, and they require, from their instability, a cargo so equally distributed, that when you want to rise for an instant, you must warn the rowers to lean to the opposite side. Without this precaution the water would necessarily enter the side pressed down. It is difficult to form an idea of the inconveniences that are suffered in such wretched vessels.

The missionary from the cataracts made the preparations for our voyage with greater energy than we wished. Lest there might not be a sufficient number of the Maco and Guahibe Indians, who are acquainted with the labyrinth of small channels and cascades of which the Raudales or cataracts are composed, two Indians were, during the night, placed in the cepo—a sort of stocks in which they were made to lie with their legs between two pieces of wood, notched and fastened together by a chain with a padlock. Early in the morning we were awakened by the cries of a young man, mercilessly beaten with a whip of manatee skin. His name was Zerepe, a very intelligent young Indian, who proved highly useful to us in the sequel, but who now refused to accompany us. He was born in the Mission of Atures; but his father was a Maco, and his mother a native of the nation of the Maypures. He had returned to the woods (al monte), and having lived some years with the unsubdued Indians, he had thus acquired the knowledge of several languages, and the missionary employed him as an interpreter. We obtained with difficulty the pardon of this young man. "Without these acts of severity," we were told, "you would want for everything. The Indians of the Raudales and the Upper Orinoco are a stronger and more laborious race than the inhabitants of the Lower Orinoco. They know that they are much sought after at Angostura. If left to their own will, they would all go down the river to sell their productions, and live in full liberty among the whites. The Missions would be totally deserted."

These reasons, I confess, appeared to me more specious than sound. Man, in order to enjoy the advantages of a social state, must no doubt sacrifice a part of his natural rights, and his original independence; but, if the sacrifice imposed on him be not compensated by the benefits of civilization, the savage, wise in his simplicity, retains the wish of returning to the forests that gave him birth. It is because the Indian of the woods is treated like a person in a state of villanage in the greater part of the Missions, because he enjoys not the fruit of his labours, that the Christian establishments on the Orinoco remain deserts. A government founded on the ruins of the liberty of the natives extinguishes the intellectual faculties, or stops their progress.

To say that the savage, like the child, can be governed only by force, is merely to establish false analogies. The Indians of the Orinoco have something infantine in the expression of their joy, and the quick succession of their emotions, but they are not great children; they are as little so as the poor labourers in the east of Europe, whom the barbarism of our feudal institutions has held in the rudest state. To consider the employment of force as the first and sole means of the civilization of the savage, is a principle as far from being true in the education of nations as in the education of youth. Whatever may be the state of weakness or degradation in our species, no faculty is entirely annihilated. The human understanding exhibits only different degrees of strength and development. The savage, like the child, compares the present with the past; he directs his actions, not according to blind instinct, but motives of interest. Reason can everywhere enlighten reason; and its progress will be retarded in proportion as the men who are called upon to bring up youth, or govern nations, substitute constraint and force for that moral influence which can alone unfold the rising faculties, calm the irritated passions, and give stability to social order.

We could not set sail before ten on the morning of the 10th. To gain something in breadth in our new canoe, a sort of lattice-work had been constructed on the stern with branches of trees, that extended on each side beyond the gunwale. Unfortunately, the toldo or roof of leaves, that covered this lattice-work, was so low that we were obliged to lie down, without seeing anything, or, if seated, to sit nearly double. The necessity of carrying the canoe across the rapids, and even from one river to another; and the fear of giving too much hold to the wind, by making the toldo higher, render this construction necessary for vessels that go up towards the Rio Negro. The toldo was intended to cover four persons, lying on the deck or lattice-work of brush-wood; but our legs reached far beyond it, and when it rained half our bodies were wet. Our couches consisted of ox-hides or tiger-skins, spread upon branches of trees, which were painfully felt through so thin a covering. The fore part of the boat was filled with Indian rowers, furnished with paddles, three feet long, in the form of spoons. They were all naked, seated two by two, and they kept time in rowing with a surprising uniformity, singing songs of a sad and monotonous character. The small cages containing our birds and our monkeys, the number of which augmented as we advanced, were hung some to the toldo and others to the bow of the boat. This was our travelling menagerie. Notwithstanding the frequent losses occasioned by accidents, and above all by the fatal effects of exposure to the sun, we had fourteen of these little animals alive at our return from the Cassiquiare. Naturalists, who wish to collect and bring living animals to Europe, might cause boats to be constructed expressly for this purpose at Angostura, or at Grand Para, the two capitals situated on the banks of the Orinoco and the Amazon, the fore-deck of which boats might be fitted up with two rows of cages sheltered from the rays of the sun. Every night, when we established our watch, our collection of animals and our instruments occupied the centre; around these were placed first our hammocks, then the hammocks of the Indians; and on the outside were the fires which are thought indispensable against the attacks of the jaguar. About sunrise the monkeys in our cages answered the cries of the monkeys of the forest. These communications between animals of the same species sympathizing with one another, though unseen, one party enjoying that liberty which the other regrets, have something melancholy and affecting.

In a canoe not three feet wide, and so incumbered, there remained no other place for the dried plants, trunks, a sextant, a dipping-needle, and the meteorological instruments, than the space below the lattice-work of branches, on which we were compelled to remain stretched the greater part of the day. If we wished to take the least object out of a trunk, or to use an instrument, it was necessary to row ashore and land. To these inconveniences were joined the torment of the mosquitos which swarmed under the toldo, and the heat radiated from the leaves of the palm-trees, the upper surface of which was continually exposed to the solar rays. We attempted every instant, but always without success, to amend our situation. While one of us hid himself under a sheet to ward off the insects, the other insisted on having green wood lighted beneath the toldo, in the hope of driving away the mosquitos by the smoke. The painful sensations of the eyes, and the increase of heat, already stifling, rendered both these contrivances alike impracticable. With some gaiety of temper, with feelings of mutual good-will, and with a vivid taste for the majestic grandeur of these vast valleys of rivers, travellers easily support evils that become habitual.

Our Indians showed us, on the right bank of the river, the place which was formerly the site of the Mission of Pararuma, founded by the Jesuits about the year 1733. The mortality occasioned by the smallpox among the Salive Indians was the principal cause of the dissolution of the mission. The few inhabitants who survived this cruel epidemic, removed to the village of Carichana. It was at Pararuma, that, according to the testimony of Father Roman, hail was seen to fall during a great storm, about the middle of the last century. This is almost the only instance of it I know in a plain that is nearly on a level with the sea; for hail falls generally, between the tropics, only at three hundred toises of elevation. If it form at an equal height over plains and table-lands, we must suppose that it melts as it falls, in passing through the lowest strata of the atmosphere, the mean temperature of which is from 27.5 to 24° of the centigrade thermometer. I acknowledge it is very difficult to explain, in the present state of meteorology, why it hails at Philadelphia, at Rome, and at Montpelier, during the hottest months, the mean temperature of which attains 25 or 26°; while the same phenomenon is not observed at Cumana, at La Guayra, and in general, in the equatorial plains. In the United States, and in the south of Europe, the heat of the plains (from 40 to 43° latitude) is nearly the same as within the tropics; and according to my researches the decrement of caloric equally varies but little. If then the absence of hail within the torrid zone, at the level of the sea, be produced by the melting of the hailstones in crossing the lower strata of the air, we must suppose that these hail-stones, at the moment of their formation, are larger in the temperate than in the torrid zone. We yet know so little of the conditions under which water congeals in a stormy cloud in our climates, that we cannot judge whether the same conditions be fulfilled on the equator above the plains. The clouds in which we hear the rattling of the hailstones against one another before they fall, and which move horizontally, have always appeared to me of little elevation; and at these small heights we may conceive that extraordinary refrigerations are caused by the dilatation of the ascending air, of which the capacity for caloric augments; by currents of cold air coming from a higher latitude, and above all, according to M. Gay Lussac, by the radiation from the upper surface of the clouds. I shall have occasion to return to this subject when speaking of the different forms under which hail and hoar-frost appear on the Andes, at two thousand and two thousand six hundred toises of height; and when examining the question whether we may consider the stratum of clouds that envelops the mountains as a horizontal continuation of the stratum which we see immediately above us in the plains.

The Orinoco, full of islands, begins to divide itself into several branches, of which the most western remain dry during the months of January and February. The total breadth of the river exceeds two thousand five hundred or three thousand toises. We perceived to the East, opposite the island of Javanavo, the mouth of the Cano Aujacoa. Between this Cano and the Rio Paruasi or Paruati, the country becomes more and more woody. A solitary rock, of extremely picturesque aspect, rises in the midst of a forest of palm-trees, not far from the Orinoco. It is a pillar of granite, a prismatic mass, the bare and steep sides of which attain nearly two hundred feet in height. Its point, which overtops the highest trees of the forest, is terminated by a shelf of rock with a horizontal and smooth surface. Other trees crown this summit, which the missionaries call the peak, or Mogote de Cocuyza. This monument of nature, in its simple grandeur recalls to mind the Cyclopean remains of antiquity. Its strongly-marked outlines, and the group of trees and shrubs by which it is crowned, stand out from the azure of the sky. It seems a forest rising above a forest.

Further on, near the mouth of the Paruasi, the Orinoco narrows. On the east is perceived a mountain with a bare top, projecting like a promontory. It is nearly three hundred feet high, and served as a fortress for the Jesuits. They had constructed there a small fort, with three batteries of cannon, and it was constantly occupied by a military detachment. We saw the cannon dismounted, and half-buried in the sand, at Carichana and at Atures. This fort of the Jesuits has been destroyed since the dissolution of their society; but the place is still called El Castillo. I find it set down, in a manuscript map, lately completed at Caracas by a member of the secular clergy, under the denomination of Trinchera del despotismo monacal.[*]

The garrison which the Jesuits maintained on this rock, was not intended merely to protect the Missions against the incursions of the Caribs: it was employed also in an offensive war, or, as they say here, in the conquest of souls (conquista de almas). The soldiers, excited by the allurement of gain, made military incursions (entradas) into the lands of the independent Indians. They killed all those who dared to make any resistance, burnt their huts, destroyed their plantations, and carried away the women, children, and old men, as prisoners. These prisoners were divided among the Missions of the Meta, the Rio Negro, and the Upper Orinoco. The most distant places were chosen,

---

[*] Intrenchmnent of monachal despotism.

that they might not be tempted to return to their native country. This violent manner of conquering souls, though prohibited by the Spanish laws, was tolerated by the civil governors, and vaunted by the superiors of the society, as beneficial to religion, and the aggrandizement of the Missions. "The voice of the Gospel is heard only," said a Jesuit of the Orinoco, very candidly, in the Cartas Edifiantes, "where the Indians have heard also the sound of fire-arms (el eco de la polvora). Mildness is a very slow measure. By chastising the natives, we facilitate their conversion." These principles, which degrade humanity, were certainly not common to all the members of a society which, in the New World, and wherever education has remained exclusively in the hands of monks, has rendered service to letters and civilization. But the entradas, the spiritual conquests with the assistance of bayonets, was an inherent vice in a system, that tended to the rapid aggrandizement of the Missions. It is pleasing to find that the same system is not followed by the Franciscan, Dominican, and Augustinian monks who now govern a vast portion of South America; and who, by the mildness or harshness of their manners, exert a powerful influence over the fate of so many thousands of natives. Military incursions are almost entirely abolished; and when they do take place, they are disavowed by the superiors of the orders. We will not decide at present, whether this amelioration of the monachal system be owing to want of activity and cold indolence; or whether it must be attributed, as we would wish to believe, to the progress of knowledge, and to feelings more elevated, and more conformable to the true spirit of Christianity.

Beyond the mouth of the Rio Paruasi, the Orinoco again narrows. Full of little islands and masses of granite rock, it presents rapids, or small cascades (remolinos), which at first sight may alarm the traveller by the continual eddies of the water, but which at no season of the year are dangerous for boats. A range of shoals, that crosses almost the whole river, bears the name of the Raudal de Marimara. We passed it without difficulty by a narrow channel, in which the water seems to boil

239

up as it issues out impetuously[*] below the Piedra de Marimara, a compact mass of granite eighty feet high, and three hundred feet in circumference, without fissures, or any trace of stratification. The river penetrates far into the land, and forms spacious bays in the rocks. One of these bays, inclosed between two promontories destitute of vegetation, is called the Port of Carichana.[†] The spot has a very wild aspect. In the evening the rocky coasts project their vast shadows over the surface of the river. The waters appear black from reflecting the image of these granitic masses, which, in the colour of their external surface, sometimes resemble coal, and sometimes lead-ore. We passed the night in the small village of Carichana, where we were received at the priest's house, or convento. It was nearly a fortnight since we had slept under a roof.

To avoid the effects of the inundations, often so fatal to health, the Mission of Carichana has been established at three quarters of a league from the river. The Indians in this Mission are of the nation of the Salives, and they have a disagreeable and nasal pronunciation. Their language, of which the Jesuit Anisson has composed a grammar still in manuscript, is, with the Caribbean, the Tamanac, the Maypure, the Ottomac, the Guahive, and the Jaruro, one of the mother-tongues most general on the Orinoco. Father Gili thinks that the Ature, the Piraoa, and the Quaqua or Mapoye, are only dialects of the Salive. My journey was much too rapid to enable me to judge of the accuracy of this opinion; but we shall soon see that, in the village of Ature, celebrated on account of its situation near the great cataracts, neither the Salive nor the Ature is now spoken, but the language of the Maypures. In the Salive of Carichana, man is called cocco; woman, gnacu; water, cagua; fire, eyussa; the earth, seke; the sky, mumeseke (earth on high); the jaguar, impii; the crocodile, cuipoo; maize, giomu; the plantain, paratuna; cassava, peibe. I may here mention one of those

---

[*] These places are called chorreros in the Spanish colonies.

[†] Piedra y puerto de Carichana.

descriptive compounds that seem to characterise the infancy of language, though they are retained in some very perfect idioms.[*] Thus, as in the Biscayan, thunder is called the noise of the cloud (odotsa); the sun bears the name, in the Salive dialect, of mume-seke-cocco, the man (cocco) of the earth (seke) above (mume).

The most ancient abode of the Salive nation appears to have been on the western banks of the Orinoco, between the Rio Vichada[†] and the Guaviare, and also between the Meta and the Rio Paute. Salives are now found not only at Carichana, but in the Missions of the province of Casanre, at Cabapuna, Guanapalo, Cabiuna, and Macuco. They are a social, mild, almost timid people; and more easy, I will not say to civilize, but to subdue, than the other tribes on the Orinoco. To escape from the dominion of the Caribs, the Salives willingly joined the first Missions of the Jesuits. Accordingly these fathers everywhere in their writings praise the docility and intelligence of that people. The Salives have a great taste for music: in the most remote times they had trumpets of baked earth, four or five feet long, with several large globular cavities communicating with one another by narrow pipes. These trumpets send forth most dismal sounds. The Jesuits have cultivated with success the natural taste of the Salives for instrumental music; and even since the destruction of the society, the missionaries of Rio Meta have continued at San Miguel de Macuco a fine church choir, and musical instruction for the Indian youth. Very lately a traveller was surprised to see the natives playing on the violin, the violoncello, the triangle, the guitar, and the flute.

We found among these Salive Indians, at Carichana, a white woman, the sister of a Jesuit of New Grenada. It is difficult to define the satisfaction that is felt when, in the midst of nations of whose language we are ignorant, we meet with a being with whom we can converse without an interpreter. Every mission

---

[*] See volume 1 chapter 1.9.
[†] The Salive mission, on the Rio Vichada, was destroyed by the Caribs.

has at least two interpreters (lenguarazes). They are Indians, a little less stupid than the rest, through whose medium the missionaries of the Orinoco, who now very rarely give themselves the trouble of studying the idioms of the country, communicate with the neophytes. These interpreters attended us in all our herborizations; but they rather understand than speak Castilian. With their indolent indifference, they answer us by chance, but always with an officious smile, "Yes, Father; no, Father," to every question addressed to them.

The vexation that arises from such a style of conversation continued for months may easily be conceived, when you wish to be enlightened upon objects in which you take the most lively interest. We were often forced to employ several interpreters at a time, and several successive translators, in order to communicate with the natives.[*]

"After leaving my Mission," said the good monk of Uruana, "you will travel like mutes." This prediction was nearly accomplished; and, not to lose the advantage we might derive from intercourse even with the rudest Indians, we sometimes preferred the language of signs. When a native perceives that you will not employ an interpreter; when you interrogate him directly, showing him the objects; he rouses himself from his habitual apathy, and manifests an extraordinary capacity to make himself comprehended. He varies his signs, pronounces his words slowly, and repeats them without being desired. The consequence conferred upon him, in suffering yourself to be instructed by him, flatters his self-love. This facility in making

---

[*] To form a just idea of the perplexity of these communications by interpreters, we may recollect that, in the expedition of Lewis and Clarke to the river Columbia, in order to converse with the Chopunnish Indians, Captain Lewis addressed one of his men in English; that man translated the question into French to Chaboneau; Chaboneau translated it to his Indian wife in Minnetaree; the woman translated it into Shoshonee to a prisoner; and the prisoner translated it into Chopunnish. It may be feared that the sense of the question was a little altered by these successive translations.

himself comprehended is particularly remarkable in the independent Indian. It cannot be doubted that direct intercourse with the natives is more instructive and more certain than the communication by interpreters, provided the questions be simplified, and repeated to several individuals under different forms. The variety of idioms spoken on the banks of the Meta, the Orinoco, the Cassiquiare, and the Rio Negro, is so prodigious, that a traveller, however great may be his talent for languages, can never hope to learn enough to make himself understood along the navigable rivers, from Angostura to the small fort of San Carlos del Rio Negro. In Peru and Quito it is sufficient to know the Quichua, or the Inca language; in Chile, the Araucan; and in Paraguay, the Guarany; in order to be understood by most of the population. But it is different in the Missions of Spanish Guiana, where nations of various races are mingled in the village. It is not even sufficient to have learned the Caribee or Carina, the Guamo, the Guahive, the Jaruro, the Ottomac, the Maypure, the Salive, the Marivitan, the Maquiritare, and the Guaica, ten dialects, of which there exist only imperfect grammars, and which have less affinity with each other than the Greek, German, and Persian languages.

The environs of the Mission of Carichana appeared to us to be delightful. The little village is situated in one of those plains covered with grass that separate all the links of the granitic mountains, from Encaramada to beyond the Cataracts of Maypures. The line of the forests is seen only in the distance. The horizon is everywhere bounded by mountains, partly wooded and of a dark tint, partly bare, with rocky summits gilded by the beams of the setting sun. What gives a peculiar character to the scenery of this country are banks of rock (laxas) nearly destitute of vegetation, and often more than eight hundred feet in circumference, yet scarcely rising a few inches above the surrounding savannahs. They now make a part of the plain. We ask ourselves with surprise, whether some extraordinary revolutions may have carried away the earth and plants; or whether the granite nucleus of our planet shows itself bare,

243

because the germs of life are not yet developed on all its points. The same phenomenon seems to be found also in the desert of Shamo, which separates Mongolia from China. Those banks of solitary rock in the desert are called tsy. I think they would be real table-lands, if the surrounding plains were stripped of the sand and mould that cover them, and which the waters have accumulated in the lowest places. On these stony flats of Carichana we observed with interest the rising vegetation in the different degrees of its development. We there found lichens cleaving the rock, and collected in crusts more or less thick; little portions of sand nourishing succulent plants; and lastly layers of black mould deposited in the hollows, formed from the decay of roots and leaves, and shaded by tufts of evergreen shrubs.

At the distance of two or three leagues from the Mission, we find, in these plains intersected by granitic hills, a vegetation no less rich than varied. On comparing the site of Carichana with that of all the villages above the Great Cataracts, we are surprised at the facility with which we traverse the country, without following the banks of the rivers, or being stopped by the thickness of the forests. M. Bonpland made several excursions on horseback, which furnished him with a rich harvest of plants. I shall mention only the paraguatan, a magnificent species of the macrocnemum, the bark of which yields a red dye;[*] the guaricamo, with a poisonous root;[†] the Jacaranda obtusifolia; and the serrape, or jape[‡] of the Salive Indians, which is the Coumarouna of Aublet, so celebrated throughout Terra Firma for its aromatic fruit. This fruit, which at Caracas is placed among linen, as in Europe it is in snuff, under the name of tonca, or Tonquin bean, is regarded as poisonous. It is a false notion, very general in the province of Cumana, that the excellent liqueur fabricated at Martinique owes its peculiar

---

[*] Macrocnemum tinctorium.

[†] Ityania coccidea.

[‡] Dipterix odorata, Willd. or Baryosma tongo of Gaertner. The jape furnishes Carichana with excellent timber.

flavour to the jape. In the Missions it is called simaruba; a name that may occasion serious mistakes, the true simaruba being a febrifuge species of the Quassia genus, found in Spanish Guiana only in the valley of Rio Caura, where the Paudacot Indians give it the name of achecchari.

I found the dip of the magnetic needle, in the great square at Carichana, 33.7° (new division). The intensity of the magnetic action was expressed by two hundred and twenty-seven oscillations in ten minutes of time; an increase of force that would seem to indicate some local attraction. Yet the blocks of the granite, blackened by the waters of the Orinoco, have no perceptible action upon the needle.

The river had risen several inches during the day on the 10th of April; this phenomenon surprised the natives so much the more, as the first swellings are almost imperceptible, and are usually followed in the month of April by a fall for some days. The Orinoco was already three feet higher than the level of the lowest waters. The natives showed us on a granite wall the traces of the great rise of the waters of late years. We found them to be forty-two feet high, which is double the mean rise of the Nile. But this measure was taken in a place where the bed of the Orinoco is singularly hemmed in by rocks, and I could only notice the marks shown me by the natives. It may easily be conceived that the effect and the height of the increase differs according to the profile of the river, the nature of the banks more or less elevated, the number of rivers flowing in that collect the pluvial waters, and the length of ground passed over. It is an unquestionable fact that at Carichana, at San Borja, at Atures, and at Maypures, wherever the river has forced its way through the mountains, you see at a hundred, sometimes at a hundred and thirty feet, above the highest present swell of the river, black bands and erosions, that indicate the ancient levels of the waters. Is then this river, which appears to us so grand and so majestic, only the feeble remains of those immense currents of fresh water which heretofore traversed the country at the east of the Andes, like arms of inland seas? What must have been the state of those low countries of Guiana that now undergo the effects of annual inundations? What immense numbers of crocodiles, manatees, and boas must have inhabited these vast spaces of land, converted alternately into marshes of stagnant water, and into barren and fissured plains! The more peaceful world which we inhabit has then succeeded to a world of tumult. The bones of mastodons and American elephants are found dispersed on the table-lands of the Andes. The megatherium inhabited the plains of Uruguay. On digging deep into the ground, in high valleys, where neither palm-trees nor arborescent ferns can grow, strata of coal are discovered, that still show vestiges of gigantic monocotyledonous plants.

There was a remote period then, in which the classes of plants were otherwise distributed, when the animals were larger, and the rivers broader and of greater depth. There end those records of nature, that it is in our power to consult. We are ignorant whether the human race, which at the time of the discovery of America scarcely formed a few feeble tribes on the east of the Cordilleras, had already descended into the plains; or whether the ancient tradition of the great waters, which is found among the nations of the Orinoco, the Erevato, and the Caura, belong to other climates, whence it has been propagated to this part of the New Continent.

On the 11th of April, we left Carichana at two in the afternoon, and found the course of the river more and more encumbered by blocks of granite rocks. We passed on the west the Cano Orupe, and then the great rock known by the name of Piedra del Tigre. The river is there so deep, that no bottom can be found with a line of twenty-two fathoms. Towards evening the weather became cloudy and gloomy. The proximity of the storm was marked by squalls alternating with dead calms. The rain was violent, and the roof of foliage, under which we lay, afforded but little shelter. Happily these showers drove away the mosquitos, at least for some time. We found ourselves before the cataract of Cariven, and the impulse of the waters was so strong, that we had great difficulty in gaining the land. We were continually driven back to the middle of the current. At length two Salive Indians, excellent swimmers, leaped into the water, and having drawn the boat to shore by means of a rope, made it fast to the Piedra de Carichana Vieja, a shelf of bare rock, on which we passed the night. The thunder continued to roll during a part of the night; the swell of the river became considerable; and we were several times afraid that our frail bark would be driven from the shore by the impetuosity of the waves.

The granitic rock on which we lay is one of those, where travellers on the Orinoco have heard from time to time, towards sunrise, subterraneous sounds, resembling those of the organ. The missionaries call these stones laxas de musica. "It is witchcraft (cosa de bruxas)," said our young Indian pilot, who could speak Spanish. We never ourselves heard these mysterious sounds, either at Carichana Vieja, or in the Upper Orinoco; but from information given us by witnesses worthy of belief, the existence of a phenomenon that seems to depend on a certain state of the atmosphere, cannot be denied. The shelves of rock are full of very narrow and deep crevices. They are heated during the day to 48 or 50°. I several times found their temperature at the surface, during the night, at 39°, the surrounding atmosphere being at 28°. It may easily be conceived, that the difference of temperature between the subterranean and the external air attains its maximum about sunrise, or at that moment which is at the same time farthest from the period of the maximum of the heat of the preceding day. May not these organ-like sounds, which are heard when a person lays his ear in contact with the stone, be the effect of a current of air that issues out through the crevices? Does not the impulse of the air against the elastic spangles of mica that intercept the crevices, contribute to modify the sounds? May we not admit that the ancient inhabitants of Egypt, in passing incessantly up and down the Nile, had made the same observation on some rock of the Thebaid; and that the music of

247

the rocks there led to the jugglery of the priests in the statue of Memnon? Perhaps, when, "the rosy-fingered Aurora rendered her son, the glorious Memnon, vocal,"[*] the voice was that of a man hidden beneath the pedestal of the statue; but the observation of the natives of the Orinoco, which we relate, seems to explain in a natural manner what gave rise to the Egyptian belief of a stone that poured forth sounds at sunrise.

Almost at the same period at which I communicated these conjectures to some of the learned of Europe, three French travellers, MM. Jomard, Jollois, and Devilliers, were led to analogous ideas. They heard, at sunrise, in a monument of granite, at the centre of the spot on which stands the palace of Karnak, a noise resembling that of a string breaking. Now this comparison is precisely that which the ancients employed in speaking of the voice of Memnon. The French travellers thought, like me, that the passage of rarefied air through the fissures of a sonorous stone might have suggested to the Egyptian priests the invention of the juggleries of the Memnomium.

We left the rock at four in the morning. The missionary had told us that we should have great difficulty in passing the rapids and the mouth of the Meta. The Indians rowed twelve hours and a half without intermission, and during all that time, they took no other nourishment than cassava and plantains. When we consider the difficulty of overcoming the force of the current, and of passing the cataracts; when we reflect on the constant employment of the muscular powers during a navigation of two months; we are equally surprised at the constitutional vigour and the abstinence of the Indians of the Orinoco and the Amazon. Amylaceous and saccharine substances, sometimes fish and the fat of turtles' eggs, supply the place of food drawn from the first two classes of the animal kingdom, those of quadrupeds and birds.

---

[*] These are the words of an inscription, which attests that sounds were heard on the 13th of the month Pachon, in the tenth year of the reign of Antoninus. See Monuments de l'Egypte Ancienne.

We found the bed of the river, to the length of six hundred toises, full of granite rocks. Here is what is called the Raudal de Cariven. We passed through channels that were not five feet broad. Our canoe was sometimes jammed between two blocks of granite. We sought to avoid these passages, into which the waters rushed with a fearful noise; but there is really little danger, in a canoe steered by a good Indian pilot. When the current is too violent to be resisted the rowers leap into the water, and fasten a rope to the point of a rock, to warp the boat along. This manoeuvre is very tedious; and we sometimes availed ourselves of it, to climb the rocks among which we were entangled. They are of all dimensions, rounded, very black, glossy like lead, and destitute of vegetation. It is an extraordinary phenomenon to see the waters of one of the largest rivers on the globe in some sort disappear. We perceived, even far from the shore, those immense blocks of granite, rising from the ground, and leaning one against another. The intervening channels in the rapids are more than twenty-five fathoms deep; and are the more difficult to be observed, as the rocks are often narrow toward their bases, and form vaults suspended over the surface of the river. We perceived no crocodiles in the raudal; these animals seem to shun the noise of cataracts.

From Cabruta to the mouth of the Rio Sinaruco, a distance of nearly two degrees of latitude, the left bank of the Orinoco is entirely uninhabited; but to the west of the Raudal de Cariven an enterprising man, Don Felix Relinchon, had assembled some Jaruro and Ottomac Indians in a small village. It is an attempt at civilization, on which the monks have had no direct influence. It is superfluous to add, that Don Felix lives at open war with the missionaries on the right bank of the Orinoco.

Proceeding up the river we arrived, at nine in the morning, before the mouth of the Meta, opposite the spot where the Mission of Santa Teresa, founded by the Jesuits, was heretofore situated.

Next to the Guaviare, the Meta is the most considerable river that flows into the Orinoco. It may be compared to the Danube, not for the length of its course, but for the volume of its waters. Its mean depth is thirty-six feet, and it sometimes reaches eighty-four. The union of these two rivers presents a very impressive spectacle. Lonely rocks rise on the eastern bank. Blocks of granite, piled upon one another, appear from afar like castles in ruins. Vast sandy shores keep the skirting of the forest at a distance from the river; but we discover amid them, in the horizon, solitary palm-trees, backed by the sky, and crowning the tops of the mountains. We passed two hours on a large rock, standing in the middle of the Orinoco, and called the Piedra de la Paciencia, or the Stone of Patience, because the canoes, in going up, are sometimes detained there two days, to extricate themselves from the whirlpool caused by this rock.

The Rio Meta, which traverses the vast plains of Casanare, and which is navigable as far as the foot of the Andes of New Grenada, will one day be of great political importance to the inhabitants of Guiana and Venezuela. From the Golfo Triste and the Boca del Drago a small fleet may go up the Orinoco and the Meta to within fifteen or twenty leagues of Santa Fe de Bogota. The flour of New Grenada may be conveyed the same way. The Meta is like a canal of communication between countries placed in the same latitude, but differing in their productions as much as France and Senegal. The Meta has its source in the union of two rivers which descend from the paramos of Chingasa and Suma Paz. The first is the Rio Negro, which, lower down, receives the Pachaquiaro; the second is the Rio de Aguas Blancas, or Umadea. The junction takes place near the port of Marayal. It is only eight or ten leagues from the Passo de la Cabulla, where you quit the Rio Negro, to the capital of Santa Fe. From the villages of Xiramena and Cabullaro to those of Guanapalo and Santa Rosalia de Cabapuna, a distance of sixty leagues, the banks of the Meta are more inhabited than those of the Orinoco. We find in this space fourteen Christian settlements, in part very populous; but from the mouths of the rivers Pauto and Casanare, for a space of more than fifty leagues, the Meta is infested by the Guahibos, a race of savages.[*]

The navigation of this river was much more active in the time of the Jesuits, and particularly during the expedition of Iturriaga, in 1756, than it is at present. Missionaries of the same order then governed the banks of the Meta and of the Orinoco. The villages of Macuco, Zurimena, and Casimena, were founded by the Jesuits, as well as those of Uruana, Encaramada, and Carichana.

These Fathers had conceived the project of forming a series of Missions from the junction of the Casanare with the Meta to that of the Meta with the Orinoco. A narrow zone of cultivated land would have crossed the vast steppes that separate the forests of Guiana from the Andes of New Grenada.

At the period of the harvest of turtles' eggs, not only the flour of Santa Fe descended the river, but the salt of Chita,[†] the cotton cloth of San Gil, and the printed counterpanes of Socorro. To give some security to the little traders who devoted themselves to this inland commerce, attacks were made from time to time from the castillo or fort of Carichana, on the Guahibos.

To keep these Guahibos in awe, the Capuchin missionaries, who succeeded the Jesuits in the government of the Missions of the Orinoco, formed the project of founding a city at the mouth of the Meta, under the name of the Villa de San Carlos. Indolence, and the dread of tertian fevers, have prevented the execution of this project; and all that has ever existed of the city of San Carlos, is a coat of arms painted on fine parchment, with an enormous cross erected on the bank of the Meta. The Guahibos, who, it is said, are some thousands in number, have

---

[*] I find the word written Guajibos, Guahivos, and Guagivos. They call themselves Gua-iva.

[†] East of Labranza Grande, and the north-west of Pore, now the capital of the province of Casanare.

become so insolent, that, at the time of our passage by Carichana, they sent word to the missionary that they would come on rafts, and burn his village. These rafts (valzas), which we had an opportunity of seeing, are scarcely three feet broad, and twelve feet long. They carry only two or three Indians; but fifteen or sixteen of these rafts are fastened to each other with the stems of the paullinia, the dolichos, and other creeping plants. It is difficult to conceive how these small craft remain tied together in passing the rapids. Many fugitives from the villages of the Casanare and the Apure have joined the Guahibos, and taught them the practice of eating beef, and preparing hides. The farms of San Vicente, Rubio, and San Antonio, have lost great numbers of their horned cattle by the incursions of the Indians, who also prevent travellers, as far as the junction of the Casanare, from sleeping on the shore in going up the Meta. It often happens, while the waters are low, that the traders of New Grenada, some of whom still visit the encampment of Pararuma, are killed by the poisoned arrows of the Guahibos.

From the mouth of the Meta, the Orinoco appeared to us to be freer of shoals and rocks. We navigated in a channel five hundred toises broad. The Indians remained rowing in the boat, without towing or pushing it forward with their arms, and wearying us with their wild cries. We passed the Canos of Uita and Endava on the west. It was night when we reached the Raudal de Tabaje. The Indians would not hazard passing the cataract; and we slept on a very incommodious spot, on the shelf of a rock, with a slope of more than eighteen degrees, and of which the crevices sheltered a swarm of bats. We heard the cries of the jaguar very near us during the whole night. They were answered by our great dog in lengthened howlings. I waited the appearance of the stars in vain: the sky was exceedingly black; and the hoarse sounds of the cascades of the Orinoco mingled with the rolling of the distant thunder.

Early in the morning of the 13th April we passed the rapids of Tabaje, and again disembarked. Father Zea, who accompanied us, desired to perform mass in the new Mission of San Borja, established two years before. We there found six houses inhabited by uncatechised Guahibos. They differ in nothing from the wild Indians. Their eyes, which are large and black, have more vivacity than those of the Indians who inhabit the ancient missions. We in vain offered them brandy; they would not even taste it. The faces of all the young girls were marked with round black spots; like the patches by which the ladies of Europe formerly imagined they set off the whiteness of their skins. The bodies of the Guahibos were not painted. Several of them had beards, of which they seemed proud; and, taking us by the chin, showed us by signs, that they were made like us. Their shape was in general slender. I was again struck, as I had been among the Salives and the Macos, with the little uniformity of features to be found among the Indians of the Orinoco. Their look is sad and gloomy; but neither stern nor ferocious. Without having any notion of the practices of the Christian religion, they behaved with the utmost decency at church. The Indians love to exhibit themselves; and will submit temporarily to any restraint or subjection, provided they are sure of drawing attention. At the moment of the consecration, they made signs to one another, to indicate beforehand that the priest was going to raise the chalice to his lips. With the exception of this gesture, they remained motionless and in imperturbable apathy.

The interest with which we examined these poor savages became perhaps the cause of the destruction of the mission. Some among them, who preferred a wandering life to the labours of agriculture, persuaded the rest to return to the plains of the Meta. They told them, that the white men would come back to San Borja, to take them away in the boats, and sell them as poitos, or slaves, at Angostura. The Guahibos awaited the news of our return from the Rio Negro by the Cassiquiare; and when they heard that we were arrived at the first great cataract, that of Atures, they all deserted, and fled to the savannahs that border the Orinoco on the west. The Jesuit Fathers had already formed a mission on this spot, and bearing the same name. No tribe is more difficult to fix to the soil than the Guahibos. They would rather feed on stale fish, scolopendras, and worms, than cultivate a little spot of ground. The other Indians say, that a Guahibo eats everything that exists, both on and under the ground.

In ascending the Orinoco more to the south, the heat, far from increasing, became more bearable. The air in the day was at 26 or 27.5°; and at night, at 23.7. The water of the Orinoco retained its habitual temperature of 27.7°. The torment of the mosquitos augmented severely, notwithstanding the decrease of heat. We never suffered so much from them as at San Borja. We could neither speak nor uncover our faces without having our mouths and noses filled with insects. We were surprised not to find the thermometer at 35 or 36°; the extreme irritation of the skin made us believe that the air was scorching. We passed the night on the beach of Guaripo. The fear of the little caribe fish prevented us from bathing. The crocodiles we had met with this day were all of an extraordinary size, from twenty-two to twenty-four feet.

Our sufferings from the zancudos made us depart at five o'clock on the morning of the 14th. There are fewer insects in the strata of air lying immediately on the river, than near the edge of the forests. We stopped to breakfast at the island of Guachaco, or Vachaco, where the granite is immediately covered by a formation of sandstone, or conglomerate. This sandstone contains fragments of quartz, and even of feldspar, cemented by indurated clay. It exhibits little veins of brown iron-ore, which separate in laminae, or plates, of one line in thickness. We had already found these plates on the shores between Encaramada and Baraguan, where the missionaries had sometimes taken them for an ore of gold, and sometimes for tin. It is probable, that this secondary formation occupied formerly a larger space. Having passed the mouth of the Rio Parueni, beyond which the Maco Indians dwell, we spent the night on the island of Panumana. I could with difficulty take the altitudes of Canopus, in order to fix the longitude of the point, near which the river suddenly turns towards the west. The island of Panumana is rich in plants. We there again found those shelves of bare rock, those tufts of melastomas, those thickets of small shrubs, the blended scenery of which had charmed us in the plains of Carichana. The mountains of the Great Cataracts bounded the horizon towards the south-east. In proportion as we advanced, the shores of the Orinoco exhibited a more imposing and picturesque aspect.

CHAPTER 2.20.

THE MOUTH OF THE RIO ANAVENI. PEAK OF UNIANA. MISSION OF ATURES. CATARACT, OR RAUDAL OF MAPARA. ISLETS OF SURUPAMANA AND UIRAPURI.

The river of the Orinoco, in running from south to north, is crossed by a chain of granitic mountains. Twice confined in its course, it turbulently breaks on the rocks, that form steps and transverse dykes. Nothing can be grander than the aspect of this spot. Neither the fall of the Tequendama, near Santa Fe de Bogota, nor the magnificent scenes of the Cordilleras, could weaken the impression produced upon my mind by the first view of the rapids of Atures and of Maypures. When the spectator is so stationed that the eye can at once take in the long succession of cataracts, the immense sheet of foam and vapours illumined by the rays of the setting sun, the whole river seems as it were suspended over its bed.

Scenes so astonishing must for ages have fixed the attention of the inhabitants of the New World. When Diego de Todaz, Alfonzo de Herrera, and the intrepid Raleigh, anchored at the mouth of the Orinoco, they were informed by the Indians of the Great Cataracts, which they themselves had never visited, and which they even confounded with cascades farther to the east. Whatever obstacles the force of vegetation under the torrid zone may throw in the way of intercourse among nations, all that relates to the course of great rivers acquires a celebrity which extends to vast distances. The Orinoco, the Amazon, and the Uruguay, traverse, like inland arms of seas, in different directions, a land covered with forests, and inhabited by tribes, part of whom are cannibals. It is not yet two hundred years since civilization and the light of a more humane religion have pursued their way along the banks of these ancient canals traced by the hand of nature; long, however, before the introduction of agriculture, before communications for the purposes of barter were established among these scattered and often hostile tribes, the knowledge of extraordinary phenomena, of falls of water, of volcanic fires, and of snows resisting all the ardent heat of summer, was propagated by a thousand fortuitous circumstances. Three hundred leagues from the coast, in the centre of South America, among nations whose excursions do not extend to three days' journey, we find an idea of the ocean, and words that denote a mass of salt water extending as far as the eye can discern. Various events, which repeatedly occur in savage life, contribute to enlarge these conceptions. In consequence of the petty wars between neighbouring tribes, a prisoner is brought into a strange country, and treated as a poito or mero, that is to say, as a slave. After being often sold, he is dragged to new wars, escapes, and returns home; he relates what he has seen, and what he has heard from those whose tongue he has been compelled to learn. As on discovering a coast, we hear of great inland animals, so, on entering the valley of a vast river, we are surprised to find that savages, who are strangers to navigation, have acquired a knowledge of distant things. In the infant state of society, the exchange of ideas precedes, to a certain point, the exchange of productions.

The two great cataracts of the Orinoco, the celebrity of which is so far-spread and so ancient, are formed by the passage of the river across the mountains of Parima. They are called by the natives Mapara and Quittuna; but the missionaries have substituted for these names those of Atures and Maypures, after the names of the tribes which were first assembled together in the nearest villages. On the coast of Caracas, the two Great Cataracts are denoted by the simple appellation of the two Raudales, or rapids; a denomination which implies that the other falls of water, even the rapids of Camiseta and of Carichana, are not considered as worthy of attention when compared with the cataracts of Atures and Maypures.

These last, situated between five and six degrees of north latitude, and a hundred leagues west of the Cordilleras of New Grenada, in the meridian of Porto Cabello, are only twelve leagues distant from each other. It is surprising that their existence was not known to D'Anville, who, in his fine map of South America, marks the inconsiderable cascades of Marimara and San Borja, by the names of the rapids of Carichana and Tabaje. The Great Cataracts divide the Christian establishments of Spanish Guiana into two unequal parts. Those situated between the Raudal of Atures and the mouth of the river are called the Missions of the Lower Orinoco; the Missions of the Upper Orinoco comprehend the villages between the Raudal of Maypures and the mountains of Duida. The course of the Lower Orinoco, if we estimate the sinuosities at one-third of the distance in a direct line, is two hundred and sixty nautical leagues: the course of the Upper Orinoco, supposing its sources to be three degrees east of Duida, includes one hundred and sixty-seven leagues.

Beyond the Great Cataracts an unknown land begins. The country is partly mountainous and partly flat, receiving at once the confluents of the Amazon and the Orinoco. From the facility of its communications with the Rio Negro and Grand Para, it appears to belong still more to Brazil than to the Spanish colonies. None of the missionaries who have described the Orinoco before me, neither Father Gumilla, Gili, nor Caulin, had passed the Raudal of Maypures. We found but three Christian establishments above the Great Cataracts, along the shores of the Orinoco, in an extent of more than a hundred leagues; and these three establishments contained scarcely six or eight white persons, that is to say, persons of European race. We cannot be surprised that such a desert region should have been at all times the land of fable and fairy visions. There, according to the statements of certain missionaries, are found races of men, some of whom have an eye in the centre of the forehead, whilst others have dogs' heads, and mouths below their stomachs. There they pretend to have found all that the ancients relate of the Garamantes, of the Arimaspes, and of the Hyperboreans. It would be an error to suppose that these simple and often rustic missionaries had themselves invented all these exaggerated fictions; they derived them in great part from the recitals of the Indians. A fondness for narration prevails in the Missions, as it does at sea, in the East, and in every place where the mind seeks amusement. A missionary, from his vocation, is not inclined to scepticism; he imprints on his memory what the natives have so often repeated to him; and, when returned to Europe, and restored to the civilized world, he finds a pleasure in creating astonishment by a recital of facts which he thinks he has collected, and by an animated description of remote things. These stories, which the Spanish colonists call tales of travellers and of monks (cuentos de viageros y frailes), increase in improbability in proportion as you increase your distance from the forests of the Orinoco, and approach the coasts inhabited by the whites. When, at Cumana, Nueva Barcelona, and other seaports which have frequent communication with the Missions, you betray any sign of incredulity, you are reduced to silence by these few words: The fathers have seen it, but far above the Great Cataracts (mas arriba de los Raudales).

On the 15th of April, we left the island of Panumana at four in the morning, two hours before sunrise. The sky was in great part obscured, and lightnings flashed over dense clouds at more than forty degrees of elevation. We were surprised at not hearing thunder; but possibly this was owing to the prodigious height of the storm? It appears to us, that in Europe the electric flashes without thunder, vaguely called heat-lightning, are seen generally nearer the horizon. Under a cloudy sky, that sent back the radiant caloric of the soil, the heat was stifling; not a breath of wind agitated the foliage of the trees. The jaguars, as usual, had crossed the arm of the Orinoco by which we were separated from the shore, and we heard their cries extremely near. During the night the Indians had advised us to quit our station in the open air, and retire to a deserted hut belonging to the conucos of the inhabitants of Atures. They had taken care to barricade the opening with planks, a precaution which seemed to us superfluous; but near the Cataracts tigers are very numerous, and two years before, in these very conucos of Panumana, an Indian returning to his hut, towards the close of the rainy season, found a tigress settled in it with her two young. These animals had inhabited the dwelling for several months; they were dislodged from it with difficulty, and it was only after an obstinate combat that the former master regained possession of his dwelling. The jaguars are fond of retiring to deserted ruins, and I believe it is more prudent in general for a solitary traveller to encamp in the open air, between two fires, than to seek shelter in uninhabited huts.

On quitting the island of Panumana, we perceived on the western bank of the river the fires of an encampment of Guahibo savages. The missionary who accompanied us caused a few musket-shots to be fired in the air, which he said would intimidate them, and shew that we were in a state to defend ourselves. The savages most likely had no canoes, and were not desirous of troubling us in the middle of the river. We passed at sunrise the mouth of the Rio Anaveni, which descends from the eastern mountains. On its banks, now deserted, Father Olmos had established, in the time of the Jesuits, a small village of Japuins or Jaruros. The heat was so excessive that we rested a long time in a woody spot, to fish with a hook and line, and it was not without some trouble that we carried away all the fish we had caught. We did not arrive till very late at the foot of the Great Cataract, in a bay called the lower harbour (puerto de abaxo); and we followed, not without difficulty, in a dark night, the narrow path that leads to the Mission of Atures, a league distant from the river. We crossed a plain covered with large blocks of granite.

The little village of San Juan Nepomuceno de los Atures was founded by the Jesuit Francisco Gonzales, in 1748. In going up the river this is the last of the Christian missions that owe their origin to the order of St. Ignatius. The more southern establishments, those of Atabapo, of Cassiquiare, and of Rio Negro, were formed by the fathers of the Observance of St. Francis. The Orinoco appears to have flowed heretofore where the village of Atures now stands, and the flat savannah that surrounds the village no doubt formed part of the river. I saw to the east of the mission a succession of rocks, which seemed to have been the ancient shore of the Orinoco. In the lapse of ages the river has been impelled westward, in consequence of the accumulations of earth, which occur more frequently on the side of the eastern mountains, that are furrowed by torrents. The cataract bears the name of Mapara,[*] as we have mentioned above; while the name of the village is derived from that of the nation of Atures, now believed to be extinct. I find on the maps of the seventeenth century, Island and Cataract of Athule; which is the word Atures written according to the pronunciation of the Tamanacs, who confound, like so many other people, the consonants l and r. This mountainous region was so little known in Europe, even in the middle of the eighteenth century, that D'Anville, in the first edition of his South America, makes a branch issue from the Orinoco, near Salto de los Atures, and fall

---

[*] I am ignorant of the etymology of this word, which I believe means only a fall of water. Gili translates into Maypure a small cascade (raudalito) by uccamatisi mapara canacapatirri. Should we not spell this word matpara? mat being a radical of the Maypure tongue, and meaning bad (Hervas, Saggio N. 29). The radical par (para) is found among American tribes more than five hundred leagues distant from each other, the Caribs, Maypures, Brazilians, and Peruvians, in the words sea, rain, water, lake. We must not confound mapara with mapaja; this last word signifies, in Maypure and Tamanac, the papaw or melon-tree, no doubt on account of the sweetness of its fruit, for mapa means in the Maypure, as well as in the Peruvian and Omagua tongues, the honey of bees. The Tamanacs call a cascade, or raudal, in general uatapurutpe; the Maypures, uca.

into the Amazon, to which branch he gives the name of Rio Negro.

Early maps, as well as Father Gumilla's work, place the Mission in latitude 1° 30'. Abbe Gili gives it 3° 50'. I found, by meridian altitudes of Canopus and a of the Southern Cross, 5° 38' 4" for the latitude; and by the chronometer 4 hours 41 minutes 17 seconds of longitude west of the meridian of Paris.

We found this small Mission in the most deplorable state. It contained, even at the time of the expedition of Solano, commonly called the expedition of the boundaries, three hundred and twenty Indians. This number had diminished, at the time of our passage by the Cataracts, to forty-seven; and the missionary assured us that this diminution became from year to year more sensible. He showed us, that in the space of thirty-two months only one marriage had been entered in the registers of the parish church. Two others had been contracted by uncatechised natives, and celebrated before the Indian Gobernador. At the first foundation of the Mission, the Atures, Maypures, Meyepures, Abanis, and Quirupas, had been assembled together. Instead of these tribes we found only Guahibos, and a few families of the nation of Macos. The Atures have almost entirely disappeared; they are no longer known, except by the tombs in the cavern of Ataruipe, which recall to mind the sepulchres of the Guanches at Teneriffe. We learned on the spot, that the Atures, as well as the Quaquas, and the Macos or Piaroas, belong to the great stock of the Salive nations; while the Maypures, the Abanis, the Parenis, and the Guaypunaves, are of the same race as the Cabres or Caveres, celebrated for their long wars with the Caribs. In this labyrinth of petty nations, divided from one another as the nations of Latium, Asia Minor, and Sogdiana, formerly were, we can trace no general relations but by following the analogy of tongues. These are the only monuments that have reached us from the early ages of the world; the only monuments, which, not being fixed to the soil, are at once moveable and lasting, and have as it were traversed time and space. They owe their duration, and the extent they occupy, much less to conquering and polished nations, than to those wandering and half-savage tribes, who, fleeing before a powerful enemy, carried along with them in their extreme wretchedness only their wives, their children, and the languages of their fathers.

Between the latitudes of 4 and 8°, the Orinoco not only separates the great forest of the Parime from the bare savannahs of the Apure, Meta, and Guaviare, but also forms the boundary between tribes of very different manners. To the westward, over treeless plains, wander the Guahibos, the Chiricoas, and the Guamos; nations, proud of their savage independence, whom it is difficult to fix to the soil, or habituate to regular labour. The Spanish missionaries characterise them well by the name of Indios andantes (errant or vagabond Indians), because they are perpetually moving from place to place. To the east of the Orinoco, between the neighbouring sources of the Caura, Cataniapo, and Ventuari, live the Macos, the Salives, the Curacicanas, Parecas, and Maquiritares, mild, tranquil tribes, addicted to agriculture, and easily subjected to the discipline of the Missions. The Indian of the plains differs from the Indian of the forests in language as well as manners and mental disposition; both have an idiom abounding in spirited and bold terms; but the language of the former is harsher, more concise, and more impassioned; that of the latter, softer, more diffuse, and fuller of ambiguous expressions.

The Mission of Atures, like most of the Missions of the Orinoco, situated between the mouths of the Apure and the Atabapo, is composed of both the classes of tribes we have just described. We there find the Indians of the forests, and the Indians heretofore nomadic[*] (Indios monteros and Indios

---

[*] I employ the word nomadic as synonymous with wandering, and not in its primitive signification. The wandering nations of America (those of the indigenous tribes, it is to be understood) are never shepherds; they live by fishing and hunting, on the fruit of a few trees, the farinaceous pith of palm-trees, etc.

llaneros, or andantes). We visited with the missionary the huts of Macos, whom the Spaniards call Piraoas, and those of the Guahibos. The first indicated more love of order, cleanliness, and ease. The independent Macos (I do not designate them by the name of savages) have their rochelas, or fixed dwellings, two or three days' journey east of Atures, toward the sources of the little river Cataniapo. They are very numerous. Like most of the natives of the woods, they cultivate, not maize, but cassava; and they live in great harmony with the Christian Indians of the mission. The harmony was established and wisely cultivated by the Franciscan monk, Bernardo Zea. This alcalde of the reduced Macos quitted the village of Atures for a few months every year, to live in the plantations which he possessed in the midst of the forests near the hamlet of the independent Macos. In consequence of this peaceful intercourse, many of the Indios monteros came and established themselves some time ago in the mission. They asked eagerly for knives, fishing hooks, and those coloured glass beads, which, notwithstanding the positive prohibition of the priests, were employed not as necklaces, but as ornaments of the guayuco (perizoma). Having obtained what they sought, they returned to the woods, weary of the regulations of the mission. Epidemic fevers, which prevailed with violence at the entrance of the rainy season, contributed greatly to this unexpected flight. In 1799 the mortality was very considerable at Carichana, on the banks of the Meta, and at the Raudal of Atures. The Indian of the forest conceives a horror of the life of the civilized man, when, I will not say any misfortune befalls his family settled in the mission, but merely any disagreeable or unforeseen accident. Natives, who were neophytes, have been known to desert for ever the Christian establishments, on account of a great drought; as if this calamity would not have reached them equally in their plantations, had they remained in their primitive independence.

The fevers which prevail during a great part of the year in the villages of Atures and Maypures, around the two Great Cataracts of the Orinoco, render these spots highly dangerous to

European travellers. They are caused by violent heats, in combination with the excessive humidity of the air, bad nutriment, and, if we may believe the natives, the pestilent exhalations rising from the bare rocks of the Raudales. These fevers of the Orinoco appeared to us to resemble those which prevail every year between New Barcelona, La Guayra, and Porto Cabello, in the vicinity of the sea; and which often degenerate into adynamic fevers. "I have had my little fever (mi calenturita) only eight months," said the good missionary of the Atures, who accompanied us to the Rio Negro; speaking of it as of an habitual evil, easy to be borne. The fits were violent, but of short duration. He was sometimes seized with them when lying along in the boat under a shelter of branches of trees, sometimes when exposed to the burning rays of the sun on an open beach. These tertian agues are attended with great debility of the muscular system; yet we find poor ecclesiastics on the Orinoco, who endure for several years these calenturitas, or tercianas: their effects are not so fatal as those which are experienced from fevers of much shorter duration in temperate climates.

I have just alluded to the noxious influence on the salubrity of the atmosphere, which is attributed by the natives, and even the missionaries, to the bare rocks. This opinion is the more worthy of attention, as it is connected with a physical phenomenon lately observed in different parts of the globe, and not yet sufficiently explained. Among the cataracts, and wherever the Orinoco, between the Missions of Carichana and of Santa Barbara, periodically washes the granitic rocks, they become smooth, black, and as if coated with plumbago. The colouring matter does not penetrate the stone, which is coarse-grained granite, containing a few solitary crystals of hornblende. Taking a general view of the primitive formation of Atures, we perceive, that, like the granite of Syene in Egypt, it is a granite with hornblende, and not a real syenite formation. Many of the layers are entirely destitute of hornblende. The black crust is 0.3 of a line in thickness; it is found chiefly on the quartzose parts. The crystals of feldspar sometimes preserve externally their reddish-white colour, and rise above the black crust. On breaking the stone with a hammer, the inside is found to be white, and without any trace of decomposition. These enormous stony masses appear sometimes in rhombs, sometimes under those hemispheric forms, peculiar to granitic rocks when they separate in blocks. They give the landscape a singularly gloomy aspect; their colour being in strong contrast with that of the foam of the river which covers them, and of the vegetation by which they are surrounded. The Indians say, that the rocks are burnt (or carbonized) by the rays of the sun. We saw them not only in the bed of the Orinoco, but in some spots as far as five hundred toises from its present shore, on heights which the waters now never reach even in their greatest swellings.

What is this brownish black crust, which gives these rocks, when they have a globular form, the appearance of meteoric stones? What idea can we form of the action of the water, which produces a deposit, or a change of colour, so extraordinary? We must observe, in the first place, that this phenomenon does not belong to the cataracts of the Orinoco alone, but is found in both hemispheres. At my return from Mexico in 1807, when I showed the granites of Atures and Maypures to M. Roziere, who had travelled over the valley of Egypt, the coasts of the Red Sea, and Mount Sinai, this learned geologist pointed out to me that the primitive rocks of the little cataracts of Syene display, like the rocks of the Orinoco, a glossy surface, of a blackish-grey, or almost leaden colour, and of which some of the fragments seem coated with tar. Recently, in the unfortunate expedition of Captain Tuckey, the English naturalists were struck with the same appearance in the yellalas (rapids and shoals) that obstruct the river Congo or Zaire. Dr. Koenig has placed in the British Museum, beside the syenites of the Congo, the granites of Atures, taken from a series of rocks which were presented by M. Bonpland and myself to the illustrious president of the Royal Society of London. "These fragments," says Mr. Koenig, "alike resemble meteoric stones; in both rocks, those of the Orinoco and of Africa, the black crust is composed, according to the analysis of Mr. Children, of the oxide of iron and manganese." Some experiments made at Mexico, conjointly with Senor del Rio, led me to think that the rocks of Atures, which blacken the paper in which they are wrapped,[*] contain, besides oxide of manganese, carbon, and supercarburetted iron. At the Orinoco, granitic masses of forty or fifty feet thick are uniformly coated with these oxides; and, however thin these crusts may appear, they must nevertheless contain pretty considerable quantities of iron and manganese, since they occupy a space of above a league square.

It must be observed that all these phenomena of coloration have hitherto appeared in the torrid zone only, in rivers that have periodical overflowings, of which the habitual temperature is from twenty-four to twenty-eight centesimal degrees, and which flow, not over gritstone or calcareous rocks, but over granite, gneiss, and hornblende rocks. Quartz and feldspar scarcely contain five or six thousandths of oxide of iron and of manganese; but in mica and hornblende these oxides, and particularly that of iron, amount, according to Klaproth and Herrmann, to fifteen or twenty parts in a hundred. The hornblende contains also some carbon, like the Lydian stone and kieselschiefer. Now, if these black crusts were formed by a slow decomposition of the granitic rock, under the double influence of humidity and the tropical sun, how is it to be conceived that these oxides are spread so uniformly over the whole surface of the stony masses, and are not more abundant round a crystal of mica or hornblende than on the feldspar and milky quartz? The ferruginous sandstones, granites, and marbles, that become cinereous and sometimes brown in damp air, have an aspect altogether different. In reflecting upon the lustre and equal thickness of the crusts, we are rather inclined to think that this

---

[*] I remarked the same phenomenon from spongy grains of platina one or two lines in length, collected at the stream-works of Taddo, in the province of Choco. Having been wrapped up in white paper during a journey of several months, they left a black stain, like that of plumbago or supercarburetted iron.

matter is deposited by the Orinoco, and that the water has penetrated even into the clefts of the rocks. Adopting this hypothesis, it may be asked whether the river holds the oxides suspended like sand and other earthy substances, or whether they are found in a state of chemical solution. The first supposition is less admissible, on account of the homogeneity of the crusts, which contain neither grains of sand, nor spangles of mica, mixed with the oxides. We must then recur to the idea of a chemical solution; and this idea is no way at variance with the phenomena daily observable in our laboratories. The waters of great rivers contain carbonic acid; and, were they even entirely pure, they would still be capable, in very great volumes, of dissolving some portions of oxide, or those metallic hydrates which are regarded as the least soluble. The mud of the Nile, which is the sediment of the matters which the river holds suspended, is destitute of manganese; but it contains, according to the analysis of M. Regnault, six parts in a hundred of oxide of iron; and its colour, at first black, changes to yellowish brown by desiccation and the contact of air. The mud consequently is not the cause of the black crusts on the rocks of Syene. Berzelius, who, at my request, examined these crusts, recognized in them, as in those of the granites of the Orinoco and River Congo, the union of iron and manganese. That celebrated chemist was of opinion that the rivers do not take up these oxides from the soil over which they flow, but that they derive them from their subterranean sources, and deposit them on the rocks in the manner of cementation, by the action of particular affinities, perhaps by that of the potash of the feldspar. A long residence at the cataracts of the Orinoco, the Nile, and the Rio Congo, and an examination of the circumstances attendant on this phenomenon of coloration, could alone lead to the complete solution of the problem we have discussed. Is this phenomenon independent of the nature of the rocks? I shall content myself with observing, in general, that neither the granitic masses remote from the ancient bed of the Orinoco, but exposed during the rainy season to the alternations of heat and moisture, nor the granitic rocks bathed by the brownish waters of the Rio Negro, assume the appearance of meteoric stones. The Indians say, that the rocks are black only where the waters are white. They ought, perhaps, to add, where the waters acquire great swiftness, and strike with force against the rocks of the banks. Cementation seems to explain why the crusts augment so little in thickness.

I know not whether it be an error, but in the Missions of the Orinoco, the neighbourhood of bare rocks, and especially of the masses that have crusts of carbon, oxide of iron, and manganese, are considered injurious to health. In the torrid zone, still more than in others, the people multiply pathogenic causes at will. They are afraid to sleep in the open air, if forced to expose the face to the rays of the full moon. They also think it dangerous to sleep on granite near the river; and many examples are cited of persons, who, after having passed the night on these black and naked rocks, have awakened in the morning with a strong paroxysm of fever. Without entirely lending faith to the assertions of the missionaries and natives, we generally avoided the laxas negras, and stretched ourselves on the beach covered with white sand, when we found no tree from which to suspend our hammocks. At Carichana, the village is intended to be destroyed, and its place changed, merely to remove it from the black rocks, or from a site where, for a space of more than ten thousand square toises, banks of bare granite form the surface. From similar motives, which must appear very chimerical to the naturalists of Europe, the Jesuits Olmo, Forneri, and Mellis, removed a village of Jaruros to three different spots, between the Raudal of Tabaje and the Rio Anaveni. I merely state these facts as they were related to me, because we are almost wholly ignorant of the nature of the gaseous mixtures which cause the insalubrity of the atmosphere. Can it be admitted that, under the influence of excessive heat and of constant humidity, the black crusts of the granitic rocks are capable of acting upon the ambient air, and producing miasmata with a triple basis of carbon, azote, and hydrogen? This I doubt. The granites of the Orinoco, it is true, often contain hornblende; and those who are accustomed to practical labour in mines are not ignorant that the most noxious exhalations rise from galleries wrought in syenitic and hornblende rocks: but in an atmosphere renewed every instant by the action of little currents of air, the effect cannot be the same as in a mine.

It is probably dangerous to sleep on the laxas negras, only because these rocks retain a very elevated temperature during the night. I have found their temperature in the day at 48°, the air in the shade being at 29.7°; during the night the thermometer on the rock indicated 36°, the air being at 26°. When the accumulation of heat in the stony masses has reached a stationary degree, these masses become at the same hours nearly of the same temperature. What they have acquired more in the day they lose at night by radiation, the force of which depends on the state of the surface of the radiating body, the interior arrangement of its particles, and, above all, on the clearness of the sky, that is, on the transparency of the atmosphere and the absence of clouds. When the declination of the sun varies very little, this luminary adds daily nearly the same quantities of heat, and the rocks are not hotter at the end than in the middle of summer. There is a certain maximum which they cannot pass, because they do not change the state of their surface, their density, or their capacity for caloric. On the shores of the Orinoco, on getting out of one's hammock during the night, and touching with the bare feet the rocky surface of the ground, the sensation of heat experienced is very remarkable. I observed pretty constantly, in putting the bulb of the thermometer in contact with the ledges of bare rocks, that the laxas negras are hotter during the day than the reddish-white granites at a distance from the river; but the latter cool during the night less rapidly than the former. It may be easily conceived that the emission and loss of caloric is more rapid in masses with black crusts than in those which abound in laminae of silvery mica. When walking between the hours of one and three in the afternoon, at Carichana, Atures, or Maypures, among those blocks of stone destitute of vegetable mould, and piled up to great heights, one feels a sensation of suffocation, as if standing before the opening of a furnace. The winds, if ever felt in those woody regions, far from bringing coolness, appear more heated when they have passed over beds of stone, and heaps of rounded blocks of granite. This augmentation of heat adds to the insalubrity of the climate.

Among the causes of the depopulation of the Raudales, I have not reckoned the small-pox, that malady which in other parts of America makes such cruel ravages that the natives, seized with dismay, burn their huts, kill their children, and renounce every kind of society. This scourge is almost unknown on the banks of the Orinoco, and should it penetrate thither, it is to be hoped that its effects may be immediately counteracted by vaccination, the blessings of which are daily felt along the coasts of Terra Firma. The causes which depopulate the Christian settlements are, the repugnance of the Indians for the regulations of the missions, insalubrity of climate, bad nourishment, want of care in the diseases of children, and the guilty practice of preventing pregnancy by the use of deleterious herbs. Among the barbarous people of Guiana, as well as those of the half-civilized islands of the South Sea, young wives are fearful of becoming mothers. If they have children, their offspring are exposed not only to the dangers of savage life, but also to other dangers arising from the strangest popular prejudices. When twins are born, false notions of propriety and family honour require that one of them should be destroyed. To bring twins into the world, say the Indians, is to be exposed to public scorn; it is to resemble rats, opossums, and the vilest animals, which bring forth a great number of young at a time. Nay, more, they affirm that two children born at the same time cannot belong to the same father. This is an axiom of physiology among the Salives; and in every zone, and in different states of society, when the vulgar seize upon an axiom, they adhere to it with more stedfastness than the better-informed men by whom it was first hazarded. To avoid the disturbance of conjugal tranquillity, the old female relations of the mother take care, that when twins are born one of them shall disappear. If a new-born infant, though not a twin, have any physical deformity, the father instantly puts it to death. They will have none but robust and well-made children, for deformities indicate some influence of the evil spirit Ioloquiamo, or the bird Tikitiki, the enemy of the human race. Sometimes children of a feeble constitution undergo the same fate. When the father is asked what is become of one of his sons, he will pretend that he has lost him by a natural death. He will disavow an action that appears to him blameable, but not criminal. "The poor boy," he will tell you, "could not follow us; we must have waited for him every moment; he has not been seen again; he did not come to sleep where we passed the night." Such is the candour and simplicity of manners—such the boasted happiness—of man in the state of nature! He kills his son to escape the ridicule of having twins, or to avoid journeying more slowly; in fact, to avoid a little inconvenience.

These acts of cruelty, I confess, are less frequent than they are believed to be; yet they occur even in the Missions, during the time when the Indians leave the village, to retire to the conucos of the neighbouring forests. It would be erroneous to attribute these actions to the state of polygamy in which the uncatechized Indians live. Polygamy no doubt diminishes the domestic happiness and internal union of families; but this practice, sanctioned by Ismaelism, does not prevent the people of the east from loving their children with tenderness. Among the Indians of the Orinoco, the father returns home only to eat, or to sleep in his hammock; he lavishes no caresses on his infants, or on his wives, whose office it is to serve him. Parental affection begins to display itself only when the son has become strong enough to take a part in hunting, fishing, and the agricultural labours of the plantations.

While our boat was unloading, we examined closely, wherever the shore could be approached, the terrific spectacle of a great river narrowed and reduced as it were to foam. I shall endeavour to paint, not the sensations we felt, but the aspect of a spot so celebrated among the scenes of the New World. The more imposing and majestic the objects we describe, the more essential it becomes to seize them in their smallest details, to fix the outline of the picture we would present to the imagination of the reader, and to describe with simplicity what characterises the great and imperishable monuments of nature.

The navigation of the Orinoco from its mouth as far as the confluence of the Anaveni, an extent of 260 leagues, is not impeded. There are shoals and eddies near Muitaco, in a cove that bears the name of the Mouth of Hell (Boca del Infierno); and there are rapids (raudalitos) near Carichana and San Borja; but in all these places the river is never entirely barred, as a channel is left by which boats can pass up and down.

In all this navigation of the Lower Orinoco travellers experience no other danger than that of the natural rafts formed by trees, which are uprooted by the river, and swept along in its great floods. Woe to the canoes that during the night strike against these rafts of wood interwoven with lianas! Covered with aquatic plants, they resemble here, as in the Mississippi, floating meadows, the chinampas or floating gardens of the Mexican lakes. The Indians, when they wish to surprise a tribe of their enemies, bring together several canoes, fasten them to each other with cords, and cover them with grass and branches, to imitate this assemblage of trunks of trees, which the Orinoco sweeps along in its middle current. The Caribs are accused of having heretofore excelled in the use of this artifice; at present the Spanish smugglers in the neighbourhood of Angostura have recourse to the same expedient to escape the vigilance of the custom-house officers.

After proceeding up the Orinoco beyond the Rio Anaveni, we find, between the mountains of Uniana and Sipapu, the Great Cataracts of Mapara and Quittuna, or, as they are more commonly called by the missionaries, the Raudales of Atures and Maypures. These bars, which extend from one bank to the other, present in general a similar aspect: they are composed of innumerable islands, dikes of rock, and blocks of granite piled on one another and covered with palm-trees. But, notwithstanding a uniformity of aspect, each of these cataracts preserves an individual character. The first, the Atures, is most easily passable when the waters are low. The Indians prefer crossing the second, the Maypures, at the time of great floods. Beyond the Maypures and the mouth of the Cano Cameji, the Orinoco is again unobstructed for the length of more than one hundred and sixty-seven leagues, or nearly to its source; that is to say, as far as the Raudalito of Guaharibos, east of the Cano Chiguire and the lofty mountains of Yumariquin.

Having visited the basins of the two rivers Orinoco and Amazon, I was singularly struck by the differences they display in their course of unequal extent. The falls of the Amazon, which is nearly nine hundred and eighty nautical leagues (twenty to a degree) in length, are pretty near its source in the first sixth of its total length, and five-sixths of its course are entirely free. We find the great falls of the Orinoco on a point far more unfavourable to navigation; if not at the half, at least much beyond the first third of its length. In both rivers it is neither the mountains, nor the different stages of flat lands lying over one another, whence they take their origin, that cause the cataracts; they are produced by other mountains, other ledges which, after a long and tranquil course, the rivers have to pass over, precipitating themselves from step to step.

The Amazon does not pierce its way through the principal chain of the Andes, as was affirmed at a period when it was gratuitously supposed that, wherever mountains are divided into parallel chains, the intermedial or central ridge must be more elevated than the others. This great river rises (and this is a point of some importance to geology) eastward of the western chain, which alone in this latitude merits the denomination of the high chain of the Andes. It is formed by the junction of the river Aguamiros with the Rio Chavinillo, which issues from the lake Llauricocha in a longitudinal valley bounded by the western and the intermedial chain of the Andes. To form an accurate idea of these hydrographical relations, it must be borne in mind that a division into three chains takes place in the colossal group or knot of the mountains of Pasco and Huanuco. The western chain, which is the loftiest, and takes the name of the Cordillera Real de Nieve, directs its course (between Huary and Caxatamba, Guamachuco and Luema, Micuipampa and Guangamarca) by the Nevados of Viuda, Pelagatos, Moyopata, and Huaylillas, and by the Paramos of Guamani and Guaringa, towards the town of Loxa. The intermedial chain separates the waters of the Upper Maranon from those of the Guallaga, and over a long space reaches only the small elevation of a thousand toises; it enters the region of perpetual snow to the south of Huanuco in the Cordillera of Sasaguanca. It stretches at first northward by Huacrachuco, Chachapoyas, Moyobamba, and the Paramo of Piscoguannuna; then it progressively lowers toward Peca, Copallin, and the Mission of Santiago, at the eastern extremity of the province of Jaen de Bracamoros. The third, or easternmost chain, skirts the right bank of the Rio Guallaga, and loses itself in the seventh degree of latitude. So long as the Amazon flows from south to north in the longitudinal valley, between two chains of unequal height (that is, from the farms of Quivilla and Guancaybamba, where the river is crossed on wooden bridges, as far as the confluence of the Rio Chinchipe), there are neither bars, nor any obstacle whatever to the navigation of boats. The falls of water begin only where the Amazon turns toward the east, crossing the intermedial chain of the Andes, which widens considerably toward the north. It meets with the first rocks of red sandstone, or ancient conglomerate, between Tambillo and the Pongo of Rentema (near which I measured the breadth, depth, and swiftness of the waters), and it leaves the rocks of red sandstone east of the famous strait of Manseriche, near the Pongo of Tayuchuc, where the hills rise no higher than forty or fifty toises above the level of its waters. The river does not reach the most easterly chain, which bounds the Pampas del Sacramento. From the hills of Tayuchuc as far as Grand Para, during a course of more than seven hundred and fifty leagues, the navigation is free from obstacles. It results from this rapid sketch, that, if the Maranon had not to pass over the hilly country between Santiago and Tomependa (which belongs to the central chain of the Andes) it would be navigable from its mouth as far as Pumpo, near Piscobamba in the province of Conchucos, forty-three leagues north of its source.

We have just seen that, in the Orinoco, as in the Amazon, the great cataracts are not found near the sources of the rivers. After a tranquil course of more than one hundred and sixty leagues from the little Raudal of Guaharibos, east of Esmeralda, as far as the mountains of Sipapu, the river, augmented by the waters of the Jao, the Ventuari, the Atabapo, and the Guaviare, suddenly changes its primitive direction from east to west, and runs from south to north: then, in crossing the land-strait[*] in the plains of Meta, meets the advanced buttresses of the Cordillera of Parima. This obstacle causes cataracts far more considerable, and presents greater impediments to navigation, than all the Pongos of the Upper Maranon, because they are proportionally nearer to the mouth of the river. These geographical details serve to prove, in the instances of the two greatest rivers of the New World, first, that it cannot be ascertained in an absolute manner that, beyond a certain number of toises, or a certain height above the level of the sea, rivers are not navigable; secondly, that the rapids are not always occasioned, as several treatises of general topography affirm, by the height of the first obstacles, by the first lines of ridges which the waters have to surmount near their sources.

The most northern of the great cataracts of the Orinoco is the only one bounded on each side by lofty mountains. The left bank of the river is generally lower, but it makes part of a plane which rises again west of Atures, towards the Peak of Uniana, a pyramid nearly three thousand feet high, and placed on a wall of

---

[*] This strait, which I have several times mentioned, is formed by the Cordilleras of the Andes of New Granada, and the Cordillera of Parima.

rock with steep slopes. The situation of this solitary peak in the plain contributes to render its aspect more imposing and majestic. Near the Mission, in the country which surrounds the cataract, the aspect of the landscape varies at every step. Within a small space we find all that is most rude and gloomy in nature, united with an open country and lovely pastoral scenery. In the physical, as in the moral world, the contrast of effects, the comparison of what is powerful and menacing with what is soft and peaceful, is a never-failing source of our pleasures and our emotions.

I shall here repeat some scattered features of a picture which I traced in another work shortly after my return to Europe.[*] The savannahs of Atures, covered with slender plants and grasses, are really meadows resembling those of Europe. They are never inundated by the rivers, and seem as if waiting to be ploughed by the hand of man. Notwithstanding their extent, these savannahs do not exhibit the monotony of our plains; they surround groups of rocks and blocks of granite piled on one another. On the very borders of these plains and this open country, glens are seen scarcely lighted by the rays of the setting sun, and hollows where the humid soil, loaded with arums, heliconias, and lianas, manifests at every step the wild fecundity of nature. Everywhere, just rising above the earth, appear those shelves of granite completely bare, which we saw at Carichana, and which I have already described. Where springs gush from the bosom of these rocks, verrucarias, psoras, and lichens are fixed on the decomposed granite, and have there accumulated mould. Little euphorbias, peperomias, and other succulent plants, have taken the place of the cryptogamous tribes; and evergreen shrubs, rhexias, and purple-flowered melastomas, form verdant isles amid desert and rocky plains. The distribution of these spots, the clusters of small trees with coriaceous and shining leaves scattered in the savannahs, the limpid rills that dig channels

---

[*] Views of Nature page 153 Bohn's edition.

across the rocks, and wind alternately through fertile places and over bare shelves of granite, all call to mind the most lovely and picturesque plantations and pleasure-grounds of Europe. We seem to recognise the industry of man, and the traces of cultivation, amid this wild scenery.

The lofty mountains that bound the horizon on every side, contribute also, by their forms and the nature of their vegetation, to give an extraordinary character to the landscape. The average height of these mountains is not more than seven or eight hundred feet above the surrounding plains. Their summits are rounded, as for the most part in granitic mountains, and covered with thick forests of the laurel-tribe. Clusters of palm-trees,[*] the leaves of which, curled like feathers, rise majestically at an angle of seventy degrees, are dispersed amid trees with horizontal branches; and their bare trunks, like columns of a hundred or a hundred and twenty feet high, shoot up into the air, and when seen in distinct relief against the azure vault of the sky, they resemble a forest planted upon another forest. When, as the moon was going down behind the mountains of Uniana, her reddish disc was hidden behind the pinnated foliage of the palm-trees, and again appeared in the aerial zone that separates the two forests, I thought myself transported for a few moments to the hermitage which Bernardin de Saint-Pierre has described as one of the most delicious scenes of the Isle of Bourbon, and I felt how much the aspect of the plants and their groupings resembled each other in the two worlds. In describing a small spot of land in an island of the Indian Ocean, the inimitable author of Paul and Virginia has sketched the vast picture of the landscape of the tropics. He knew how to paint nature, not because he had studied it scientifically, but because he felt it in all its harmonious analogies of forms, colours, and interior powers.

East of the Atures, near these rounded mountains crowned, as it were, by two superimposed forests of laurels and palms,

---

[*] El cucurito.

other mountains of a very different aspect arise. Their ridge is bristled with pointed rocks, towering like pillars above the summits of the trees and shrubs. These effects are common to all granitic table-lands, at the Harz, in the metalliferous mountains of Bohemia, in Galicia, on the limit of the two Castiles, or wherever a granite of new formation appears above the ground. The rocks, which are at distances from each other, are composed of blocks piled together, or divided into regular and horizontal beds. On the summits of those situated near the Orinoco, flamingos, soldados,* and other fishing-birds perch, and look like men posted as sentinels. This resemblance is so striking, that the inhabitants of Angostura, soon after the foundation of their city, were one day alarmed by the sudden appearance of soldados and garzas, on a mountain towards the south. They believed they were menaced with an attack of Indios monteros (wild Indians called mountaineers); and the people were not perfectly tranquilized, till they saw the birds soaring in the air, and continuing their migration towards the mouths of the Orinoco.

The fine vegetation of the mountains spreads over the plains, wherever the rock is covered with mould, We generally find that this black mould, mixed with fibrous vegetable matter, is separated from the granitic rock by a layer of white sand. The missionary assured us that verdure of perpetual freshness prevails in the vicinity of the cataracts, produced by the quantity of vapour which the river, broken into torrents and cascades for the length of three or four thousand toises, diffuses in the air.

We had not heard thunder more than once or twice at Atures, and the vegetation everywhere displayed that vigorous aspect, that brilliancy of colour, seen on the coast only at the end of the rainy season. The old trees were decorated with beautiful orchideas,† yellow bannisterias, blue-flowered bignonias,

---

* The soldado (soldier) is a large species of heron.
† Cymbidium violaceum, Habenaria angustifolia, etc.

peperomias, arums, and pothoses. A single trunk displays a greater variety of vegetable forms than are contained within an extensive space of ground in our countries. Close to the parasite plants peculiar to very hot climates we observed, not without surprise, in the centre of the torrid zone, and near the level of the sea, mosses resembling in every respect those of Europe. We gathered, near the Great Cataract of Atures, that fine specimen of Grimmia* with fontinalis leaves, which has so much fixed the attention of botanists. It is suspended to the branches of the loftiest trees. Of the phaenerogamous plants, those which prevail in the woody spots are the mimosa, ficus, and laurinea. This fact is the more characteristic as, according to the observations of Mr. Brown, the laurineae appear to be almost entirely wanting on the opposite continent, in the equinoctial part of Africa. Plants that love humidity adorn the scenery surrounding the cataracts. We there find in the plains groups of heliconias and other scitamineae with large and glossy leaves, bamboos, and the three palm-trees, the murichi, jagua, and vadgiai, each of which forms a separate group. The murichi, or mauritia with scaly fruits, is the celebrated sago-tree of the Guaraon Indians. It has palmate leaves, and has no relation to the palm-trees with pinnate and curled leaves; to the jagua, which appears to be a species of the cocoa-tree; or to the vadgiai or cucurito, which may be assimilated to the fine species Oreodoxa. The cucurito, which is the palm most prevalent around the cataracts of the Atures and Maypures, is remarkable for its stateliness. Its leaves, or rather its palms, crown a trunk of eighty or one hundred feet high; their direction is almost perpendicular when young, as well as at their full growth, the points only being incurvated. They look like plumes of the most soft and verdant green. The

---

* Grimmia fontinaloides. See Hooker's Musci Exotici, 1818 tab. 2. The learned author of the Monography of the Jungermanniae (Mr. Jackson Hooker), with noble disinterestedness, published at his own expense, in London, the whole collection of cryptogamous plants, brought by Bonpland and Humboldt from the equinoctial regions of America.

cucurito, the pirijao, the fruit of which resembles the apricot, the Oreodoxa regia or palma real of the island of Cuba, and the ceroxylon of the high Andes, are the most majestic of all the palm-trees we saw in the New World. As we advance toward the temperate zone, the plants of this family decrease in size and beauty. What a difference between the species we have just mentioned, and the date-tree of the East, which unfortunately has become to the landscape painters of Europe the type of a group of palm-trees!

It is not suprising that persons who have travelled only in the north of Africa, in Sicily, or in Spain, cannot conceive that, of all large trees, the palm is the most grand and beautiful in form. Incomplete analogies prevent Europeans from having a just idea of the aspect of the torrid zone. All the world knows, for instance, that this zone is embellished by the contrasts exhibited in the foliage of the trees, and particularly by the great number of those with pinnate leaves. The ash, the service-tree, the inga, the acacia of the United States, the gleditsia, the tamarind, the mimosa, the desmanthus, have all pinnate leaves, with foliolae more or less long, slender, tough, and shining. But can a group of ash-trees, of service-trees, or of sumach, recall the picturesque effect of tamarinds or mimosas, when the azure of the sky appears through their small, slender, and delicately pinnated leaves? These considerations are more important than they may at first seem. The forms of plants determine the physiognomy of nature; and this physiognomy influences the moral dispositions of nations. Every type comprehends species, which, while exhibiting the same general appearance, differ in the varied development of the similar organs. The palm-trees, the scitamineae, the malvaceae, the trees with pinnate leaves, do not all display the same picturesque beauties; and generally the most beautiful species of each type, in plants as in animals, belong to the equinoctial zone.

The proteaceae,[*] crotons, agaves, and the great tribe of the cactuses, which inhabit exclusively the New World, disappear gradually, as we ascend the Orinoco above the Apure and the Meta. It is, however, the shade and humidity, rather than the distance from the coast, which oppose the migration of the cactuses southward. We found forests of them mingled with crotons, covering a great space of arid land to the east of the Andes, in the province of Bracamoros, towards the Upper Maranon. The arborescent ferns seem to fail entirely near the cataracts of the Orinoco; we found no species as far as San Fernando de Atabapo, that is, to the confluence of the Orinoco and the Guaviare.

Having now examined the vicinity of the Atures, it remains for me to speak of the rapids themselves, which occur in a part of the valley where the bed of the river, deeply ingulfed, has almost inaccessible banks. It was only in a very few spots that we could enter the Orinoco to bathe, between the two cataracts, in coves where the waters have eddies of little velocity. Persons who have dwelt in the Alps, the Pyrenees, or even the Cordilleras, so celebrated for the fractures and the vestiges of destruction which they display at every step, can scarcely picture to themselves, from a mere narration, the state of the bed of the river. It is traversed, in an extent of more than five miles, by innumerable dikes of rock, forming so many natural dams, so many barriers resembling those of the Dnieper, which the ancients designated by the name of phragmoi. The space between the rocky dikes of the Orinoco is filled with islands of different dimensions; some hilly, divided into several peaks, and two or three hundred toises in length, others small, low, and like mere shoals. These islands divide the river into a number of torrents, which boil up as they break against the rocks. The jaguas and cucuritos with plumy leaves, with which all the islands are covered, seem like groves of palm-trees rising from

---

[*] Rhopalas, which characterise the vegetation of the Llanos.

the foamy surface of the waters. The Indians, whose task it is to pass the boats empty over the raudales, distinguish every shelf, and every rock, by a particular name. On entering from the south you find first the Leap of the Toucan (Salto del Piapoco); and between the islands of Avaguri and Javariveni is the Raudal of Javariveni, where, on our return from Rio Negro, we passed some hours amid the rapids, waiting for our boat. A great part of the river appeared dry. Blocks of granite are heaped together, as in the moraines which the glaciers of Switzerland drive before them. The river is ingulfed in caverns; and in one of these caverns we heard the water roll at once over our heads and beneath our feet. The Orinoco seems divided into a multitude of arms or torrents, each of which seeks to force a passage through the rocks. We were struck with the little water to be seen in the bed of the river, the frequency of subterraneous falls, and the tumult of the waters breaking on the rocks in foam.

Cuncta fremunt undis; ac multo murmure montis
Spumeus invictis canescit fluctibus amnis.[*]

Having passed the Raudal of Javariveni (I name here only the principal falls) we come to the Raudal of Canucari, formed by a ledge of rocks uniting the islands of Surupamana and Uirapuri. When the dikes, or natural dams, are only two or three feet high, the Indians venture to descend them in boats. In going up the river, they swim on before, and if, after many vain efforts, they succeed in fixing a rope to one of the points of rock that crown the dike, they then, by means of that rope, draw the bark to the top of the raudal. The bark, during this arduous task, often fills with water; at other times it is stove against the rocks, and the Indians, their bodies bruised and bleeding, extricate themselves with difficulty from the whirlpools, and reach, by swimming, the nearest island. When the steps or rocky barriers are very high, and entirely bar the river, light boats are carried on shore, and with the help of branches of trees placed under

---

[*] Lucan, Pharsalia lib 10 v 132.

them to serve as rollers, they are drawn as far as the place where the river again becomes navigable. This operation is seldom necessary when the water is high. We cannot speak of the cataracts of the Orinoco without recalling to mind the manner heretofore employed for descending the cataracts of the Nile, of which Seneca has left us a description probably more poetical than accurate. I shall cite the passage, which traces with fidelity what may be seen every day at Atures, Maypures, and in some pongos of the Amazon. "Two men embark in a small boat; one steers, and the other empties it as it fills with water. Long buffeted by the rapids, the whirlpools, and the contrary currents, they pass through the narrowest channels, avoid the shoals, and rush down the whole river, guiding the course of the boat in its accelerated fall." (Nat. Quaest. lib 4 cap 2 edit. Elzev. tome 2 page 609.)

In hydrographic descriptions of countries, the vague names of cataracts, cascades, falls, and rapids,[*] denoting those tumultuous movements of water which arise from very different circumstances, are generally confounded with one another. Sometimes a whole river precipitating itself from a great height, and by one single fall, renders navigation impossible. Such is the majestic fall of the Rio Tequendama, which I have represented in my Views of the Cordilleras; such are the falls of Niagara and of the Rhine, much less remarkable for their elevation, than for the mass of water they contain. Sometimes stony dikes of small height succeed each other at great distances, and form distinct falls; such are the cachoeiras of the Rio Negro and the Rio Madeira, the saltos of the Rio Cauca, and the greater part of the pongos that are found in the Upper Maranon, from the confluence of the Chinchipe to the village of San Borja. The highest and most formidable of these pongos, which are descended on rafts, that of Mayasi, is however only three feet in height. Sometimes small rocky dikes are so near each other that

---

[*] The corresponding terms in use among the people of South America, are saltos, chorros, pongos, cachoeiras, and raudales.

they form for several miles an uninterrupted succession of cascades and whirlpools (chorros and remolinos); these are properly what are called rapids (raudales). Such are the yellalas, or rapids of the River Zaire,[*] or Congo, which Captain Tuckey has recently made known to us; the rapids of the Orange River in Africa, above Pella; and the falls of the Missouri, which are four leagues in length, where the river issues from the Rocky Mountains. Such also are the cataracts of Atures and Maypures; the only cataracts which, situated in the equinoctial region of the New World, are adorned with the noble growth of palm-trees. At all seasons they exhibit the aspect of cascades, and present the greatest obstacles to the navigation of the Orinoco, while the rapids of the Ohio and of Upper Egypt are scarcely visible at the period of floods. A solitary cataract, like Niagara, or the cascade of Terni, affords a grand but single picture, varying only as the observer changes his place. Rapids, on the contrary, especially when adorned with large trees, embellish a landscape during a length of several leagues. Sometimes the tumultuous movement of the waters is caused only by extraordinary contractions of the beds of the rivers. Such is the angostura of Carare, in the river Magdalena, a strait that impedes communication between Santa Fe de Bogota and the coast of Carthagena; and such is the pongo of Manseriche, in the Upper Maranon.

The Orinoco, the Rio Negro, and almost all the confluents of the Amazon and the Maranon, have falls or rapids, either because they cross the mountains where they take rise, or because they meet other mountains in their course. If the

---

[*] Voyage to explore the River Zaire, 1818, pages 152, 327, 340. What the inhabitants of Upper Egypt and Nubia call chellal in the Nile, is called yellala in the River Congo. This analogy between words signifying rapids is remarkable, on account of the enormous distance of the yellalas of the Congo from the chellal and djenadel of the Nile. Did the word chellal penetrate with the Moors into the west of Africa? If, with Burckhardt, we consider the origin of this word as Arabic (Travels in Nubia, 1819), it must be derived from the root challa, to disperse, which forms chelil, water falling through a narrow channel.

Amazon, from the pongo of Manseriche (or, to speak with more precision, from the pongo of Tayuchuc) as far as its mouth, a space of more than seven hundred and fifty leagues, exhibit no tumultuous movement of the waters, the river owes this advantage to the uniform direction of its course. It flows from west to east in a vast plain, forming a longitudinal valley between the mountains of Parima and the great mass of the mountains of Brazil.

I was surprised to find by actual measurement that the rapids of the Orinoco, the roar of which is heard at the distance of more than a league, and which are so eminently picturesque from the varied appearance of the waters, the palm-trees and the rocks, have not probably, on their whole length, a height of more than twenty-eight feet perpendicular. In reflecting on this, we find that it is a great deal for rapids, while it would be very little for a single cataract. The Yellalas of the Rio Congo, in the contracted part of the river from Banza Noki as far as Banza Inga, furnish, between the upper and lower levels, a much more considerable difference; but Mr. Barrow observes, that among the great number of these rapids there is one fall, which alone is thirty feet high. On the other hand, the famous pongos of the river Amazon, so dangerous to go up, the falls of Rentema, of Escurrebragas, and of Mayasi, are but a few feet in perpendicular height. Those who are engaged in hydraulic works know the effect that a bar of eighteen or twenty inches' height produces in a great river. The whirling and tumultuous movement of the water does not depend solely on the greatness of partial falls; what determines the force and impetuosity is the nearness of these falls, the steepness of the rocky ledges, the returning sheets of water which strike against and surmount each other, the form of the islands and shoals, the direction of the counter-currents, and the contraction and sinuosity of the channels through which the waters force a passage between two adjacent levels. In two rivers equally large, that of which the falls have least height may sometimes present the greatest dangers and the most impetuous movements.

It is probable that the river Orinoco loses part of its waters in the cataracts, not only by increased evaporation, caused by the dispersion of minute drops in the atmosphere, but still more by filtrations into the subterraneous cavities. These losses, however, are not very perceptible when we compare the mass of waters entering into the raudal with that which issues out near the mouth of the Rio Anaveni. It was by a similar comparison that the existence of subterraneous cavities in the yellalas or rapids of the river Congo was discovered. The pongo of Manseriche, which ought rather to be called a strait than a fall, ingulfs, in a manner not yet sufficiently explored, a part of the waters and all the floating wood of the Upper Maranon.

The spectator, seated on the bank of the Orinoco, with his eyes fixed on those rocky dikes, is naturally led to inquire whether, in the lapse of ages, the falls change their form or height. I am not much inclined to believe in such effects of the shock of water against blocks of granite, and in the erosion of siliceous matter. The holes narrowed toward the bottom, the funnels that are discovered in the raudales, as well as near so many other cascades in Europe, are owing only to the friction of the sand, and the movement of quartz pebbles. We saw many such, whirled perpetually by the current at the bottom of the funnels, and contributing to enlarge them in every direction. The pongos of the river Amazon are easily destroyed, because the rocky dikes are not granite, but a conglomerate, or red sandstone with large fragments. A part of the pongo of Rentama was broken down eighty years ago, and the course of the waters being interrupted by a new bar, the bed of the river remained dry for some hours, to the great astonishment of the inhabitants of the village of Payaya, seven leagues below the pongo. The Indians of Atures assert (and in this their testimony is contrary to the opinion of Caulin) that the rocks of the raudal preserve the same aspect; but that the partial torrents into which the great river divides itself as it passes through the heaped blocks of granite, change their direction, and carry sometimes more, sometimes less water towards one or the other bank; but the causes of these changes may be very remote from the cataracts, for in the rivers that spread life over the surface of the globe, as in the arteries by which it is diffused through organized bodies, all the movements are propagated to great distances. Oscillations, that at first seem partial, react on the whole liquid mass contained in the trunk as well as in its numerous ramifications.

Some of the Missionaries in their writings have alleged that the inhabitants of Atures and Maypures have been struck with deafness by the noise of the Great Cataracts, but this is untrue. When the noise is heard in the plain that surrounds the mission, at the distance of more than a league, you seem to be near a coast skirted by reefs and breakers. The noise is three times as loud by night as by day, and gives an inexpressible charm to these solitary scenes. What can be the cause of this increased intensity of sound, in a desert where nothing seems to interrupt the silence of nature? The velocity of the propagation of sound, far from augmenting, decreases with the lowering of the temperature. The intensity diminishes in air agitated by a wind which is contrary to the direction of the sound; it diminishes also by dilatation of the air, and is weaker in the higher than in the lower regions of the atmosphere, where the number of particles of air in motion is greater in the same radius. The intensity is the same in dry air, and in air mingled with vapours; but it is feebler in carbonic acid gas than in mixtures of azote and oxygen. From these facts, which are all we know with any certainty, it is difficult to explain a phenomenon observed near every cascade in Europe, and which, long before our arrival in the village of Atures, had struck the missionary and the Indians.

It may be thought that, even in places not inhabited by man, the hum of insects, the song of birds, the rustling of leaves agitated by the feeblest winds, occasion during the day a confused noise, which we perceive the less because it is uniform, and constantly strikes the ear. Now this noise, however slightly perceptible it may be, may diminish the intensity of a louder noise; and this diminution may cease if during the calm of the night the song of birds, the hum of insects, and the action of the wind upon the leaves be interrupted. But this reasoning, even admitting its justness, can scarcely be applied to the forests of the Orinoco, where the air is constantly filled by an innumerable quantity of mosquitos, where the hum of insects is much louder by night than by day, and where the breeze, if ever it be felt, blows only after sunset.

I rather think that the presence of the sun acts upon the propagation and intensity of sound by the obstacles met in currents of air of different density, and by the partial undulations of the atmosphere arising from the unequal heating of different parts of the soil. In calm air, whether dry or mingled with vesicular vapours equally distributed, sound-waves are propagated without difficulty. But when the air is crossed in every direction by small currents of hotter air, the sonorous undulation is divided into two undulations where the density of the medium changes abruptly; partial echoes are formed that weaken the sound, because one of the streams comes back upon itself; and those divisions of undulations take place of which M. Poisson has developed the theory with great sagacity.[*] It is not therefore the movement of the particles of air from below to above in the ascending current, or the small oblique currents that we consider as opposing by a shock the propagation of the sonorous undulations. A shock given to the surface of a liquid will form circles around the centre of percussion, even when the liquid is agitated. Several kinds of undulations may cross each other in water, as in air, without being disturbed in their propagation: little movements may, as it were, ride over each other, and the real cause of the less intensity of sound during the day appears to be the interpretation of homogeneity in the elastic medium. During the day there is a sudden interruption of density wherever small streamlets of air of a high temperature rise over

[*] Annales de Chimie tome 7 page 293.

parts of the soil unequally heated. The sonorous undulations are divided, as the rays of light are refracted and form the mirage wherever strata of air of unequal density are contiguous. The propagation of sound is altered when a stratum of hydrogen gas is made to rise in a tube closed at one end above a stratum of atmospheric air; and M. Biot has well explained, by the interposition of bubbles of carbonic acid gas, why a glass filled with champagne is not sonorous so long as that gas is evolved, and passing through the strata of the liquid.

In support of these ideas, I might almost rest on the authority of an ancient philosopher, whom the moderns do not esteem in proportion to his merits, though the most distinguished zoologists have long rendered ample justice to the sagacity of his observations. "Why," says Aristotle in his curious book of Problems, "why is sound better heard during the night? Because there is more calmness on account of the absence of caloric (of the hottest).* This absence renders every thing calmer, for the sun is the principle of all movement." Aristotle had no doubt a vague presentiment of the cause of the phenomenon; but he attributes to the motion of the atmosphere, and the shock of the particles of air, that which seems to be rather owing to abrupt changes of density in the contiguous strata of air.

On the 16th of April, towards evening, we received tidings that in less than six hours our boat had passed the rapids, and had arrived in good condition in a cove called el Puerto de arriba, or the Port of the Expedition. We were shown in the little church of Atures some remains of the ancient wealth of the

---

* I have placed in a parenthesis, a literal version of the term employed by Aristotle, to express in reality what we now term the matter of heat. Theodore of Gaza, in his Latin translation, expresses in the shape of a doubt what Aristotle positively asserts. I may here remark, that, notwithstanding the imperfect state of science among the ancients, the works of the Stagirite contain more ingenious observations than those of many later philosophers. It is in vain we look in Aristoxenes (De Musica), in Theophylactus Simocatta (De Quaestionibus physicis), or in the 5th Book of the Quest. Nat. of Seneca, for an explanation of the nocturnal augmentation of sound.

Jesuits. A silver lamp of considerable weight lay on the ground half-buried in the sand. Such an object, it is true, would nowhere tempt the cupidity of a savage; yet I may here remark, to the honor of the natives of the Orinoco, that they are not addicted to stealing, like the less savage tribes of the islands in the Pacific. The former have a great respect for property; they do not even attempt to steal provision, hooks, or hatchets. At Maypures and Atures, locks on doors are unknown: they will be introduced only when whites and men of mixed race establish themselves in the missions.

The Indians of Atures are mild and moderate, and accustomed, from the effects of their idleness, to the greatest privations. Formerly, being excited to labour by the Jesuits, they did not want for food. The fathers cultivated maize, French beans (frijoles), and other European vegetables; they even planted sweet oranges and tamarinds round the villages; and they possessed twenty or thirty thousand head of cows and horses, in the savannahs of Atures and Carichana. They had at their service a great number of slaves and servants (peones), to tend their herds. Nothing is now cultivated but a little cassava, and a few plantains. Such however is the fertility of the soil, that at Atures I counted on a single branch of a musa one hundred and eight fruits, four or five of which would almost have sufficed for a man's daily food. The culture of maize is entirely neglected, and the horses and cows have entirely disappeared. Near the raudal, a part of the village still bears the name of Passo del ganado (ford of the cattle), while the descendants of those very Indians whom the Jesuits had assembled in a mission, speak of horned cattle as of animals of a race now lost. In going up the Orinoco, toward San Carlos del Rio Negro, we saw the last cow at Carichana. The Fathers of the Observance, who now govern these vast countries, did not immediately succeed the Jesuits. During an interregnum of eighteen years, the missions were visited only from time to time, and by Capuchin monks. The agents of the secular government, under the title of Royal Commissioners, managed the hatos or farms of the Jesuits with culpable negligence. They killed the cattle for the sake of selling the hides. Many heifers were devoured by the jaguars, and a great number perished in consequence of wounds made by the bats of the raudales, which, though smaller, are far bolder than the bats of the Llanos. At the time of the expedition of the boundaries, horses from Encaramada, Carichana, and Atures, were conveyed as far as San Jose de Maravitanos, where, on the banks of the Rio Negro, the Portuguese could only procure them, after a long passage, and of a very inferior quality, by the rivers Amazon and Grand Para. Since the year 1795, the cattle of the Jesuits have entirely disappeared. There now remain as monuments of the ancient cultivation of these countries, and the active industry of the first missionaries, only a few trunks of the orange and tamarind, in the savannahs, surrounded by wild trees.

The tigers, or jaguars, which are less dangerous for the cattle than the bats, come into the village at Atures, and devour the swine of the poor Indians. The missionary related to us a striking instance of the familiarity of these animals, usually so ferocious. Some months before our arrival, a jaguar, which was thought to be young, though of a large size, had wounded a child in playing with him. The facts of this case, which were verified to us on the spot, are not without interest in the history of the manners of animals. Two Indian children, a boy and a girl, about eight and nine years of age, were seated on the grass near the village of Atures, in the middle of a savannah, which we several times traversed. At two o'clock in the afternoon, a jaguar issued from the forest, and approached the children, bounding around them; sometimes he hid himself in the high grass, sometimes he sprang forward, his back bent, his head hung down, in the manner of our cats. The little boy, ignorant of his danger, seemed to be sensible of it only when the jaguar with one of his paws gave him some blows on the head. These blows, at first slight, became ruder and ruder; the claws of the jaguar wounded the child, and the blood flowed freely. The little girl then took a branch of a tree, struck the animal, and it fled from her. The Indians ran up at the cries of the children, and saw the jaguar, which then bounded off without making the least show of resistance.

The little boy was brought to us, who appeared lively and intelligent. The claw of the jaguar had torn away the skin from the lower part of the forehead, and there was a second scar at the top of the head. This was a singular fit of playfulness in an animal which, though not difficult to be tamed in our menageries, nevertheless shows itself always wild and ferocious in its natural state. If we admit that, being sure of its prey, it played with the little Indian as our cats play with birds whose wings have been clipped, how shall we explain the patience of a jaguar of large size, which finds itself attacked by a girl? If the jaguar were not pressed by hunger, why did it approach the children at all? There is something mysterious in the affections and hatreds of animals. We have known lions kill three or four dogs that were put into their den, and instantly caress a fifth, which, less timid, took the king of animals by the mane. These are instincts of which we know not the secret.

We have mentioned that domestic pigs are attacked by the jaguars. There are in these countries, besides the common swine of European race, several species of peccaries, or pigs with lumbar glands, two of which only are known to the naturalists of Europe. The Indians call the little peccary (Dicotiles torquatus, Cuv.), in the Maypure tongue, chacharo; while they give the name of apida to a species of pig which they say has no pouch, is larger, and of a dark brown colour, with the belly and lower jaw white. The chacharo, reared in the houses, becomes tame like our sheep and goats. It reminds us, by the gentleness of its manners, of the curious analogies which anatomists have observed between the peccaries and the ruminating animals. The apida, which is domesticated like our swine in Europe, wanders in large herds composed of several hundreds. The presence of these herds is announced from afar, not only by their hoarse gruntings, but above all by the impetuosity with which they break down the shrubs in their way. M. Bonpland, in an herborizing excursion, warned by his Indian guide to hide himself behind the trunk of a tree, saw a number of these peccaries (cochinos or puercos del monte) pass close by him. The herd marched in a close body, the males proceeding first; and each sow was accompanied by her young. The flesh of the chacharo is flabby, and not very agreeable; it affords, however, a plentiful nourishment to the natives, who kill these animals with small lances tied to cords. We were assured at Atures, that the tiger dreads being surrounded in the forests by these herds of wild pigs; and that, to avoid being stifled, he tries to save himself by climbing up a tree. Is this a hunter's tale, or a fact that has really been observed? In several parts of America the hunters believe in the existence of a javali, or native boar with tusks curved outwardly. I never saw one, but this animal is mentioned in the works of the Spanish missionaries, a source too much neglected by zoologists; for amidst much incorrectness and extravagance, they contain many curious local observations.

Among the monkeys which we saw at the mission of the Atures, we found one new species, of the tribe of sais and sajous, which the Creoles vulgarly call machis. It is the Guvapavi with grey hair and a bluish face. It has the orbits of the eyes and the forehead as white as snow, a peculiarity which at first sight distinguishes it from the Simia capucina, the Simia apella, the Simia trepida, and the other weeping monkeys hitherto so confusedly described. This little animal is as gentle as it is ugly. A monkey of this species, which was kept in the courtyard of the missionary, would frequently mount on the back of a pig, and in this manner traverse the savannahs. We have also seen it upon the back of a large cat, which had been brought up with it in Father Zea's house.

It was among the cataracts that we began to hear of the hairy man of the woods, called salvaje, that carries off women, constructs huts, and sometimes eats human flesh. The Tamanacs call it achi, and the Maypures vasitri, or great devil. The natives and the missionaries have no doubt of the existence of this man-shaped monkey, of which they entertain a singular dread. Father Gili gravely relates the history of a lady in the town of San Carlos, in the Llanos of Venezuela, who much praised the gentle character and attentions of the man of the woods. She is stated to have lived several years with one in great domestic harmony, and only requested some hunters to take her back, because she and her children (a little hairy also) were weary of living far from the church and the sacraments. The same author, notwithstanding his credulity, acknowledges that he never knew an Indian who asserted positively that he had seen the salvaje with his own eyes. This wild legend, which the missionaries, the European planters, and the negroes of Africa, have no doubt embellished with many features taken from the description of the manners of the orang-otang,[*] the gibbon, the jocko or chimpanzee, and the pongo, followed us, during five years, from the northern to the southern hemisphere. We were everywhere blamed, in the most cultivated class of society, for being the only persons to doubt the existence of the great anthropomorphous monkey of America. There are certain regions where this belief is particularly prevalent among the people; such are the banks of the Upper Orinoco, the valley of Upar near the lake of Maracaybo, the mountains of Santa Martha and of Merida, the provinces of Quixos, and the banks of the Amazon near Tomependa. In all these places, so distant one from the other, it is asserted that the salvaje is easily recognized by the traces of its feet, the toes of which are turned backward. But if there exist a monkey of a large size in the New Continent, how has it happened that for three centuries no man worthy of belief has

---

[*] Simia satyrus. We must not believe, notwithstanding the assertions of almost all zoological writers, that the word orang-otang is applied exclusively in the Malay language to the Simia satyrus of Borneo. This expression, on the contrary, means any very large monkey, that resembles man in figure. Marsden's History of Sumatra 3rd edition page 117. Modern zoologists have arbitrarily appropriated provincial names to certain species; and by continuing to prefer these names, strangely disfigured in their orthography, to the Latin systematic names, the confusion of the nomenclature has been increased.

been able to procure the skin of one? Several hypotheses present themselves to the mind, in order to explain the source of so ancient an error or belief. Has the famous capuchin monkey of Esmeralda (Simia chiropotes), with its long canine teeth, and physiognomy much more like man's[*] than that of the orang-otang, given rise to the fable of the salvaje? It is not so large indeed as the coaita (Simia paniscus); but when seen at the top of a tree, and the head only visible, it might easily be taken for a human being. It may be also (and this opinion appears to me the most probable) that the man of the woods was one of those large bears, the footsteps of which resemble those of a man, and which are believed in every country to attack women. The animal killed in my time at the foot of the mountains of Merida, and sent, by the name of salvaje, to Colonel Ungaro, the governor of the province of Varinas, was in fact a bear with black and smooth fur. Our fellow-traveller, Don Nicolas Soto, had examined it closely. Did the strange idea of a plantigrade animal, the toes of which are placed as if it walked backward, take its origin from the habit of the real savages of the woods, the Indians of the weakest and most timid tribes, of deceiving their enemies, when they enter a forest, or cross a sandy shore, by covering the traces of their feet with sand, or walking backward?

Though I have expressed my doubts of the existence of an unknown species of large monkey in a continent which appears entirely destitute of quadrumanous animals of the family of the orangs, cynocephali, mandrils, and pongos; yet it should be remembered that almost all matters of popular belief, even those most absurd in appearance, rest on real facts, but facts ill observed. In treating them with disdain, the traces of a discovery may often be lost, in natural philosophy as well as in zoology. We will not then admit, with a Spanish author, that the fable of the man of the woods was invented by the artifice of Indian women, who pretended to have been carried off, when they had

---

[*] The whole of the features—the expression of the physiognomy; but not the forehead.

been long absent unknown to their husbands. Travellers who may hereafter visit the missions of the Orinoco will do well to follow up our researches on the salvaje or great devil of the woods; and examine whether it be some unknown species of bear, or some very rare monkey analogous to the Simia chiropotes, or Simia satanas, which may have given rise to such singular tales.

After having spent two days near the cataract of Atures, we were not sorry when our boat was reladen, and we were enabled to leave a spot where the temperature of the air is generally by day twenty-nine degrees, and by night twenty-six degrees, of the centigrade thermometer. This temperature seemed to us to be still much more elevated, from the feeling of heat which we experienced. The want of concordance between the instruments and the sensations must be attributed to the continual irritation of the skin excited by the mosquitos. An atmosphere filled with venomous insects always appears to be more heated than it is in reality. We were horribly tormented in the day by mosquitos and the jejen, a small venomous fly (simulium), and at night by the zancudos, a large species of gnat, dreaded even by the natives. Our hands began to swell considerably, and this swelling increased daily till our arrival on the banks of the Temi. The means that are employed to escape from these little plagues are very extraordinary. The good missionary Bernardo Zea, who passed his life tormented by mosquitos, had constructed near the church, on a scaffolding of trunks of palm-trees, a small apartment, in which we breathed more freely. To this we went up in the evening, by means of a ladder, to dry our plants and write our journal. The missionary had justly observed, that the insects abounded more particularly in the lowest strata of the atmosphere, that which reaches from the ground to the height of twelve or fifteen feet. At Maypures the Indians quit the village at night, to go and sleep on the little islets in the midst of the cataracts. There they enjoy some rest; the mosquitoes appearing to shun air loaded with vapours. We found everywhere fewer in the middle of the river than near its banks; and thus less is suffered in descending the Orinoco than in going up in a boat.

Persons who have not navigated the great rivers of equinoctial America, for instance, the Orinoco and the Magdalena, can scarcely conceive how, at every instant, without intermission, you may be tormented by insects flying in the air; and how the multitude of these little animals may render vast regions almost uninhabitable. Whatever fortitude be exercised to endure pain without complaint, whatever interest may be felt in the objects of scientific research, it is impossible not to be constantly disturbed by the mosquitos, zancudos, jejens, and tempraneros, that cover the face and hands, pierce the clothes with their long needle-formed suckers, and getting into the mouth and nostrils, occasion coughing and sneezing whenever any attempt is made to speak in the open air. In the missions of the Orinoco, in the villages on the banks of the river, surrounded by immense forests, the plaga de las moscas, or the plague of the mosquitos, affords an inexhaustible subject of conversation. When two persons meet in the morning, the first questions they address to each other are: How did you find the zancudos during the night? How are we to-day for the mosquitos?[*] These questions remind us of a Chinese form of politeness, which indicates the ancient state of the country where it took birth. Salutations were made heretofore in the Celestial empire in the following words, vou-to-hou, Have you been incommoded in the night by the serpents?

The geographical distribution of the insects of the family of tipulae presents very remarkable phenomena. It does not appear to depend solely on heat of climate, excess of humidity, or the thickness of forests, but on local circumstances that are difficult to characterise. It may be observed that the plague of mosquitos and zancudos is not so general in the torrid zone as is commonly

---

[*] Que le han parecido los zancudos de noche? Como stamos hoy de mosquitos?

believed. On the table-lands elevated more than four hundred toises above the level of the ocean, in the very dry plains remote from the beds of great rivers (for instance, at Cumana and Calabozo), there are not sensibly more gnats than in the most populous parts of Europe. They are perceived to augment enormously at Nueva Barcelona, and more to the west, on the coast that extends towards Cape Codera. Between the little harbour of Higuerote and the mouth of the Rio Unare, the wretched inhabitants are accustomed to stretch themselves on the ground, and pass the night buried in the sand three or four inches deep, leaving out the head only, which they cover with a handkerchief. You suffer from the sting of insects, but in a manner easy to bear, in descending the Orinoco from Cabruta towards Angostura, and in going up from Cabruta towards Uruana, between the latitudes of 7 and 8°. But beyond the mouth of the Rio Arauca, after having passed the strait of Baraguan, the scene suddenly changes. From this spot the traveller may bid farewell to repose. If he have any poetical remembrance of Dante, he may easily imagine he has entered the citta dolente, and he will seem to read on the granite rocks of Baraguan these lines of the Inferno:

Noi sem venuti al luogo, ov' i' t'ho detto
Che tu vedrai le genti dolorose.

The lower strata of air, from the surface of the ground to the height of fifteen or twenty feet, are absolutely filled with venomous insects. If in an obscure spot, for instance in the grottos of the cataracts formed by superincumbent blocks of granite, you direct your eyes toward the opening enlightened by the sun, you see clouds of mosquitos more or less thick. At the mission of San Borja, the suffering from mosquitos is greater than at Carichana; but in the Raudales, at Atures, and above all at Maypures, this suffering may be said to attain its maximum. I doubt whether there be a country upon earth where man is exposed to more cruel torments in the rainy season. Having passed the fifth degree of latitude, you are somewhat less stung;

but on the Upper Orinoco the stings are more painful, because the heat and the absolute want of wind render the air more burning and more irritating in its contact with the skin.

"How comfortable must people be in the moon!" said a Salive Indian to Father Gumilla; "she looks so beautiful and so clear, that she must be free from mosquitos." These words, which denote the infancy of a people, are very remarkable. The satellite of the earth appears to all savage nations the abode of the blessed, the country of abundance. The Esquimaux, who counts among his riches a plank or trunk of a tree, thrown by the currents on a coast destitute of vegetation, sees in the moon plains covered with forests; the Indian of the forests of Orinoco there beholds open savannahs, where the inhabitants are never stung by mosquitos.

After proceeding further to the south, where the system of yellowish-brown waters commences,[*] on the banks of the Atabapo, the Tuni, the Tuamini, and the Rio Negro, we enjoyed an unexpected repose. These rivers, like the Orinoco, cross thick forests, but the tipulary insects, as well as the crocodiles, shun the proximity of the black waters. Possibly these waters, which are a little colder, and chemically different from the white waters, are adverse to the larvae of tipulary insects and gnats, which may be considered as real aquatic animals. Some small rivers, the colour of which is deep blue, or yellowish-brown (as the Toparo, the Mataveni, and the Zama), are exceptions to the almost general rule of the absence of mosquitos over the black waters. These three rivers swarm with them; and the Indians themselves fixed our attention on the problematic causes of this phenomenon. In going down the Rio Negro, we breathed freely at Maroa, Daripe, and San Carlos, villages situated on the boundaries of Brazil. But this improvement of our situation was of short continuance; our sufferings recommenced as soon as we entered the Cassiquiare. At Esmeralda, at the eastern extremity

---

[*] Generally called black waters, aguas negras.

of the Upper Orinoco, where ends the known world of the Spaniards, the clouds of mosquitos are almost as thick as at the Great Cataracts. At Mandavaca we found an old missionary, who told us with an air of sadness, that he had had his twenty years of mosquitos in America[*]. He desired us to look at his legs, that we might be able to tell one day, beyond sea (por alla), what the poor monks suffer in the forests of Cassiquiare. Every sting leaving a small darkish brown point, his legs were so speckled that it was difficult to recognize the whiteness of his skin through the spots of coagulated blood. If the insects of the genus Simulium abound in the Cassiquiare, which has white waters, the culices or zancudos are so much the more rare; you scarcely find any there; while on the rivers of black waters, in the Atabapo and the Rio, there are generally some zancudos and no mosquitos.

I have just shown, from my own observations, how much the geographical distribution of venomous insects varies in this labyrinth of rivers with white and black waters. It were to be wished that a learned entomologist could study on the spot the specific differences of these noxious insects,[†] which in the torrid zone, in spite of their minute size, act an important point in the economy of nature. What appeared to us very remarkable, and is a fact known to all the missionaries, is, that the different species do not associate together, and that at different hours of the day you are stung by distinct species. Every time that the scene changes, and, to use the simple expression of the missionaries, other insects mount guard, you have a few minutes, often a quarter of an hour, of repose. The insects that disappear have not their places instantly supplied by their successors. From half-past-six in the morning till five in the afternoon, the air is filled with mosquitos; which have not, as some travellers have stated,

---

[*] "Yo tengo mis veinte anos de mosquitos."
[†] The mosquito bovo or tenbiguai; the melero, which always settles upon the eyes; the tempranero, or putchiki; the jejen; the gnat rivau, the great zancudo, or matchaki; the cafafi, etc.

the form of our gnats,[*] but that of a small fly. They are simuliums of the family Nemocera of the system of Latreille. Their sting is as painful as that of the genus Stomox. It leaves a little reddish brown spot, which is extravased and coagulated blood, where their proboscis has pierced the skin. An hour before sunset a species of small gnats, called tempraneros,[†] because they appear also at sunrise, take the place of the mosquitos. Their presence scarcely lasts an hour and a half; they disappear between six and seven in the evening, or, as they say here, after the Angelus (a la oracion). After a few minutes' repose, you feel yourself stung by zancudos, another species of gnat with very long legs. The zancudo, the proboscis of which contains a sharp-pointed sucker, causes the most acute pain, and a swelling that remains several weeks. Its hum resembles that of the European gnat, but is louder and more prolonged. The Indians pretend to distinguish the zancudos and the tempraneros by their song; the latter are real twilight insects, while the zancudos are most frequently nocturnal insects, and disappear toward sunrise.

In our way from Carthagena to Santa Fe de Bogota, we observed that between Mompox and Honda, in the valley of the Rio Magdalena, the zancudos darkened the air from eight in the evening till midnight; that towards midnight they diminished in number, and were hidden for three or four hours; and lastly that they returned in crowds, about four in the morning. What is the cause of these alternations of motion and rest? Are these animals fatigued by long flight? It is rare on the Orinoco to see real gnats by day; while at the Rio Magdalena we were stung night and

---

[*] Culex pipiens. This difference between mosquito (little fly, simulium) and zancudo (gnat, culex) exists in all the Spanish colonies. The word zancudo signifies long legs, qui tiene las zancas largas. The mosquitos of the Orinoco are the moustiques; the zancudos are the maringouins of French travellers.

[†] Which appear at an early hour (temprano). Some persons say, that the zancudo is the same as the tempranero, which returns at night, after hiding itself for some time. I have doubts of this identity of the species; the pain caused by the sting of the two insects appeared to me different.

day, except from noon till about two o'clock. The zancudos of the two rivers are no doubt of different species.

We have seen that the insects of the tropics everywhere follow a certain standard in the periods at which they alternately arrive and disappear. At fixed and invariable hours, in the same season, and the same latitude, the air is peopled with new inhabitants, and in a zone where the barometer becomes a clock,[*] where everything proceeds with such admirable regularity, we might guess blindfold the hour of the day or night, by the hum of the insects, and by their stings, the pain of which differs according to the nature of the poison that each species deposits in the wound.

At a period when the geography of animals and of plants had not yet been studied, the analogous species of different climates were often confounded. It was believed that the pines and ranunculuses, the stags, the rats, and the tipulary insects of the north of Europe, were to be found in Japan, on the ridge of the Andes, and at the Straits of Magellan. Justly celebrated naturalists have thought that the zancudo of the torrid zone was the gnat of our marshes, become more vigorous, more voracious, and more noxious, under the influence of a burning climate. This is a very erroneous opinion. I carefully examined and described upon the spot those zancudos, the stings of which are most tormenting. In the rivers Magdalena and Guayaquil alone there are five distinct species.

The culices of South America have generally the wings, corslet, and legs of an azure colour, ringed and variegated with a mixture of spots of metallic lustre. Here as in Europe, the males, which are distinguished by their feathered antennae, are extremely rare; you are seldom stung except by females. The preponderance of this sex explains the immense increase of the species, each female laying several hundred eggs. In going up

---

[*] By the extreme regularity of the horary variations of the atmospheric pressure.

303

one of the great rivers of America, it is observed, that the appearance of a new species of culex denotes the proximity of a new stream flowing in. I shall mention an instance of this curious phenomenon. The Culex lineatus, which belongs to the Cano Tamalamec, is only perceived in the valley of the Rio Grande de la Magdalena, at a league north of the junction of the two rivers; it goes up, but scarcely ever descends the Rio Grande. It is thus, that, on a principal vein, the appearance of a new substance in the gangue indicates to the miner the neighbourhood of a secondary vein that joins the first.

On recapitulating the observations here recorded, we see, that within the tropics, the mosquitos and zancudos do not rise on the slope of the Cordilleras[*] toward the temperate region, where the mean heat is below 19 or 20°; and that, with few exceptions, they shun the black waters, and dry and unwooded spots.[†] The atmosphere swarms with them much more in the Upper than in the Lower Orinoco, because in the former the river is surrounded with thick forests on its banks, and the skirts of the forests are not separated from the river by a barren and extensive beach. The mosquitos diminish on the New Continent with the diminution of the water, and the destruction of the woods; but the effects of these changes are as slow as the progress of cultivation. The towns of Angostura, Nueva Barcelona, and Mompox, where from the want of police, the streets, the great squares, and the interior of court-yards are overgrown with brushwood, are sadly celebrated for the abundance of zancudos.

---

[*] The culex pipiens of Europe does not, like the culex of the torrid zone, shun mountainous places. Giesecke suffered from these insects in Greenland, at Disco, in latitude 70°. They are found in Lapland in summer, at three or four hundred toises high, and at a temperature of 11 or 12°.

[†] Trifling modifications in the waters, or in the air, often appear to prevent the development of the mosquitos. Mr. Bowdich remarks that there are none at Coomassie, in the kingdom of the Ashantees, though the town is surrounded by marshes, and though the thermometer keeps up between seventeen and twenty-eight centesimal degrees, day and night.

People born in the country, whether whites, mulattoes, negroes, or Indians, all suffer from the sting of these insects. But as cold does not render the north of Europe uninhabitable, so the mosquitos do not prevent men from dwelling in the countries where they abound, provided that, by their situation and government, they afford resources for agriculture and industry. The inhabitants pass their lives in complaining of the insufferable torment of the mosquitos, yet, notwithstanding these continual complaints, they seek, and even with a sort of predilection, the commercial towns of Mompox, Santa Marta, and Rio de la Hacha. Such is the force of habit in evils which we suffer every hour of the day, that the three missions of San Borja, Atures, and Esmeralda, where, to make use of an hyperbolical expression of the monks, there are more mosquitos than air,[*] would no doubt become flourishing towns, if the Orinoco afforded planters the same advantages for the exchange of produce, as the Ohio and the Lower Mississippi.

It is a curious fact, that the whites born in the torrid zone may walk barefoot with impunity, in the same apartment where a European recently landed is exposed to the attack of the nigua or chegoe (Pulex penetrans). This animal, almost invisible to the eye, gets under the toe-nails, and there acquires the size of a small pea, by the quick increase of its eggs, which are placed in a bag under the belly of the insect. The nigua therefore distinguishes what the most delicate chemical analysis could not distinguish, the cellular membrane and blood of a European from those of a creole white. The mosquitos, on the contrary, attack equally the natives and the Europeans; but the effects of the sting are different in the two races of men. The same venomous liquid, deposited in the skin of a copper-coloured man of Indian race, and in that of a white man newly landed, causes no swelling in the former, while in the latter it produces hard blisters, greatly inflamed, and painful for several days; so

---

[*] Mas moscas que aire.

different is the action on the epidermis, according to the degree of irritability of the organs in different races and different individuals!

I shall here recite several facts, which prove that the Indians, and in general all the people of colour, at the moment of being stung, suffer like the whites, although perhaps with less intensity of pain. In the day-time, and even when labouring at the oar, the natives, in order to chase the insects, are continually giving one another smart slaps with the palm of the hand. They even strike themselves and their comrades mechanically during their sleep. The violence of their blows reminds one of the Persian tale of the bear that tried to kill with his paw the insects on the forehead of his sleeping master. Near Maypures we saw some young Indians seated in a circle and rubbing cruelly each others' backs with the bark of trees dried at the fire. Indian women were occupied, with a degree of patience of which the copper-coloured race alone are capable, in extracting, by means of a sharp bone, the little mass of coagulated blood that forms the centre of every sting, and gives the skin a speckled appearance. One of the most barbarous nations of the Orinoco, that of the Ottomacs, is acquainted with the use of mosquito-curtains (mosquiteros) woven from the fibres of the moriche palm-tree. At Higuerote, on the coast of Caracas, the copper-coloured people sleep buried in the sand. In the villages of the Rio Magdalena the Indians often invited us to stretch ourselves as they did on ox-skins, near the church, in the middle of the plaza grande, where they had assembled all the cows in the neighbourhood. The proximity of cattle gives some repose to man. The Indians of the Upper Orinoco and the Cassiquiare, seeing that M. Bonpland could not prepare his herbal, owing to the continual torment of the mosquitos, invited him to enter their ovens (hornitos). Thus they call little chambers, without doors or windows, into which they creep horizontally through a very low opening. When they have driven away the insects by means of a fire of wet brushwood, which emits a great deal of smoke, they close the opening of the oven. The absence of the mosquitos is purchased dearly enough by the excessive heat of the stagnated air, and the smoke of a torch of copal, which lights the oven during your stay in it. M. Bonpland, with courage and patience well worthy of praise, dried hundreds of plants, shut up in these hornitos of the Indians.

These precautions of the Indians sufficiently prove that, notwithstanding the different organization of the epidermis, the copper-coloured man, like the white man, suffers from the stings of insects; but the former seems to feel less pain, and the sting is not followed by those swellings which, during several weeks, heighten the irritability of the skin, and throw persons of a delicate constitution into that feverish state which always accompanies eruptive maladies. Whites born in equinoctial America, and Europeans who have long sojourned in the Missions, on the borders of forests and great rivers, suffer much more than the Indians, but infinitely less than Europeans newly arrived. It is not, therefore, as some travellers assert, the thickness of the skin that renders the sting more or less painful at the moment when it is received; nor is it owing to the particular organization of the integuments, that in the Indians the sting is followed by less of swelling and inflammatory symptoms; it is on the nervous irritability of the epidermis that the acuteness and duration of the pain depend. This irritability is augmented by very warm clothing, by the use of alcoholic liquors, by the habit of scratching the wounds, and lastly, (and this physiological observation is the result of my own experience,) that of baths repeated at too short intervals. In places where the absence of crocodiles permits people to enter a river, M. Bonpland and myself observed that the immoderate use of baths, while it moderated the pain of old stings of zancudos, rendered us more sensible to new stings. By bathing more than twice a day, the skin is brought into a state of nervous irritability, of which no idea can be formed in Europe. It would seem as if all feeling were carried toward the integuments.

As the mosquitos and gnats pass two-thirds of their lives in the water, it is not surprising that these noxious insects become less numerous in proportion as you recede from the banks of the great rivers which intersect the forests. They seem to prefer the spots where their metamorphosis took place, and where they go to deposit their eggs. In fact the wild Indians (Indios monteros) experience the greater difficulty in accustoming themselves to the life of the missions, as they suffer in the Christian establishments a torment which they scarcely know in their own inland dwellings. The natives at Maypures, Atures, and Esmeralda, have been seen fleeing to the woods, or, as they say, al monte, solely from the dread of mosquitos. Unfortunately, all the Missions of the Orinoco have been established too near the banks of the river. At Esmeralda the inhabitants assured us that if the village were situated in one of the five plains surrounding the high mountains of Duida and Maraguaca, they should breathe freely, and enjoy some repose. The great cloud of mosquitos (la nube de moscas) to use the expression of the monks, is suspended only over the Orinoco and its tributary streams, and is dissipated in proportion as you remove from the rivers. We should form a very inaccurate idea of Guiana and Brazil, were we to judge of that great forest four hundred leagues wide, lying between the sources of the Madeira and the Lower Orinoco, from the valleys of the rivers by which it is crossed.

I learned that the little insects of the family of the nemocerae migrate from time to time like the alouate monkeys, which live in society. In certain spots, at the commencement of the rainy season, different species appear, the sting of which has not yet been felt. We were informed at the Rio Magdalena, that at Simiti no other culex than the jejen was formerly known; and it was then possible to enjoy a tranquil night's rest, for the jejen is not a nocturnal insect. Since the year 1801, the great blue-winged gnat (Culex cyanopterus) has appeared in such numbers, that the poor inhabitants of Simiti know not how to procure an undisturbed sleep. In the marshy channels (esteros) of the isle of Baru, near

Carthagena, is found a little white fly called cafafi. It is scarcely visible to the naked eye, and causes very painful swellings. The toldos or cottons used for mosquito-curtains, are wetted to prevent the cafafi penetrating through the interstices left by the crossing threads. This insect, happily rare elsewhere, goes up in January, by the channel (dique) of Mahates, as far as Morales. When we went to this village in the month of May, we found there cimuliae and zancudos, but no jejens.

The insects most troublesome at Orinoco, or as the Creoles say, the most ferocious (los mas feroces), are those of the great cataracts of Esmeralda and Mandavaca. On the Rio Magdalena the Culex cyanopterus is dreaded, particularly at Mompox, Chiloa, and Tamalameca. At these places this insect is larger and stronger, and its legs blacker. It is difficult to avoid smiling on hearing the missionaries dispute about the size and voracity of the mosquitos at different parts of the same river. In a region the inhabitants of which are ignorant of all that is passing in the rest of the world, this is the favourite subject of conversation. "How I pity your situation!" said the missionary of the Raudales to the missionary of Cassiquiare, at our departure; "you are alone, like me, in this country of tigers and monkeys; with you fish is still more rare, and the heat more violent; but as for my mosquitos (mias moscas) I can boast that with one of mine I would beat three of yours."

This voracity of insects in certain spots, the fury with which they attack man,[*] the activity of the venom varying in the same species, are very remarkable facts; which find their analogy, however, in the classes of large animals. The crocodile of Angostura pursues men, while at Nueva Barcelona you may bathe tranquilly in the Rio Neveri amidst these carnivorous

---

[*] This voracity, this appetite for blood, seems surprising in little insects, that live on vegetable juices, and in a country almost entirely uninhabited. "What would these animals eat, if we did not pass this way?" say the Creoles, in going through countries where there are only crocodiles covered with a scaly skin, and hairy monkeys.

reptiles. The jaguars of Maturin, Cumanacoa, and the isthmus of Panama, are timid in comparison of those of the Upper Orinoco. The Indians well know that the monkeys of some valleys are easily tamed, while others of the same species, caught elsewhere, will rather die of hunger than submit to slavery.[*]

The common people in America have framed systems respecting the salubrity of climates and pathological phenomena, as well as the learned of Europe; and their systems, like ours, are diametrically opposed to each other, according to the provinces into which the New Continent is divided. At the Rio Magdalena the frequency of mosquitos is regarded as troublesome, but salutary. These animals, say the inhabitants, give us slight bleedings, and preserve us, in a country excessively hot, from the scarlet fever, and other inflammatory diseases. But at the Orinoco, the banks of which are very insalubrious, the sick blame the mosquitos for all their sufferings. It is unnecessary to refute the fallacy of the popular belief that the action of the mosquitos is salutary by its local bleedings. In Europe the inhabitants of marshy countries are not ignorant that the insects irritate the epidermis, and stimulate its functions by the venom which they deposit in the wounds they make. Far from diminishing the inflammatory state of the skin, the stings increase it.

The frequency of gnats and mosquitos characterises unhealthy climates only so far as the development and multiplication of these insects depend on the same causes that give rise to miasmata. These noxious animals love a fertile soil

---

[*] I might have added the example of the scorpion of Cumana, which it is very difficult to distinguish from that of the island of Trinidad, Jamaica, Carthagena, and Guayaquil; yet the former is not more to be feared than the Scorpio europaeus (of the south of France), while the latter produces consequences far more alarming than the Scorpio occitanus (of Spain and Barbary). At Carthagena and Guayaquil, the sting of the scorpion (alacran) instantly causes the loss of speech. Sometimes a singular torpor of the tongue is observed for fifteen or sixteen hours. The patient, when stung in the legs, stammers as if he had been struck with apoplexy.

covered with plants, stagnant waters, and a humid air never agitated by the wind; they prefer to an open country those shades, that softened day, that tempered degree of light, heat, and moisture which, while it favours the action of chemical affinities, accelerates the putrefaction of organised substances. May not the mosquitos themselves increase the insalubrity of the atmosphere? When we reflect that to the height of three or four toises a cubic foot of air is often peopled by a million of winged insects,[*] which contain a caustic and venomous liquid; when we recollect that several species of culex are 1.8 lines long from the head to the extremity of the corslet (without reckoning the legs); lastly, when we consider that in this swarm of mosquitos and gnats, diffused in the atmosphere like smoke, there is a great number of dead insects raised by the force of the ascending air, or by that of the lateral currents which are caused by the unequal heating of the soil, we are led to inquire whether the presence of so many animal substances in the air must not occasion particular miasmata. I think that these substances act on the atmosphere differently from sand and dust; but it will be prudent to affirm nothing positively on this subject. Chemistry has not yet unveiled the numerous mysteries of the insalubrity of the air; it has only taught us that we are ignorant of many things with which a few years ago we believed we were acquainted.

Daily experience appears in a certain degree to prove the fact that at the Orinoco, Cassiquiare, Rio Caura, and wherever the air is very unhealthy, the sting of the mosquito augments the disposition of the organs to receive the impression of miasmata. When you are exposed day and night, during whole months, to the torment of insects, the continual irritation of the skin causes febrile commotions; and, from the sympathy existing between the dermoid and the gastric systems, injures the functions of the stomach. Digestion first becomes difficult, the cutaneous inflammation excites profuse perspirations, an unquenchable

---

[*] It is sufficient to mention, that the cubic foot contains 2,985,984 cubic lines.

311

thirst succeeds, and, in persons of a feeble constitution, increasing impatience is succeeded by depression of mind, during which all the pathogenic causes act with increased violence. It is neither the dangers of navigating in small boats, the savage Indians, nor the serpents, crocodiles, or jaguars, that make Spaniards dread a voyage on the Orinoco; it is, as they say with simplicity, "el sudar y las moscas," (the perspiration and the flies). We have reason to believe that mankind, as they change the surface of the soil, will succeed in altering by degrees the constitution of the atmosphere. The insects will diminish when the old trees of the forest have disappeared; when, in those countries now desert, the rivers are seen bordered with cottages, and the plains covered with pastures and harvests.

Whoever has lived long in countries infested by mosquitos will be convinced, as we were, that there exists no remedy for the torment of these insects. The Indians, covered with anoto, bolar earth, or turtle oil, are not protected from their attacks. It is doubtful whether the painting even relieves: it certainly does not prevent the evil. Europeans, recently arrived at the Orinoco, the Rio Magdalena, the river Guayaquil, or Rio Chagres (I mention the four rivers where the insects are most to be dreaded) at first obtain some relief by covering their faces and hands, but they soon feel it difficult to endure the heat, are weary of being condemned to complete inactivity, and finish with leaving the face and hands uncovered. Persons who would renounce all kind of occupation during the navigation of these rivers, might bring some particular garment from Europe in the form of a bag, under which they could remain covered, opening it only every half-hour. This bag should be distended by whalebone hoops, for a close mask and gloves would be perfectly insupportable. Sleeping on the ground, on skins, or in hammocks, we could not make use of mosquito-curtains (toldos) while on the Orinoco. The toldo is useful only where it forms a tent so well closed around the bed that there is not the smallest opening by which a gnat can pass. This is difficult to accomplish; and often when you succeed (for instance, in going up the Rio Magdalena, where you travel with some degree of convenience), you are forced, in order to avoid being suffocated by the heat, to come out from beneath your toldo, and walk about in the open air. A feeble wind, smoke, and powerful smells, scarcely afford any relief in places where the insects are very numerous and very voracious. It is erroneously affirmed that these little animals fly from the peculiar smell emitted by the crocodile. We were fearfully stung at Bataillez, in the road from Carthagena to Honda, while we were dissecting a crocodile eleven feet long, the smell of which infested all the surrounding atmosphere. The Indians much commend the fumes of burnt cow-dung. When the wind is very strong, and accompanied by rain, the mosquitos disappear for some time: they sting most cruelly at the approach of a storm, particularly when the electric explosions are not followed by heavy showers.

Anything waved about the head and the hands contributes to chase away the insects. "The more you stir yourself, the less you will be stung," say the missionaries. The zancudo makes a buzzing before it settles; but, when it has assumed confidence, when it has once begun to fix its sucker, and distend itself, you may touch its wings without its being frightened. It remains the whole time with its two hind legs raised; and, if left to suck to satiety, no swelling takes place, and no pain is left behind. We often repeated this experiment on ourselves in the valley of the Rio Magdalena. It may be asked whether the insect deposits the stimulating liquid only at the moment of its flight, when it is driven away, or whether it draws the liquid up again when left to suck undisturbed. I incline to this latter opinion; for on quietly presenting the back of my hand to the Culex cyanopterus, I observed that the pain, though violent in the beginning, diminishes in proportion as the insect continues to suck, and ceases altogether when it voluntarily flies away. I also wounded my skin with a pin, and rubbed the pricks with bruised mosquitos, and no swelling ensued. The irritating liquid, in which chemists have not yet recognized any acid properties, is contained, as in the ant and other hymenopterous insects, in

313

particular glands; and is probably too much diluted, and consequently too much weakened, if the skin be rubbed with the whole of the bruised insect.

I have thrown together at the close of this chapter all we learned during the course of our travels on phenomena which naturalists have hitherto singularly neglected, though they exercise a great influence on the welfare of the inhabitants, the salubrity of the climate, and the establishment of new colonies on the rivers of equinoctial America. I might justly have incurred the charge of having treated this subject too much in detail, were it not connected with general physiological views. Our imagination is struck only by what is great; but the lover of natural philosophy should reflect equally on little things. We have just seen that winged insects, collected in society, and concealing in their sucker a liquid that irritates the skin, are capable of rendering vast countries almost uninhabitable. Other insects equally small, the termites (comejen),[*] create obstacles to the progress of civilization, in several hot and temperate parts of the equinoctial zone, that are difficult to be surmounted. They devour paper, pasteboard, and parchment with frightful rapidity, utterly destroying records and libraries. Whole provinces of Spanish America do not possess one written document that dates a hundred years back. What improvement can the civilization of nations acquire if nothing link the present with the past; if the depositories of human knowledge must be repeatedly renewed; if the records of genius and reason cannot be transmitted to posterity?

In proportion as you ascend the table-land of the Andes these evils disappear. Man breathes a fresh and pure air. Insects no more disturb the labours of the day or the slumbers of the night. Documents can be collected in archives without our having to complain of the voracity of the termites. Mosquitos are no longer feared at a height of two hundred toises; and the termites, still

---

[*] Literally, the eaters or the devourers.

very frequent at three hundred toises of elevation,[*] become very rare at Mexico, Santa Fe de Bogota, and Quito. In these great capitals, situated on the back of the Cordilleras, we find libraries and archives, augmented from day to day by the enlightened zeal of the inhabitants. These circumstances, combined with others, insure a moral preponderance to the Alpine region over the lower regions of the torrid zone. If we admit, agreeably to the ancient traditions collected in both the old and new worlds, that at the time of the catastrophe which preceded the renewal of our species, man descended from the mountains into the plains, we may admit, with still greater confidence, that these mountains, the cradle of so many various nations, will for ever remain the centre of human civilization in the torrid zone. From these fertile and temperate table-lands, from these islets scattered in the aerial ocean, knowledge and the blessings of social institutions will be spread over those vast forests extending along the foot of the Andes, now inhabited only by savage tribes whom the very wealth of nature has retained in indolence and barbarism.

---

[*] There are some at Popayan (height 910 toises; mean temperature 18.7°), but they are species that gnaw wood only.

CHAPTER 2.21.

RAUDAL OF GARCITA. MAYPURES. CATARACTS OF QUITUNA. MOUTH OF THE VICHADA AND THE ZAMA. ROCK OF ARICAGUA. SIQUITA.

We directed our course to the Puerto de arriba, above the cataract of Atures, opposite the mouth of the Rio Cataniapo, where our boat was to be ready for us. In the narrow path that leads to the embarcadero we beheld for the last time the peak of Uniana. It appeared like a cloud rising above the horizon of the plains. The Guahibos wander at the foot of the mountains, and extend their course as far as the banks of the Vichada. We were shown at a distance, on the right of the river, the rocks that surround the cavern of Ataruipe; but we had not time to visit that cemetery of the destroyed tribe of the Atures. Father Zea had repeatedly described to us this extraordinary cavern, the skeletons painted with anoto, the large vases of baked earth, in which the bones of separate families appear to be collected; and many other curious objects, which we proposed to examine on our return from the Rio Negro. "You will scarcely believe," said the missionaries, "that these skeletons, these painted vases, things which we believed were unknown to the rest of the world, have brought trouble upon me and my neighbour, the missionary of Carichana. You have seen the misery in which I live in the raudales. Though devoured by mosquitos, and often in want of plantains and cassava, yet I have found envious people even in this country! A white man, who inhabits the pastures between the Meta and the Apure, denounced me recently in the Audencia of Caracas, as concealing a treasure I had discovered, jointly with the missionary of Carichana, amid the tombs of the Indians. It is asserted that the Jesuits of Santa Fe de Bogota were apprised beforehand of the destruction of their company; and that, in order to save the riches they possessed in money and precious vases, they sent them, either by the Rio Meta or the Vichada, to the Orinoco, with orders to have them hidden in the islets amid the raudales. These treasures I am supposed to have appropriated unknown to my superiors. The Audencia of Caracas brought a complaint before the governor of Guiana, and we were ordered to appear in person. We uselessly performed a journey of one hundred and fifty leagues; and, although we declared that we had found in the cavern only human bones, and dried bats and polecats, commissioners were gravely nominated to come hither and search on the spot for the supposed treasures of the Jesuits. We shall wait long for these commissioners. When they have gone up the Orinoco as far as San Borja, the fear of the mosquitos will prevent them from going farther. The cloud of flies which envelopes us in the raudales is a good defence."

The account given by the missionary was entirely conformable to what we afterwards learned at Angostura from the governor himself. Fortuitous circumstances had given rise to the strangest suspicions. In the caverns where the mummies and skeletons of the nation of the Atures are found, even in the midst of the cataracts, and in the most inaccessible islets, the Indians long ago discovered boxes bound with iron, containing various European tools, remnants of clothes, rosaries, and glass trinkets. These objects are thought to have belonged to Portuguese traders of the Rio Negro and Grand Para, who, before the establishment of the Jesuits on the banks of the Orinoco, went up to Atures by the portages and interior communications of rivers, to trade with the natives. It is supposed that these men sunk beneath the epidemic maladies so common in the raudales, and that their chests became the property of the Indians, the wealthiest of whom were usually buried with all they possessed most valuable during their lives. From these very uncertain traditions the tale of hidden treasures has been fabricated. As in the Andes of Quito every ruined building, not excepting the foundations of the pyramids erected by the French savans for the measurement of the meridian, is regarded as Inga pilca,* that is, the work of the Inca; so on the Orinoco every hidden treasure can belong only to the Jesuits, an order which, no doubt, governed the missions better than the Capuchins and the monks of the Observance, but whose riches and success in the civilization of the Indians have been much exaggerated. When the Jesuits of Santa Fe were arrested, those heaps of piastres, those emeralds of Muzo, those bars of gold of Choco, which the enemies of the company supposed they possessed, were not found in their dwellings. I can cite a respectable testimony, which proves incontestibly, that the viceroy of New Granada had not warned the Jesuits of Santa Fe of the danger with which they were menaced. Don Vicente Orosco, an engineer officer in the Spanish army, related to me that, being arrived at Angostura, with Don Manuel Centurion, to arrest the missionaries of Carichana, he met an Indian boat that was going down the Rio Meta. The boat being manned with Indians who could speak none of the tongues of the country, gave rise to suspicions. After useless researches, a bottle was at length discovered, containing a letter, in which the Superior of the company residing at Santa Fe informed the missionaries of the Orinoco of the persecutions to which the Jesuits were exposed in New Grenada. This letter recommended no measure of precaution; it was short, without ambiguity, and respectful towards the government, whose orders were executed with useless and unreasonable severity.

Eight Indians of Atures had conducted our boat through the raudales, and seemed well satisfied with the slight recompence we gave them. They gain little by this employment; and in order to give a just idea of the poverty and want of commerce in the missions of the Orinoco, I shall observe that during three years,

* Pilca (properly in Quichua pirca), wall of the Inca.

with the exception of the boats sent annually to Angostura by the commander of San Carlos de Rio Negro, to fetch the pay of the soldiers, the missionary had seen but five canoes of the Upper Orinoco pass the cataract, which were bound for the harvest of turtles' eggs, and eight boats laden with merchandize.

About eleven on the morning of the 17th of April we reached our boat. Father Zea caused to be embarked, with our instruments, the small store of provisions he had been able to procure for the voyage, on which he was to accompany us; these provisions consisted of a few bunches of plantains, some cassava, and fowls. Leaving the embarcadero, we immediately passed the mouth of the Cataniapo, a small river, the banks of which are inhabited by the Macos, or Piaroas, who belong to the great family of the Salive nations.

Besides the Piaroas of Cataniapo, who pierce their ears, and wear as ear-ornaments the teeth of caymans and peccaries, three other tribes of Macos are known: one, on the Ventuari, above the Rio Mariata; the second, on the Padamo, north of the mountains of Maraguaca; and the third, near the Guaharibos, towards the sources of the Orinoco, above the Rio Gehette. This last tribe bears the name of Macos-Macos. I collected the following words from a young Maco of the banks of the Cataniapo, whom we met near the embarcadero, and who wore in his ears, instead of a tusk of the peccary, a large wooden cylinder.[*]

> Planta(in, *Paruru* , (in Tamanac, also *paruru).*
>
> Cassava, *Elente* , (in Maco, *cahig).*
>
> Maize, *Niarne.*,
>
> The sun, *Jama* , (in Salive, *mume-seke-cocco).*
>
> The moon, *Jama* , (in Salive, *vexio).*
>
> Water, *Ahia* , (in Salive, *cagua).*
>
> One, *Nianti.*

---

[*] This custom is observed among the Cabres, the Maypures, and the Pevas of the Amazon. These last, described by La Condamine, stretch their ears by weights of a considerable size.

319

Two, *Tajus.*

Three, *Percotahuja.*

Four, *Imontegroa.*

The young man could not reckon as far as five, which certainly is no proof that the word five does not exist in the Maco tongue. I know not whether this tongue be a dialect of the Salive, as is pretty generally asserted; for idioms derived from one another, sometimes furnish words utterly different for the most common and most important things.[*] But in discussions on mother-tongues and derivative languages, it is not the sounds, the roots only, that are decisive; but rather the interior structure and grammatical forms. In the American idioms, which are notwithstanding rich, the moon is commonly enough called the sun of night or even the sun of sleep; but the moon and sun very rarely bear the same name, as among the Macos. I know only a few examples in the most northerly part of America, among the Woccons, the Ojibbeways, the Muskogulges, and the Mohawks.[†] Our missionary asserted that jama, in Maco, indicated at the same time the Supreme Being, and the great orbs of night and day; while many other American tongues, for instance the Tamanac, and the Caribbee, have distinct words to denote God, the Moon, and the Sun. We shall soon see how anxious the missionaries of the Orinoco are not to employ, in their translations of the prayers of the church, the native words which denote the Divinity, the Creator (Amanene), the Great Spirit who animates all nature. They choose rather to Indianize the Spanish word Dios, converting it, according to the differences of pronunciation, and the genius of the different dialects, into Dioso, Tiosu, or Piosu.

When we again embarked on the Orinoco, we found the river free from shoals. After a few hours we passed the Raudal of

---

[*] The great family of the Esthonian (or Tschoudi) languages, and of the Samoiede languages, affords numerous examples of these differences.

[†] Nipia-kisathwa in the Shawanese (the idiom of Canada), from nippi, to sleep, and kisathwa, the sun.

Garcita, the rapids of which are easy of ascent, when the waters are high. To the eastward is seen a small chain of mountains called the chain of Cumadaminari, consisting of gneiss, and not of stratified granite. We were struck with a succession of great holes at more than one hundred and eighty feet above the present level of the Orinoco, yet which, notwithstanding, appear to be the effects of the erosion of the waters. We shall see hereafter, that this phenomenon occurs again nearly at the same height, both in the rocks that border the cataracts of Maypures, and fifty leagues to the east, near the mouth of the Rio Jao. We slept in the open air, on the left bank of the river, below the island of Tomo. The night was beautiful and serene, but the torment of the mosquitos was so great near the ground, that I could not succeed in levelling the artificial horizon; consequently I lost the opportunity of making an observation.

On the 18th we set out at three in the morning, to be more sure of arriving before the close of the day at the cataract known by the name of the Raudal de los Guahibos. We stopped at the mouth of the Rio Tomo. The Indians went on shore, to prepare their food, and take some repose. When we reached the foot of the raudal, it was near five in the afternoon. It was extremely difficult to go up the current against a mass of water, precipitated from a bank of gneiss several feet high. An Indian threw himself into the water, to reach, by swimming, the rock that divides the cataract into two parts. A rope was fastened to the point of this rock, and when the canoe was hauled near enough, our instruments, our dry plants, and the provision we had collected at Atures, were landed in the raudal itself. We remarked with surprise, that the natural damn over which the river is precipitated, presents a dry space of considerable extent; where we stopped to see the boat go up.

The rock of gneiss exhibits circular holes, the largest of which are four feet deep, and eighteen inches wide. These funnels contain quartz pebbles, and appear to have been formed by the friction of masses rolled along by the impulse of the waters. Our situation, in the midst of the cataract, was singular enough, but unattended by the smallest danger. The missionary, who accompanied us, had his fever-fit on him. In order to quench the thirst by which he was tormented, the idea suggested itself to us of preparing a refreshing beverage for him in one of the excavations of the rock. We had taken on board at Atures an Indian basket called a mapire, filled with sugar, limes, and those grenadillas, or fruits of the passion-flower, to which the Spaniards give the name of parchas. As we were absolutely destitute of large vessels for holding and mixing liquids, we poured the water of the river, by means of a calabash, into one of the holes of the rock: to this we added sugar and lime-juice. In a few minutes we had an excellent beverage, which is almost a refinement of luxury, in that wild spot; but our wants rendered us every day more and more ingenious.

After an hour of expectation, we saw the boat arrive above the raudal, and we were soon ready to depart. After quitting the rock, our passage was not exempt from danger. The river is eight hundred toises broad, and must be crossed obliquely, above the cataract, at the point where the waters, impelled by the slope of their bed, rush with extreme violence toward the ledge from which they are precipitated. We were overtaken by a storm, accompanied happily by no wind, but the rain fell in torrents. After rowing for twenty minutes, the pilot declared that, far from gaining upon the current, we were again approaching the raudal. These moments of uncertainty appeared to us very long: the Indians spoke only in whispers, as they do always when they think their situation perilous. They redoubled their efforts, and we arrived at nightfall, without any accident, in the port of Maypures.

Storms within the tropics are as short as they are violent. The lightning had fallen twice near our boat, and had no doubt struck the surface of the water. I mention this phenomenon, because it is pretty generally believed in those countries that the clouds, the surface of which is charged with electricity, are at so great a height that the lightning reaches the ground more rarely than in Europe. The night was extremely dark, and we could not in less than two hours reach the village of Maypures. We were wet to the skin. In proportion as the rain ceased, the zancudos reappeared, with that voracity which tipulary insects always display immediately after a storm. My fellow-travellers were uncertain whether it would be best to stop in the port or proceed on our way on foot, in spite of the darkness of the night. Father Zea was determined to reach his home. He had given directions for the construction of a large house of two stories, which was to be begun by the Indians of the mission. "You will there find," said he gravely, "the same conveniences as in the open air; I have neither a bench nor a table, but you will not suffer so much from the flies, which are less troublesome in the mission than on the banks of the river." We followed the counsel if the missionary, who caused torches of copal to be lighted. These torches are tubes made of bark, three inches in diameter, and filled with copal resin. We walked at first over beds of rock, which were bare and slippery, and then we entered a thick grove of palm trees. We were twice obliged to pass a stream on trunks of trees hewn down. The torches had already ceased to give light. Being formed on a strange principle, the woody substance which resembles the wick surrounding the resin, they emit more smoke than light, and are easily extinguished. The Indian pilot, who expressed himself with some facility in Spanish, told us of snakes, water-serpents, and tigers, by which we might be attacked. Such conversations may be expected as matters of course, by persons who travel at night with the natives. By intimidating the European traveller, the Indians imagine they render themselves more necessary, and gain the confidence of the stranger. The rudest inhabitant of the missions fully understands the deceptions which everywhere arise from the relations between men of unequal fortune and civilization. Under the absolute and sometimes vexatious government of the monks, the Indian seeks to ameliorate his condition by those little artifices which are the weapons of physical and intellectual weakness.

Having arrived during the night at San Jose de Maypures we were forcibly struck by the solitude of the place; the Indians were plunged in profound sleep, and nothing was heard but the cries of nocturnal birds, and the distant sound of the cataract. In the calm of the night, amid the deep repose of nature, the monotonous sound of a fall of water has in it something sad and solemn. We remained three days at Maypures, a small village founded by Don Jose Solano at the time of the expedition of the boundaries, the situation of which is more picturesque, it might be said still more admirable, than that of Atures.

The raudal of Maypures, called by the Indians Quituna, is formed, as all cataracts are, by the resistance which the river encounters in its way across a ridge of rocks, or a chain of mountains. The lofty mountains of Cunavami and Calitamini, between the sources of the rivers Cataniapo and Ventuari, stretch toward the west in a chain of granitic hills. From this chain flow three small rivers, which embrace in some sort the cataract of Maypures. There are, on the eastern bank, the Sanariapo, and on the western, the Cameji and the Toparo. Opposite the village of Maypures, the mountains fall back in an arch, and, like a rocky coast, form a gulf open to the south-east. The irruption of the river is effected between the mouths of the Toparo and the Sanariapo, at the western extremity of this majestic amphitheatre.

The waters of the Orinoco now roll at the foot of the eastern chain of the mountains, and have receded from the west, where, in a deep valley, the ancient shore is easily recognized. A savannah, scarcely raised thirty feet above the mean level of the river, extends from this valley as far as the cataracts. There the small church of Maypures has been constructed. It is built of trunks of palm-trees, and is surrounded by seven or eight huts. The dry valley, which runs in a straight line from south to north, from the Cameji to the Toparo, is filled with granitic and solitary mounds, all resembling those found in the shape of islands and shoals in the present bed of the river. I was struck with this analogy of form, on comparing the rocks of Keri and Oco, situated in the descrted bed of the river, west of Maypures, with the islets of Ouivitari and Caminitamini, which rise like old castles amid the cataracts to the east of the mission. The geological aspect of these scenes, the insular form of the elevations farthest from the present shore of the Orinoco, the cavities which the waves appear to have hollowed in the rock Oco, and which are precisely on the same level (twenty-five or thirty toises high) as the excavations perceived opposite to them in the isle of Ouivitari; all these appearances prove that the whole of this bay, now dry, was formerly covered by water. Those waters probably formed a lake, the northern dike preventing their running out: but, when this dike was broken down, the savannah that surrounds the mission appeared at first like a very low island, bounded by two arms of the same river. It may be supposed that the Orinoco continued for some time to fill the ravine, which we shall call the valley of Keri, because it contains the rock of that name; and that the waters retired wholly toward the eastern chain, leaving dry the western arm of the river, only as they gradually diminished. Coloured stripes, which no doubt owe their black tint to the oxides of iron and manganese, seem to justify this conjecture. They are found on all the stones, far from the mission, and indicate the former abode of the waters. In going up the river, all merchandise is discharged at the confluence of the Rio Toparo and the Orinoco. The boats are entrusted to the natives, who have so perfect a knowledge of the raudal, that they have a particular name for every step. They conduct the boats as far as the mouth of the Cameji, where the danger is considered as past.

I will here describe the cataract of Quituna or Maypures as it appeared at the two periods when I examined it, in going down and up the river. It is formed, like that of Mapara or Atures, by an archipelago of islands, which, to the length of three thousand toises, fill the bed of the river, and by rocky dikes, which join the islands together. The most remarkable of these dikes, or natural dams, are Purimarimi, Manimi, and the Leap of the Sardine (Salto de la Sardina). I name them in the order in which I saw them in succession from south to north. The last of these three stages is near nine feet high, and forms by its breadth a magnificent cascade. I must here repeat, however, that the turbulent shock of the precipitated and broken waters depends not so much on the absolute height of each step or dike, as upon the multitude of counter-currents, the grouping of the islands and shoals, that lie at the foot of the raudalitos or partial cascades, and the contraction of the channels, which often do not leave a free navigable passage of twenty or thirty feet. The eastern part of the cataract of Maypures is much more dangerous than the western; and therefore the Indian pilots prefer the left bank of the river to conduct the boats down or up. Unfortunately, in the season of low waters, this bank remains partly dry, and recourse must be had to the process of portage; that is, the boats are obliged to be dragged on cylinders, or round logs.

To command a comprehensive view of these stupendous scenes, the spectator must be stationed on the little mountain of Manimi, a granitic ridge, which rises from the savannah, north of the church of the mission, and is itself only a continuation of the ridges of which the raudalito of Manimi is composed. We often visited this mountain, for we were never weary of gazing on this astonishing spectacle. From the summit of the rock is descried a sheet of foam, extending the length of a whole mile. Enormous masses of stone, black as iron, issue from its bosom. Some are paps grouped in pairs, like basaltic hills; others resemble towers, fortified castles, and ruined buildings. Their gloomy tint contrasts with the silvery splendour of the foam. Every rock, every islet is covered with vigorous trees, collected in clusters. At the foot of those paps, far as the eye can reach, a thick vapour is suspended over the river, and through this whitish fog the tops of the lofty palm-trees shoot up. What name shall we give to these majestic plants? I suppose them to be the vadgiai, a new species of the genus Oreodoxa, the trunk of which is more than eighty feet high. The feathery leaves of this palm-tree have a brilliant lustre, and rise almost straight toward the sky. At every hour of the day the sheet of foam displays different aspects. Sometimes the hilly islands and the palm-trees project their broad shadows; sometimes the rays of the setting sun are refracted in the cloud that hangs over the cataract, and coloured arcs are formed which vanish and appear alternately.

Such is the character of the landscape discovered from the top of the mountain of Manimi, which no traveller has yet described. I do not hesitate to repeat, that neither time, nor the view of the Cordilleras, nor any abode in the temperate valleys of Mexico, has effaced from my mind the powerful impression of the aspect of the cataracts. When I read a description of those places in India that are embellished by running waters and a vigorous vegetation, my imagination retraces a sea of foam and palm-trees, the tops of which rise above a stratum of vapour. The majestic scenes of nature, like the sublime works of poetry and the arts, leave remembrances that are incessantly awakening, and which, through the whole of life, mingle with all our feelings of what is grand and beautiful.

The calm of the atmosphere, and the tumultuous movement of the waters, produce a contrast peculiar to this zone. Here no breath of wind ever agitates the foliage, no cloud veils the splendour of the azure vault of heaven; a great mass of light is diffused in the air, on the earth strewn with plants with glossy leaves, and on the bed of the river, which extends as far as the eye can reach. This appearance surprises the traveller born in the north of Europe. The idea of wild scenery, of a torrent rushing from rock to rock, is linked in his imagination with that of a climate where the noise of the tempest is mingled with the sound of the cataract; and where, in a gloomy and misty day, sweeping clouds seem to descend into the valley, and to rest upon the tops of the pines. The landscape of the tropics in the low regions of the continents has a peculiar physiognomy, something of greatness and repose, which it preserves even where one of the elements is struggling with invincible obstacles. Near the equator, hurricanes and tempests belong to islands only, to deserts destitute of plants, and to those spots where parts of the atmosphere repose upon surfaces from which the radiation of heat is very unequal.

The mountain of Manimi forms the eastern limit of a plain which furnishes for the history of vegetation, that is, for its progressive development in bare and desert places, the same phenomena which we have described above in speaking of the raudal of Atures. During the rainy season, the waters heap vegetable earth upon the granitic rock, the bare shelves of which extend horizontally. These islands of mould, decorated with beautiful and odoriferous plants, resemble the blocks of granite covered with flowers, which the inhabitants of the Alps call gardens or courtils, and which pierce the glaciers of Switzerland.

In a place where we had bathed the day before, at the foot of the rock of Manimi, the Indians killed a serpent seven feet and a half long. The Macos called it a camudu. Its back displayed, upon a yellow ground, transverse bands, partly black, and partly inclining to a brown green: under the belly the bands were blue, and united in rhombic spots. This animal, which is not venomous, is said by the natives to attain more than fifteen feet in length. I thought at first, that the camudu was a boa; but I saw with surprise, that the scales beneath the tail were divided into two rows. It was therefore a viper (coluber); perhaps a python of the New Continent: I say perhaps, for great naturalists appear to admit that all the pythons belong to the Old, and all the boas to the New World. As the boa of Pliny was a serpent of Africa and of the south of Europe, it would have been well if the boas of America had been named pythons, and the pythons of India been called boas. The first notions of an enormous reptile capable of seizing man, and even the great quadrupeds, came to us from India and the coast of Guinea. However indifferent names may be, we can scarcely admit the idea, that the hemisphere in which Virgil described the agonies of Laocoon (a fable which the

Greeks of Asia borrowed from much more southern nations) does not possess the boa-constrictor. I will not augment the confusion of zoological nomenclature by proposing new changes, and shall confine myself to observing that at least the missionaries and the latinized Indians of the missions, if not the planters of Guiana, clearly distinguish the traga-venados (real boas, with simple anal plates) from the culebras de agua, or water-snakes, like the camudu (pythons with double anal scales). The traga-venados have no transverse bands on the back, but a chain of rhombic or hexagonal spots. Some species prefer the driest places; others love the water, as the pythons, or culebras de agua.

Advancing towards the west, we find the hills or islets in the deserted branch of the Orinoco crowned with the same palm-trees that rise on the rocks of the cataracts. One of these hills, called Keri, is celebrated in the country on account of a white spot which shines from afar, and in which the natives profess to see the image of the full moon. I could not climb this steep rock, but I believe the white spot to be a large nodule of quartz, formed by the union of several of those veins so common in granites passing into gneiss. Opposite Keri, or the Rock of the Moon, on the twin mountain Ouivitari, which is an islet in the midst of the cataracts, the Indians point out with mysterious awe a similar white spot. It has the form of a disc; and they say this is the image of the sun (Camosi). Perhaps the geographical situation of these two objects has contributed to their having received these names. Keri is on the side of the setting, Camosi on that of the rising sun. Languages being the most ancient historical monuments of nations, some learned men have been singularly struck by the analogy between the American word camosi and camosch, which seems to have signified originally, the sun, in one of the Semitic dialects. This analogy has given rise to hypotheses which appear to me at least very problematical. The god of the Moabites, Chemosh, or Camosch, who has so wearied the patience of the learned; Apollo Chomens, cited by Strabo and by Ammianus Marcellinus;

329

Belphegor; Amun or Hamon; and Adonis: all, without doubt, represent the sun in the winter solstice; but what can we conclude from a solitary and fortuitous resemblance of sounds in languages that have nothing besides in common?

The Maypure tongue is still spoken at Atures, although the mission is inhabited only by Guahibos and Macos. At Maypures the Guareken and Pareni tongues only are now spoken. From the Rio Anaveni, which falls into the Orinoco north of Atures, as far as beyond Jao, and to the mouth of the Guaviare (between the fourth and sixth degrees of latitude), we everywhere find rivers, the termination of which, veni,[*] recalls to mind the extent to which the Maypure tongue heretofore prevailed. Veni, or weni, signifies water, or a river. The words camosi and keri, which we have just cited, are of the idiom of the Pareni Indians,[†] who, I think I have heard from the natives, lived originally on the banks of the Mataveni.[‡] The Abbe Gili considers the Pareni as a simple dialect of the Maypure. This question cannot be solved by a comparison of the roots merely. Being totally ignorant of the grammatical structure of the Pareni, I can raise but feeble doubts against the opinion of the Italian missionary. The Pareni is perhaps a mixture of two tongues that belong to different families; like the Maquiritari, which is composed of the Maypure and the Caribbee; or, to cite an example better known, the modern Persian, which is allied at the same time to the Sanscrit and to the Semitic tongues. The following are Pareni words, which I carefully compared with Maypure words.[§]

---

[*] Anaveni, Mataveni, Maraveni, etc.

[†] Or Parenas, who must not be confounded either with the Paravenes of the Rio Caura (Caulin page 69), or with the Parecas, whose language belongs to the great family of the Tamanac tongues. A young Indian of Maypures, who called himself a Paragini, answered my questions almost in the same words that M. Bonpland heard from a Pareni. I have indicated the differences in the table, see below.

[‡] South of the Rio Zama. We slept in the open air near the mouth of the Mataveni on the 28th day of May, in our return from the Rio Negro.

[§] The words of the Maypure language have been taken from the works of

TABLE OF PARENI AND MAYPURE WORDS COMPARED.

English Word	Pareni Word	Maypure Word
The sun	Camosi	Kie (Kiepurig).
The moon	Keri	Kejapi (Cagijapi).
A star	Ouipo	Urrupu.
The devil	Amethami	Vasuri.
Water	Oneui (ut)	Oueni.
Fire	Casi	Catti.
Lightning	Eno	Eno-ima.[*]
The head	Ossipo	Nuchibucu.[†]
The hair	Nomao.	
The eyes	Nopurizi	Nupuriki.
The nose	Nosivi	Nukirri.
The mouth	Nonoma	Nunumacu.
The teeth	Nasi	Nati.
The tongue	Notate	Nuare.
The ear	Notasine	Nuakini.
The cheek	Nocaco.	
The neck	Nono	Noinu.
The arm	Nocano	Nuana.
The hand	Nucavi	Nucapi.
The breast	Notoroni.	
The back	Notoli.	
The thigh	Nocazo.	
The nipples	Nocini.	
The foot	Nocizi	Nukii.
The toes	Nociziriani.	
The calf of the leg	Nocavua.	
A crocodile	Cazuiti	Amana.
A fish	Cimasi	Timaki.

Gili and Hervas. I collected the words placed between parentheses from a young Maco Indian, who understood the Maypure language.

[*] I am ignorant of what ima signifies in this compound word. Eno means in Maypure the sky and thunder. Ina signifies mother.

[†] The syllables no and nu, joined to the words that designate parts of the body, might have been suppressed; they answer to the possessive pronoun my.

331

English Word	Pareni Word	Maypure Word
Maize	Cana	Jomuki.
Plantain	Paratana (Teot)*	Arata.
Cacao	Cacavua†.	
Tobacco	Jema	Jema.
Pimento	(Pumake).	
Mimosa inga	(Caraba).	
Cecropia peltata	(Jocovi).	
Agaric	(Cajuli).	
Agaric	Puziana (Pagiana)	Papeta (Popetas).
Agaric	Sinapa (Achinafe)	Avanume (Avanome).
Agaric	Meteuba (Meuteufafa)	Apekiva (Pejiiveji).
Agaric	Puriana vacavi	(Jaliva).
Agaric	Puriana vacavi uschanite.	
Agaric	Puriassima vacavi	(Javiji).

This comparison seems to prove that the analogies observed in the roots of the Pareni and the Maypure tongues are not to be neglected; they are, however, scarcely more frequent than those that have been observed between the Maypure of the Upper Orinoco and the language of the Moxos, which is spoken on the banks of the Marmora, from 15 to 20° of south latitude. The Parenis have in their pronunciation the English th, or tsa of the

---

* We may be surprised to find the word teot denote the eminently nutritive substance that supplies the place of corn (the gift of a beneficent divinity), and on which the subsistence of man within the tropics depends. I may here mention, that the word Teo, or Teot, which in Aztec signifies God (Teotl, properly Teo, for tl is only a termination), is found in the language of the Betoi of the Rio Meta. The name of the moon, in this language so remarkable for the complication of its grammatical structure, is Teo-ro. The name of the sun is Teo-umasoi. The particle ro designates a woman, umasoi a man. Among the Betoi, the Maypures, and so many other nations of both continents, the moon is believed to be the wife of the sun. But what is this root Teo? It appears to me very doubtful, that Teo-ro should signify God-woman, for Memelu is the name of the All-powerful Being in the Betoi langnage.

† Has this word been introduced from a communication with Europeans? It is almost identical with the Mexican (Aztec) word cacava.

Arabians, as I clearly heard in the word Amethami (devil, evil spirit). I need not again notice the origin of the word camosi. Solitary resemblances of sounds are as little proof of communication between nations as the dissimilitude of a few roots furnishes evidence against the affiliation of the German from the Persian and the Greek. It is remarkable, however, that the names of the sun and moon are sometimes found to be identical in languages, the grammatical construction of which is entirely different; I may cite as examples the Guarany and the Omagua,[*] languages of nations formerly very powerful. It may be conceived that, with the worship of the stars and of the powers of nature, words which have a relation to these objects might pass from one idiom to another. I showed the constellation of the Southern Cross to a Pareni Indian, who covered the lantern while I was taking the circum-meridian heights of the stars; and he called it Bahumehi, a name which the caribe fish, or serra salme, also bears in Pareni. He was ignorant of the name of the belt of Orion; but a Poignave Indian,[†] who knew the constellations better, assured me that in his tongue the belt of Orion bore the name of Fuebot; he called the moon Zenquerot. These two words have a very peculiar character for words of American origin. As the names of the constellations may have been transmitted to immense distances from one nation to another, these Poignave words have fixed the attention of the learned, who have imagined they recognize the Phoenician and Moabite tongues in the word camosi of the Pareni. Fuebot and zenquerot seem to remind us of the Phoenician words mot (clay), ardod (oak-tree), ephod, etc. But what can we conclude from simple terminations which are most frequently foreign to the

---

[*] Sun and Moon, in Guarany, Quarasi and Jasi; in Omagua, Huarassi and Jase. I shall give, farther on, these same words in the principal languages of the old and new worlds. See note below.

[†] At the Orinoco the Puignaves, or Poignaves, are distinguished from the Guipunaves (Uipunavi). The latter, on account of their language, are considered as belonging to the Maypure and Cabre nations; yet water is called in Poignave, as well as in Maypure, oueni.

roots? In Hebrew the feminine plurals terminate also in oth. I noted entire phrases in Poignave; but the young man whom I interrogated spoke so quick that I could not seize the division of the words, and should have mixed them confusedly together had I attempted to write them down.[*]

The Mission near the raudal of Maypures was very considerable in the time of the Jesuits, when it reckoned six hundred inhabitants, among whom were several families of whites. Under the government of the Fathers of the Observance the population was reduced to less than sixty. It must be observed that in this part of South America cultivation has been diminishing for half a century, while beyond the forests, in the provinces near the sea, we find villages that contain from two or three thousand Indians. The inhabitants of Maypures are a mild, temperate people, and distinguished by great cleanliness. The savages of the Orinoco for the most part have not that inordinate fondness for strong liquors which prevails in North America. It is true that the Ottomacs, the Jaruros, the Achaguas, and the Caribs, are often intoxicated by the immoderate use of chiza and many other fermented liquors, which they know how to prepare with cassava, maize, and the saccharine fruit of the palm-tree; but travellers have as usual generalized what belongs only to the manners of some tribes. We were frequently unable to prevail upon the Guahibos, or the Maco-Piroas, to taste brandy while they were labouring for us, and seemed exhausted by fatigue. It will require a longer residence of Europeans in these countries to spread there the vices that are already common among the Indians on the coast. In the huts of the natives of Maypures we found an appearance of order and neatness, rarely met with in the houses of the missionaries.

---

[*] For a curious example of this, see the speech of Artabanes in Aristophanes (Acharn. act 1 scene 3) where a Greek has attempted to give a Persian oration. See also Gibbon's Roman Empire chapter 53 note 54, for a curious example of the way in which foreign languages have been disfigured when it has been attempted to represent them in a totally different tongue.

These natives cultivate plantains and cassava, but no maize. Cassava, made into thin cakes, is the bread of the country. Like the greater part of the Indians of the Orinoco, the inhabitants of Maypures have beverages which may be considered nourishing; one of these, much celebrated in that country, is furnished by a palm-tree which grows wild in the vicinity of the mission on the banks of the Auvana. This tree is the seje: I estimated the number of flowers on one cluster at forty-four thousand; and that of the fruit, of which the greater part fall without ripening, at eight thousand. The fruit is a small fleshy drupe. It is immersed for a few minutes in boiling water, to separate the kernel from the parenchymatous part of the sarcocarp, which has a sweet taste, and is pounded and bruised in a large vessel filled with water. The infusion yields a yellowish liquor, which tastes like milk of almonds. Sometimes papelon (unrefined sugar) is added. The missionary told us that the natives become visibly fatter during the two or three months in which they drink this seje, into which they dip their cakes of cassava. The piaches, or Indian jugglers, go into the forests, and sound the botuto (the sacred trumpet) under the seje palm-trees, to force the tree, they say, to yield an ample produce the following year. The people pay for this operation, as the Mongols, the Arabs, and nations still nearer to us, pay the chamans, the marabouts, and other classes of priests, to drive away the white ants and the locusts by mystic words or prayers, or to procure a cessation of continued rain, and invert the order of the seasons.

"I have a manufacture of pottery in my village," said Father Zea, when accompanying us on a visit to an Indian family, who were occupied in baking, by a fire of brushwood, in the open air, large earthen vessels, two feet and a half high. This branch of manufacture is peculiar to the various tribes of the great family of Maypures, and they appear to have followed it from time immemorial. In every part of the forests, far from any human habitation, on digging the earth, fragments of pottery and delf are found. The taste for this kind of manufacture seems to have been common heretofore to the natives of both North and South

America. To the north of Mexico, on the banks of the Rio Gila, among the ruins of an Aztec city; in the United States, near the tumuli of the Miamis; in Florida, and in every place where any traces of ancient civilization are found, the soil covers fragments of painted pottery; and the extreme resemblance of the ornaments they display is striking. Savage nations, and those civilized people[*] who are condemned by their political and religious institutions always to imitate themselves, strive, as if by instinct, to perpetuate the same forms, to preserve a peculiar type or style, and to follow the methods and processes which were employed by their ancestors. In North America, fragments of delf ware have been discovered in places where there exist lines of fortification, and the walls of towns constructed by some unknown nation, now entirely extinct. The paintings on these fragments have a great similitude to those which are executed in our days on earthenware by the natives of Louisiana and Florida. Thus too, the Indians of Maypures often painted before our eyes the same ornaments as those we had observed in the cavern of Ataruipe, on the vases containing human bones. They were grecques, meanders, and figures of crocodiles, of monkeys, and of a large quadruped which I could not recognize, though it had always the same squat form. I might hazard the hypothesis that it belongs to another country, and that the type had been brought thither in the great migration of the American nations from the north-west to the south and south-east; but I am rather inclined to believe that the figure is intended to represent a tapir, and that the deformed image of a native animal has become by degrees one of the types that has been preserved.

The Maypures execute with the greatest skill grecques, or ornaments formed by straight lines variously combined, similar to those that we find on the vases of Magna Grecia, on the Mexican edifices at Mitla, and in the works of so many nations

---

[*] The Hindoos, the Tibetians, the Chinese, the ancient Egyptians, the Aztecs, the Peruvians; with whom the tendency toward civilization in a body has prevented the free development of the faculties of individuals.

who, without communication with each other, find alike a sensible pleasure in the symmetric repetition of the same forms. Arabesques, meanders, and grecques, please our eyes, because the elements of which their series is composed, follow in rhythmic order. The eye finds in this order, in the periodical return of the same forms, what the ear distinguishes in the cadenced succession of sounds and concords. Can we then admit a doubt that the feeling of rhythm manifests itself in man at the first dawn of civilization, and in the rudest essays of poetry and song?

Among the natives of Maypures, the making of pottery is an occupation principally confined to the women. They purify the clay by repeated washings, form it into cylinders, and mould the largest vases with their hands. The American Indian is unacquainted with the potter's wheel, which was familiar to the nations of the east in the remotest antiquity. We may be surprised that the missionaries have not introduced this simple and useful machine among the natives of the Orinoco, yet we must recollect that three centuries have not sufficed to make it known among the Indians of the peninsula of Araya, opposite the port of Cumana. The colours used by the Maypures are the oxides of iron and manganese, and particularly the yellow and red ochres that are found in the hollows of sandstone. Sometimes the fecula of the Bignonia chica is employed, after the pottery has been exposed to a feeble fire. This painting is covered with a varnish of algarobo, which is the transparent resin of the Hymenaea courbaril. The large vessels in which the chiza is preserved are called ciamacu, the smallest bear the name of mucra, from which word the Spaniards of the coast have framed murcura. Not only the Maypures, but also the Guaypunaves, the Caribs, the Ottomacs, and even the Guamos, are distinguished at the Orinoco as makers of painted pottery, and this manufacture extended formerly towards the banks of the Amazon. Orellana was struck with the painted ornaments on the ware of the Omaguas, who in his time were a populous commercial nation.

337

The following facts throw some light on the history of American civilization. In the United States, west of the Allegheny mountains, particularly between the Ohio and the great lakes of Canada, on digging the earth, fragments of painted pottery, mingled with brass tools, are constantly found. This mixture may well surprise us in a country where, on the first arrival of Europeans, the natives were ignorant of the use of metals. In the forests of South America, which extend from the equator as far as the eighth degree of north latitude, from the foot of the Andes to the Atlantic, this painted pottery is discovered in the most desert places, but it is found accompanied by hatchets of jade and other hard stones, skilfully perforated. No metallic tools or ornaments have ever been discovered; though in the mountains on the shore, and at the back of the Cordilleras, the art of melting gold and copper, and of mixing the latter metal with tin to make cutting instruments, was known. How can we account for these contrasts between the temperate and the torrid zone? The Incas of Peru had pushed their conquests and their religious wars as far as the banks of the Napo and the Amazon, where their language extended over a small space of land; but the civilization of the Peruvians, of the inhabitants of Quito, and of the Muyscas of New Grenada, never appears to have had any sensible influence on the moral state of the nations of Guiana. It must be observed further, that in North America, between the Ohio, Miami, and the Lakes, an unknown people, whom systematic authors would make the descendants of the Toltecs and Aztecs, constructed walls of earth and sometimes of stone without mortar,[*] from ten to fifteen feet high, and seven or eight thousand feet long. These singular circumvallations sometimes enclosed a hundred and fifty acres of ground. In the plains of the Orinoco, as in those of Marietta, the Miami, and the Ohio, the centre of an ancient civilization is

---

[*] Of siliceous limestone, at Pique, on the Great Miami; of sandstone at Creek Point, ten leagues from Chillakothe, where the wall is fifteen hundred toises long.

found in the west on the back of the mountains; but the Orinoco, and the countries lying between that great river and the Amazon, appear never to have been inhabited by nations whose constructions have resisted the ravages of time. Though symbolical figures are found engraved on the hardest rocks, yet further south than eight degrees of latitude, no tumulus, no circumvallation, no dike of earth similar to those that exist farther north in the plains of Varinas and Canagua, has been found. Such is the contrast that may be observed between the eastern parts of North and South America, those parts which extend from the table-land of Cundinamarca[*] and the mountains of Cayenne towards the Atlantic, and those which stretch from the Andes of New Spain towards the Alleghenies. Nations advanced in civilization, of which we discover traces on the banks of lake Teguyo and in the Casas grandes of the Rio Gila, might have sent some tribes eastward into the open countries of the Missouri and the Ohio, where the climate differs little from that of New Mexico; but in South America, where the great flux of nations has continued from north to south, those who had long enjoyed the mild temperature of the back of the equinoctial Cordilleras no doubt dreaded a descent into burning plains bristled with forests, and inundated by the periodical swellings of rivers. It is easy to conceive how much the force of vegetation, and the nature of the soil and climate, within the torrid zone, embarrassed the natives in regard to migration in numerous bodies, prevented settlements requiring an extensive space, and perpetuated the misery and barbarism of solitary hordes.

The feeble civilization introduced in our days by the Spanish monks pursues a retrograde course. Father Gili relates that, at the time of the expedition to the boundaries, agriculture began to make some progress on the banks of the Orinoco; and that cattle, especially goats, had multiplied considerably at Maypures. We

---

[*] This is the ancient name of the empire of the Zaques, founded by Bochica or Idacanzas, the high priest of Iraca, in New Grenada.

found no goats, either in the mission or in any other village of the Orinoco; they had all been devoured by the tigers. The black and white breeds of pigs only, the latter of which are called French pigs (puercos franceses), because they are believed to have come from the Caribbee Islands, have resisted the pursuit of wild beasts. We saw with much pleasure guacamayas, or tame macaws, round the huts of the Indians, and flying to the fields like our pigeons. This bird is the largest and most majestic species of parrot with naked cheeks that we found in our travels. It is called in Marativitan, cahuei. Including the tail, it is two feet three inches long. We had observed it also on the banks of the Atabapo, the Temi, and the Rio Negro. The flesh of the cahuei, which is frequently eaten, is black and somewhat tough. These macaws, whose plumage glows with vivid tints of purple, blue, and yellow, are a great ornament to the Indian farm-yards; they do not yield in beauty to the peacock, the golden pheasant, the pauxi, or the alector. The practice of rearing parrots, birds of a family so different from the gallinaceous tribes, was remarked by Columbus. When he discovered America he saw macaws, or large parrots, which served as food to the natives of the Caribbee Islands, instead of fowls.

A majestic tree, more than sixty feet high, which the planters call fruta de burro, grows in the vicinity of the little village of Maypures. It is a new species of the unona, and has the stateliness of the Uvaria zeylanica of Aublet. Its branches are straight, and rise in a pyramid, nearly like the poplar of the Mississippi, erroneously called the Lombardy poplar. The tree is celebrated for its aromatic fruit, the infusion of which is a powerful febrifuge. The poor missionaries of the Orinoco, who are afflicted with tertian fevers during a great part of the year, seldom travel without a little bag filled with frutas de burro. I have already observed that between the tropics, the use of aromatics, for instance very strong coffee, the Croton cascarilla, or the pericarp of the Unona xylopioides, is generally preferred to that of the astringent bark of cinchona, or of Bonplandia trifolatia, which is the Angostura bark. The people of America have the most inveterate prejudice against the employment of different kinds of cinchona; and in the very countries where this valuable remedy grows, they try (to use their own phrase) to cut off the fever, by infusions of Scoparia dulcis, and hot lemonade prepared with sugar and the small wild lime, the rind of which is equally oily and aromatic.

The weather was unfavourable for astronomical observations. I obtained, however, on the 20th of April, a good series of corresponding altitudes of the sun, according to which the chronometer gave 70° 37' 33" for the longitude of the mission of Maypures; the latitude was found, by a star observed towards the north, to be 5° 13' 57"; and by a star observed towards the south, 5° 13' 7". The error of the most recent maps is half a degree of longitude and half a degree of latitude. It would be difficult to relate the trouble and torments which these nocturnal observations cost us. Nowhere is a denser cloud of mosquitos to be found. It formed, as it were, a particular stratum some feet above the ground, and it thickened as we brought lights to illumine our artificial horizon. The inhabitants of Maypures, for the most part, quit the village to sleep in the islets amid the cataracts, where the number of insects is less; others make a fire of brushwood in their huts, and suspend their hammocks in the midst of the smoke.

We spent two days and a half in the little village of Maypures, on the banks of the great Upper Cataract, and on the 21st April we embarked in the canoe we had obtained from the missionary of Carichana. It was much damaged by the shoals it had struck against, and the carelessness of the Indians; but still greater dangers awaited it. It was to be dragged over land, across an isthmus of thirty-six thousand feet; from the Rio Tuamini to the Rio Negro, to go up by the Cassiquiare to the Orinoco, and to repass the two raudales.

When the traveller has passed the Great Cataracts, he feels as if he were in a new world, and had overstepped the barriers which nature seems to have raised between the civilized

341

countries of the coast and the savage and unknown interior. Towards the east, in the bluish distance, we saw for the last time the high chain of the Cunavami mountains. Its long, horizontal ridge reminded us of the Mesa of the Brigantine, near Cumana; but it terminates by a truncated summit. The Peak of Calitamini (the name given to this summit) glows at sunset as with a reddish fire. This appearance is every day the same. No one ever approached this mountain, the height of which does not exceed six hundred toises. I believe this splendour, commonly reddish but sometimes silvery, to be a reflection produced by large plates of talc, or by gneiss passing into mica-slate. The whole of this country contains granitic rocks, on which here and there, in little plains, an argillaceous grit-stone immediately reposes, containing fragments of quartz and of brown iron-ore.

In going to the embarcadero, we caught on the trunk of a hevea[*] a new species of tree-frog, remarkable for its beautiful colours; it had a yellow belly, the back and head of a fine velvety purple, and a very narrow stripe of white from the point of the nose to the hinder extremities. This frog was two inches long, and allied to the Rana tinctoria, the blood of which, it is asserted, introduced into the skin of a parrot, in places where the feathers have been plucked out, occasions the growth of frizzled feathers of a yellow or red colour. The Indians showed us on the way, what is no doubt very curious in that country, traces of cartwheels in the rock. They spoke, as of an unknown animal, of those beasts with large horns, which, at the time of the expedition to the boundaries, drew the boats through the valley of Keri, from the Rio Toparo to the Rio Cameji, to avoid the cataracts, and save the trouble of unloading the merchandize. I believe these poor inhabitants of Maypures would now be as much astonished at the sight of an ox of the Spanish breed, as the Romans were at the sight of the Lucanian oxen, as they called the elephants of the army of Pyrrhus.

---

[*] One of those trees whose milk yields caoutchouc.

We embarked at Puerto de Arriba, and passed the Raudal de Cameji with some difficulty. This passage is reputed to be dangerous when the water is very high; but we found the surface of the river beyond the raudal as smooth as glass. We passed the night in a rocky island called Piedra Raton, which is three-quarters of a league long, and displays that singular aspect of rising vegetation, those clusters of shrubs, scattered over a bare and rocky soil, of which we have often spoken.

On the 22nd of April we departed an hour and a half before sunrise. The morning was humid but delicious; not a breath of wind was felt; for south of Atures and Maypures a perpetual calm prevails. On the banks of the Rio Negro and the Cassiquiare, at the foot of Cerro Duida, and at the mission of Santa Barbara, we never heard that rustling of the leaves which has such a peculiar charm in very hot climates. The windings of rivers, the shelter of mountains, the thickness of the forests, and the almost continual rains, at one or two degrees of latitude north of the equator, contribute no doubt to this phenomenon, which is peculiar to the missions of the Orinoco.

In that part of the valley of the Amazon which is south of the equator, but at the same distance from it, as the places just mentioned, a strong wind always rises two hours after mid-day. This wind blows constantly against the stream, and is felt only in the bed of the river. Below San Borja it is an easterly wind; at Tomependa I found it between north and north-north-east; it is still the same breeze, the wind of the rotation of the globe, but modified by slight local circumstances. By favour of this general breeze you may go up the Amazon under sail, from Grand Para as far as Tefe, a distance of seven hundred and fifty leagues. In the province of Jaen de Bracamoros, at the foot of the western declivity of the Cordilleras, this Atlantic breeze rises sometimes to a tempest.

It is highly probable that the great salubrity of the Amazon is owing to this constant breeze. In the stagnant air of the Upper Orinoco the chemical affinities act more powerfully, and more

343

deleterious miasmata are formed. The insalubrity of the climate would be the same on the woody banks of the Amazon, if that river, running like the Niger from west to east, did not follow in its immense length the same direction, which is that of the trade-winds. The valley of the Amazon is closed only at its western extremity, where it approaches the Cordilleras of the Andes. Towards the east, where the sea-breeze strikes the New Continent, the shore is raised but a few feet above the level of the Atlantic. The Upper Orinoco first runs from east to west, and then from north to south. Where its course is nearly parallel to that of the Amazon, a very hilly country (the group of the mountains of Parima and of Dutch and French Guiana) separates it from the Atlantic, and prevents the wind of rotation from reaching Esmeralda. This wind begins to be powerfully felt only from the confluence of the Apure, where the Lower Orinoco runs from west to east in a vast plain open towards the Atlantic, and therefore the climate of this part of the river is less noxious than that of the Upper Orinoco.

In order to add a third point of comparison, I may mention the valley of the Rio Magdalena, which, like the Amazon, has one direction only, but unfortunately, instead of being that of the breeze, it is from south to north. Situated in the region of the trade-winds, the Rio Magdalena has the stagnant air of the Upper Orinoco. From the canal of Mahates as far as Honda, particularly south of the town of Mompox, we never felt the wind blow but at the approach of the evening storms. When, on the contrary, you proceed up the river beyond Honda, you find the atmosphere often agitated. The strong winds that are ingulfed in the valley of Neiva are noted for their excessive heat. We may be at first surprised to perceive that the calm ceases as we approach the lofty mountains in the upper course of the river, but this astonishment ends when we recollect that the dry and burning winds of the Llanos de Neiva are the effect of descending currents. The columns of cold air rush from the top of the Nevados of Quindiu and of Guanacas into the valley, driving before them the lower strata of the atmosphere. Everywhere the unequal heating of the soil, and the proximity of mountains covered with perpetual snow, cause partial currents within the tropics, as well as in the temperate zone. The violent winds of Neiva are not the effect of a repercussion of the trade-winds; they rise where those winds cannot penetrate; and if the mountains of the Upper Orinoco, the tops of which are generally crowned with trees, were more elevated, they would produce the same impetuous movements in the atmosphere as we observe in the Cordilleras of Peru, of Abyssinia, and of Thibet. The intimate connection that exists between the direction of rivers, the height and disposition of the adjacent mountains, the movements of the atmosphere, and the salubrity of the climate, are subjects well worthy of attention. The study of the surface and the inequalities of the soil would indeed be irksome and useless were it not connected with more general considerations.

At the distance of six miles from the island of Piedra Raton we passed, first, on the east, the mouth of the Rio Sipapo, called Tipapu by the Indians; and then, on the west, the mouth of the Rio Vichada. Near the latter are some rocks covered by the water, that form a small cascade or raudalito. The Rio Sipapo, which Father Gili went up in 1757, and which he says is twice as broad as the Tiber, comes from a considerable chain of mountains, which in its southern part bears the name of the river, and joins the group of Calitamini and of Cunavami. Next to the Peak of Duida, which rises above the mission of Esmeralda, the Cerros of Sipapo appeared to me the most lofty of the whole Cordillera of Parima. They form an immense wall of rocks, shooting up abruptly from the plain, its craggy ridge of running from south-south-east to north-north-west. I believe these crags, these indentations, which equally occur in the sandstone of Montserrat in Catalonia,[*] are owing to blocks of granite heaped together. The Cerros de Sipapo wear a different aspect every hour of the day. At sunrise the thick vegetation with which these

---

[*] From them the name of Montserrat is derived, Monte Serrato signifying a mountain ridged or jagged like a saw.

mountains are clothed is tinged with that dark green inclining to brown, which is peculiar to a region where trees with coriaceous leaves prevail. Broad and strong shadows are projected on the neighbouring plain, and form a contrast with the vivid light diffused over the ground, in the air, and on the surface of the waters. But towards noon, when the sun reaches its zenith, these strong shadows gradually disappear, and the whole group is veiled by an aerial vapour of a much deeper azure than that of the lower regions of the celestial vault. These vapours, circulating around the rocky ridge, soften its outline, temper the effects of the light, and give the landscape that aspect of calmness and repose which in nature, as in the works of Claude Lorraine and Poussin, arises from the harmony of forms and colours.

Cruzero, the powerful chief of the Guaypunaves, long resided behind the mountains of Sipapo, after having quitted with his warlike horde the plains between the Rio Inirida and the Chamochiquini. The Indians told us that the forests which cover the Sipapo abound in the climbing plant called vehuco de maimure. This species of liana is celebrated among the Indians, and serves for making baskets and weaving mats. The forests of Sipapo are altogether unknown, and there the missionaries place the nation of the Rayas,* whose mouths are believed to be in their navels.

An old Indian, whom we met at Carichana, and who boasted of having often eaten human flesh, had seen these acephali "with his own eyes." These absurd fables are spread as far as the Llanos, where you are not always permitted to doubt the existence of the Raya Indians. In every zone intolerance

---

* Rays, on account of the pretended analogy with the fish of this name, the mouth of which seems as if forced downwards below the body. This singular legend has been spread far and wide over the earth. Shakespeare has described Othello as recounting marvellous tales:

"of cannibals that do each other eat: Of Anthropophagi, and men whose heads Do grow beneath their shoulders."

accompanies credulity; and it might be said that the fictions of ancient geographers had passed from one hemisphere to the other, did we not know that the most fantastic productions of the imagination, like the works of nature, furnish everywhere a certain analogy of aspect and of form.

We landed at the mouth of the Rio Vichada or Visata to examine the plants of that part of the country. The scenery is very singular. The forest is thin, and an innumerable quantity of small rocks rise from the plain. These form massy prisms, ruined pillars, and solitary towers fifteen or twenty feet high. Some are shaded by the trees of the forest, others have their summits crowned with palms. These rocks are of granite passing into gneiss. At the confluence of the Vichada the rocks of granite, and what is still more remarkable, the soil itself, are covered with moss and lichens. These latter resemble the Cladonia pyxidata and the Lichen rangiferinus, so common in the north of Europe. We could scarcely persuade ourselves that we were elevated less than one hundred toises above the level of the sea, in the fifth degree of latitude, in the centre of the torrid zone, which has so long been thought to be destitute of cryptogamous plants. The mean temperature of this shady and humid spot probably exceeds twenty-six degrees of the centigrade thermometer. Reflecting on the small quantity of rain which had hitherto fallen, we were surprised at the beautiful verdure of the forests. This peculiarity characterises the valley of the Upper Orinoco; on the coast of Caracas, and in the Llanos, the trees in winter (in the season called summer in South America, north of the equator) are stripped of their leaves, and the ground is covered only with yellow and withered grass. Between the solitary rocks just described arise some high plants of columnar cactus (Cactus septemangularis), a very rare appearance south of the cataracts of Atures and Maypures.

Amid this picturesque scene M. Bonpland was fortunate enough to find several specimens of Laurus cinnamomoides, a very aromatic species of cinnamon, known at the Orinoco by the

347

names of varimacu and of canelilla.* This valuable production is found also in the valley of the Rio Caura, as well as near Esmeralda, and eastward of the Great Cataracts. The Jesuit Francisco de Olmo appears to have been the first who discovered the canelilla, which he did in the country of the Piaroas, near the sources of the Cataniapo. The missionary Gili, who did not advance so far as the regions I am now describing, seems to confound the varimacu, or guarimacu, with the myristica, or nutmeg-tree of America. These barks and aromatic fruits, the cinnamon, the nutmeg, the Myrtus pimenta, and the Laurus pucheri, would have become important objects of trade, if Europe, at the period of the discovery of the New World, had not already been accustomed to the spices and aromatics of India. The cinnamon of the Orinoco, and that of the Andaquies missions, are, however, less aromatic than the cinnamon of Ceylon, and would still be so even if dried and prepared by similar processes.

Every hemisphere produces plants of a different species; and it is not by the diversity of climates that we can attempt to explain why equinoctial Africa has no laurels, and the New World no heaths; why calceolariae are found wild only in the southern hemisphere; why the birds of the East Indies glow with colours less splendid than those of the hot parts of America; finally, why the tiger is peculiar to Asia, and the ornithorynchus to Australia. In the vegetable as well as in the animal kingdom, the causes of the distribution of the species are among the mysteries which natural philosophy cannot solve. The attempts made to explain the distribution of various species on the globe by the sole influence of climate, take their date from a period when physical geography was still in its infancy; when, recurring incessantly to pretended contrasts between the two worlds, it was imagined that the whole of Africa and of America resembled the deserts of Egypt and the marshes of Cayenne. At

---

* The diminutive of the Spanish word canela, which signifies cinnamon.

present, when men judge of the state of things not from one type arbitrarily chosen, but from positive knowledge, it is ascertained that the two continents, in their immense extent, contain countries that are altogether analogous. There are regions of America as barren and burning as the interior of Africa. Those islands which produce the spices of India are scarcely remarkable for their dryness; and it is not on account of the humidity of the climate, as has been affirmed in recent works, that the New Continent is deprived of those fine species of lauriniae and myristicae, which are found united in one little corner of the earth in the archipelago of India. For some years past cinnamon has been cultivated with success in several parts of the New Continent; and a zone that produces the coumarouna, the vanilla, the pucheri, the pine-apple, the pimento, the balsam of tolu, the Myroxylon peruvianum, the croton, the citroma, the pejoa, the incienso of the Silla of Caracas, the quereme, the pancratium, and so many majestic liliaceous plants, cannot be considered as destitute of aromatics. Besides, a dry air favours the development of the aromatic or exciting properties, only in certain species of plants. The most inveterate poisons are produced in the most humid zone of America; and it is precisely under the influence of the long rains of the tropics that the American pimento (Capsicum baccatum), the fruit of which is often as caustic and fiery as Indian pepper, vegetates best. From all these considerations it follows, first, that the New Continent possesses spices, aromatics, and very active vegetable poisons, peculiar to itself, and differing specifically from those of the Old World; secondly, that the primitive distribution of species in the torrid zone cannot be explained by the influence of climate solely, or by the distribution of temperature, which we observe in the present state of our planet; but that this difference of climates leads us to perceive why a given type of organization develops itself more vigorously in such or such local circumstances. We can conceive that a small number of the families of plants, for instance the musaceae and the palms, cannot belong to very cold regions, on account of their internal

349

structure, and the importance of certain organs; but we cannot explain why no one of the family of the Melastomaceae vegetates north of the parallel of the thirtieth degree of latitude, or why no rose-tree belongs to the southern hemisphere. Analogy of climates is often found in the two continents, without identity of productions.

The Rio Vichada, which has a small raudal at its confluence with the Orinoco, appeared to me, next to the Meta and the Guaviare, to be the most considerable river coming from the west. During the last forty years no European has navigated the Vichada. I could learn nothing of its sources; they rise, I believe, with those of the Tomo, in the plains that extend to the south of Casimena. Fugitive Indians of Santa Rosalia de Cabapuna, a village situate on the banks of the Meta, have arrived even recently, by the Rio Vichada, at the cataract of Maypures; which sufficiently proves that the sources of this river are not very distant from the Meta. Father Gumilla has preserved the names of several German and Spanish Jesuits, who in 1734 fell victims to their zeal for religion, by the hands of the Caribs on the now desert banks of the Vichada.

Having passed the Cano Pirajavi on the east, and then a small river on the west, which issues, as the Indians say, from a lake called Nao, we rested for the night on the shore of the Orinoco, at the mouth of the Zama, a very considerable river, but as little known as the Vichada. Notwithstanding the black waters of the Zama, we suffered greatly from insects. The night was beautiful, without a breath of wind in the lower regions of the atmosphere, but towards two in the morning we saw thick clouds crossing the zenith rapidly from east to west. When, declining toward the horizon, they traversed the great nebulae of Sagittarius and the Ship, they appeared of a dark blue. The light of the nebulae is never more splendid than when they are in part covered by sweeping clouds. We observe the same phenomenon in Europe in the Milky Way, in the aurora borealis when it beams with a silvery light; and at the rising and setting of the sun in that part of the sky that is whitened[*] from causes which philosophers have not yet sufficiently explained.

The vast tract of country lying between the Meta, the Vichada, and the Guaviare, is altogether unknown a league from the banks; but it is believed to be inhabited by wild Indians of the tribe of Chiricoas, who fortunately build no boats. Formerly, when the Caribs, and their enemies the Cabres, traversed these regions with their little fleets of rafts and canoes, it would have been imprudent to have passed the night near the mouth of a river running from the west. The little settlements of the Europeans having now caused the independent Indians to retire from the banks of the Upper Orinoco, the solitude of these regions is such, that from Carichana to Javita, and from Esmeralda to San Fernando de Atabapo, during a course of one hundred and eighty leagues, we did not meet a single boat.

At the mouth of the Rio Zama we approach a class of rivers, that merits great attention. The Zama, the Mataveni, the Atabapo, the Tuamini, the Temi, and the Guainia, are aguas negras, that is, their waters, seen in a large body, appear brown like coffee, or of a greenish black. These waters, notwithstanding, are most beautiful, clear, and agreeable to the taste. I have observed above, that the crocodiles, and, if not the zancudos, at least the mosquitos, generally shun the black waters. The people assert too, that these waters do not colour the rocks; and that the white rivers have black borders, while the black rivers have white. In fact, the shores of the Guainia, known to Europeans by the name of the Rio Negro, frequently exhibit masses of quartz issuing from granite, and of a dazzling whiteness. The waters of the Mataveni, when examined in a glass, are pretty white; those of the Atabapo retain a slight tinge of yellowish-brown. When the least breath of wind agitates the surface of these black rivers they appear of a fine grass-green, like the lakes of Switzerland. In the shade, the Zama, the

---

[*] The dawn: in French aube (alba, albente coelo.)

Atabapo, and the Guainia, are as dark as coffee-grounds. These phenomena are so striking, that the Indians everywhere distinguish the waters by the terms black and white. The former have often served me for an artificial horizon; they reflect the image of the stars with admirable clearness.

The colour of the waters of springs, rivers, and lakes, ranks among those physical problems which it is difficult, if not impossible, to solve by direct experiments. The tints of reflected light are generally very different from the tints of transmitted light; particularly when the transmission takes place through a great portion of fluid. If there were no absorption of rays, the transmitted light would be of a colour corresponding with that of the reflected light; and in general we judge imperfectly of transmitted light, by filling with water a shallow glass with a narrow aperture. In a river, the colour of the reflected light comes to us always from the interior strata of the fluid, and not from the upper stratum.

Some celebrated naturalists, who have examined the purest waters of the glaciers, and those which flow from mountains covered with perpetual snow, where the earth is destitute of the relics of vegetation, have thought that the proper colour of water might be blue, or green. Nothing, in fact, proves, that water is by nature white; and we must always admit the presence of a colouring principle, when water viewed by reflection is coloured. In the rivers that contain a colouring principle, that principle is generally so little in quantity, that it eludes all chemical research. The tints of the ocean seem often to depend neither on the nature of the bottom, nor on the reflection of the sky on the clouds. Sir Humphrey Davy was of opinion that the tints of different seas may very likely be owing to different proportions of iodine.

On consulting the geographers of antiquity, we find that the Greeks had noticed the blue waters of Thermopylae, the red waters of Joppa, and the black waters of the hot-baths of Astyra, opposite Lesbos. Some rivers, the Rhone for instance, near

Geneva, have a decidedly blue colour. It is said, that the snow-waters of the Alps are sometimes of a dark emerald green. Several lakes of Savoy and of Peru have a brown colour approaching black. Most of these phenomena of coloration are observed in waters that are believed to be the purest; and it is rather from reasonings founded on analogy, than from any direct analysis, that we may throw any light on so uncertain a matter. In the vast system of rivers near the mouth of the Rio Zama, a fact which appears to me remarkable is, that the black waters are principally restricted to the equatorial regions. They begin about five degrees of north latitude; and abound thence to beyond the equator as far as about two degrees of south latitude. The mouth of the Rio Negro is indeed in the latitude of 3° 9'; but in this interval the black and white waters are so singularly mingled in the forests and the savannahs, that we know not to what cause the coloration must be attributed. The waters of the Cassiquiare, which fall into the Rio Negro, are as white as those of the Orinoco, from which it issues. Of two tributary streams of the Cassiquiare very near each other, the Siapa and the Pacimony, one is white, the other black.

When the Indians are interrogated respecting the causes of these strange colorations, they answer, as questions in natural philosophy or physiology are sometimes answered in Europe, by repeating the fact in other terms. If you address yourself to the missionaries, they reply, as if they had the most convincing proofs of the fact, that the waters are coloured by washing the roots of the sarsaparilla. The Smilaceae no doubt abound on the banks of the Rio Negro, the Pacimony, and the Cababury; their roots, macerated in the water, yield an extractive matter, that is brown, bitter, and mucilaginous; but how many tufts of smilax have we seen in places, where the waters were entirely white. In the marshy forest which we traversed, to convey our canoe from the Rio Tuamini to the Cano Pimichin and the Rio Negro, why, in the same soil, did we ford alternately rivulets of black and white water? Why did we find no river white near its springs, and black in the lower part of its course? I know not whether the

Rio Negro preserves its yellowish brown colour as far as its mouth, notwithstanding the great quantity of white water it receives from the Cassiquiare and the Rio Blanco.

Although, on account of the abundance of rain, vegetation is more vigorous close to the equator than eight or ten degrees north or south, it cannot be affirmed, that the rivers with black waters rise principally in the most shady and thickest forests. On the contrary, a great number of the aguas negras come from the open savannahs that extend from the Meta beyond the Guaviare towards the Caqueta. In a journey which I made with Senor Montufar from the port of Guayaquil to the Bodegas de Babaojo, at the period of the great inundations, I was struck by the analogy of colour displayed by the vast savannahs of the Invernadero del Garzal and of the Lagartero, as well as by the Rio Negro and the Atabapo. These savannahs, partly inundated during three months, are composed of paspalum, eriochloa, and several species of cyperaceae. We sailed on waters that were from four to five feet deep; their temperature was by day from 33 to 34° of the centigrade thermometer; they exhaled a strong smell of sulphuretted hydrogen, to which no doubt some rotten plants of arum and heliconia, that swam on the surface of the pools, contributed. The waters of the Lagartero were of a golden yellow by transmitted, and coffee-brown by reflected light. They are no doubt coloured by a carburet of hydrogen. An analogous phenomenon is observed in the dunghill-waters prepared by our gardeners, and in the waters that issue from bogs. May we not also admit, that it is a mixture of carbon and hydrogen, an extractive vegetable matter, that colours the black rivers, the Atabapo, the Zama, the Mataveni, and the Guainia? The frequency of the equatorial rains contributes no doubt to this coloration by filtration through a thick mass of grasses. I suggest these ideas only in the form of a doubt. The colouring principle seems to be in little abundance; for I observed that the waters of the Guainia or Rio Negro, when subjected to ebullition, do not become brown like other fluids charged with carburets of hydrogen.

It is also very remarkable, that this phenomenon of black waters, which might be supposed to belong only to the low regions of the torrid zone, is found also, though rarely, on the table-lands of the Andes. The town of Cuenca in the kingdom of Quito, is surrounded by three small rivers, the Machangara, the Rio del Matadero, and the Yanuncai; of which the two former are white, and the waters of the last are black (aguas negras). These waters, like those of the Atabapo, are of a coffee-colour by reflection, and pale yellow by transmission. They are very clear, and the inhabitants of Cuenca, who drink them in preference to any other, attribute their colour to the sarsaparilla, which it is said grows abundantly on the banks of the Rio Yanuncai.

We left the mouth of the Zama at five in the morning of the 23rd of April. The river continued to be skirted on both sides by a thick forest. The mountains on the east seemed gradually to retire farther back. We passed first the mouth of the Rio Mataveni, and afterward an islet of a very singular form; a square granitic rock that rises in the middle of the water. It is called by the missionaries El Castillito, or the Little Castle. Black bands seem to indicate, that the highest swellings of the Orinoco do not rise at this place above eight feet; and that the great swellings observed lower down are owing to the tributary streams which flow into it north of the raudales of Atures and Maypures. We passed the night on the right bank opposite the mouth of the Rio Siucurivapu, near a rock called Aricagua. During the night an innumerable quantity of bats issued from the clefts of the rock, and hovered around our hammocks.

On the 24th a violent rain obliged us early to return to our boat. We departed at two o'clock, after having lost some books, which we could not find in the darkness of the night, on the rock of Aricagua. The river runs straight from south to north; its banks are low, and shaded on both sides by thick forests. We passed the mouths of the Ucata, the Arapa, and the Caranaveni. About four in the afternoon we landed at the Conucos de Siquita,

355

the Indian plantations of the mission of San Fernando. The good people wished to detain us among them, but we continued to go up against the current, which ran at the rate of five feet a second, according to a measurement I made by observing the time that a floating body took to go down a given distance. We entered the mouth of the Guaviare on a dark night, passed the point where the Rio Atabapo joins the Guaviare, and arrived at the mission after midnight. We were lodged as usual at the Convent, that is, in the house of the missionary, who, though much surprised at our unexpected visit, nevertheless received us with the kindest hospitality.

NOTE.

If, in the philosophical study of the structure of languages, the analogy of a few roots acquires value only when they can be geographically connected together, neither is the want of resemblance in roots any very strong proof against the common origin of nations. In the different dialects of the Totonac language (that of one of the most ancient tribes of Mexico) the sun and the moon have names which custom has rendered entirely different. This difference is found among the Caribs between the language of men and women; a phenomenon that probably arises from the circumstance that, among prisoners, men were oftener put to death than women. Females introduced by degrees words of a foreign language into the Caribbee; and, as the girls followed the occupations of the women much more than the boys, a language was formed peculiar to the women. I shall record in this note the names of the sun and moon in a great number of American and Asiatic idioms, again reminding the reader of the uncertainty of all judgments founded merely on the comparison of solitary words.

TABLE OF NAMES OF THE SUN AND THE MOON.

IN THE NEW WORLD:

Language	Name of the Sun	Name of the Moon
Eastern Esquimaux (Greenland)	Ajut, kaumat, sakanach	Anningat, kaumei, tatcok.
Western Esquimaux (Kadjak)	Tschingugak, madschak	Igaluk, tangeik.
Ojibbeway	Kissis	Debicot.
Delaware	Natatane	Keyshocof.
Nootka	Opulszthl	Omulszthl.
Otomi	Hindi	Zana.
Aztec or Mexican	Tonatiuh	Meztli.
Cora	Taica	Maitsaca.
Huasteca	Aquicha	Aytz.
Muysca	Zuhe (sua)	Chia.
Yaruro	*ditto*	Goppe.
Caribbee and Tamanac	Veiou (hueiou)	Nouno (nonum).
Maypure	Kie	Kejapi.
Lule	Inni	Allit.
Vilela	Olo	Copi.
Moxo	Sachi	Cohe.
Chiquito	Suus	Copi.
Guarani	Quarasi	Jasi.
Tupi (Brasil)	Coaracy	Iacy.
Peruvian (Quichua)	Inti	Quilla.
Araucan (Chili)	Antu	Cuyen.

IN THE OLD WORLD:

Language	Name of the Sun	Name of the Moon
Mongol	Nara (naran)	Sara (saran).
Mantchou	Choun	Bia.
Tschaghatai	Koun	Ay.
Ossete (of Caucasus)	Khourr	Mai.
Tibetan	Niyma	Rdjawa.
Chinese	Jy	Yue.
Japanese	Fi	Tsouki.

357

Language	Name of the Sun	Name of the Moon
Sanscrit	Surya, aryama, mitra, aditya, arka, hamsa	Tschandra, tschandrama, soma, masi.
Persian	Chor, chorschid, afitab	Mah.
Zend	Houere.	
Pehlvi	Schemschia, zabzoba, kokma	Kokma.
Phoenician	Schemesih.	
Hebrew	Schemesch	Yarea.
Aramean or Chaldean	Schimscha	Yarha.
Syrian	Schemscho	Yarho.
Arabic	Schams	Kamar.
Ethiopian	Tzabay	Warha.

The American words are written according to the Spanish orthography. I would not change the orthography of the Nootka word onulszth, taken from Cook's Voyages, to show how much Volney's idea of introducing an uniform notation of sounds is worthy of attention, if not applied to the languages of the East written without vowels. In onulszth there are four signs for one single consonant. We have already seen that American nations, speaking languages of a very different structure, call the sun by the same name; that the moon is sometimes called sleeping sun, sun of night, light of night; and that sometimes the two orbs have the same denomination. These examples are taken from the Guarany, the Omagua, Shawanese, Miami, Maco, and Ojibbeway idioms. Thus in the Old World, the sun and moon are denoted in Arabic by niryn, the luminaries; thus, in Persian, the most common words, afitab and chorschid, are compounds. By the migration of tribes from Asia to America, and from America to Asia, a certain number of roots have passed from one language into others; and these roots have been transported, like the fragments of a shipwreck, far from the coast, into the islands. (Sun, in New England, kone; in Tschagatai, koun; in Yakout, kouini. Star, in Huastec, ot; in Mongol, oddon; in Aztec, citlal, citl; in Persian, sitareh. House, in Aztec, calli; in Wogoul, kualla or kolla. Water, in Aztec, atel (itels, a river, in Vilela); in Mongol, Tscheremiss, and Tschouvass, atl, atelch, etel, or idel. Stone, in Caribbee, tebou; in the Lesgian of Caucasus, teb; in Aztec, tepetl; in Turkish, tepe. Food, in Quichua, micunnan; in Malay, macannon. Boat, in Haitian, canoa; in Ayno, cahani; in Greenlandish, kayak; in Turkish, kayik; in Samoyiede, kayouk; in the Germanic tongues, kahn.) But we must distinguish from these foreign elements what belongs fundamentally to the American idioms themselves. Such is the effect of time, and communication among nations, that the mixture with an heterogenous language has not only an influence upon roots, but most frequently ends by modifying and denaturalizing grammatical forms. "When a language resists a regular analysis," observes William von Humboldt, in his considerations on the Mexican, Cora, Totonac, and Tarahumar tongues, "we may suspect some mixture, some foreign influence; for the faculties of man, which are, as we may say, reflected in the structure of languages, and in their grammatical forms, act constantly in a regular and uniform manner."

CHAPTER 2.22.

SAN FERNANDO DE ATABAPO. SAN BALTHASAR. THE RIVERS TEMI AND TUAMINI. JAVITA. PORTAGE FROM THE TUAMINI TO THE RIO NEGRO.

During the night, we had left, almost unperceived, the waters of the Orinoco; and at sunrise found ourselves as if transported to a new country, on the banks of a river the name of which we had scarcely ever heard pronounced, and which was to conduct us, by the portage of Pimichin, to the Rio Negro, on the frontiers of Brazil. "You will go up," said the president of the missions, who resides at San Fernando, "first the Atabapo, then the Temi, and finally, the Tuamini. When the force of the current of black waters hinders you from advancing, you will be conducted out of the bed of the river through forests, which you will find inundated. Two monks only are settled in those desert places, between the Orinoco and the Rio Negro; but at Javita you will be furnished with the means of having your canoe drawn over land in the course of four days to Cano Pimichin. If it be not broken to pieces you will descend the Rio Negro without any obstacle (from north-west to south-east) as far as the little fort of San Carlos; you will go up the Cassiquiare (from south to north), and then return to San Fernando in a month, descending the Upper Orinoco from east to west." Such was the plan traced for our passage, and we carried it into effect without danger, though not without some suffering, in the space of thirty-three days. The Orinoco runs from its source, or at least from Esmeralda, as far as San Fernando de Atabapo, from east to west; from San Fernando, (where the junction of the Guaviare and the Atabapo takes place,) as far as the mouth of the Rio Apure, it flows from south to north, forming the Great Cataracts; and from the mouth of the Apure as far as Angostura and the coast of the Atlantic its direction is from west to east. In the first part of its course, where the river flows from east to west, it forms that celebrated bifurcation so often disputed by geographers, of which I was the first enabled to determine the situation by astronomical observations. One arm of the Orinoco, (the Cassiquiare,) running from north to south, falls into the Guainia, or Rio Negro, which, in its turn, joins the Maranon, or river Amazon. The most natural way, therefore, to go from Angostura to Grand Para, would be to ascend the Orinoco as far as Esmeralda, and then to go down the Cassiquiare, the Rio Negro, and the Amazon; but, as the Rio Negro in the upper part of its course approaches very near the sources of some rivers that fall into the Orinoco near San Fernando de Atabapo (where the Orinoco abruptly changes its direction from east to west to take that from south to north), the passage up that part of the river between San Fernando and Esmeralda, in order to reach the Rio Negro, may be avoided. Leaving the Orinoco near the mission of San Fernando, the traveller proceeds up the little black rivers (the Atabapo, the Temi, and the Tuamini), and the boats are carried across an isthmus six thousand toises broad, to the banks of a stream (the Cano Pimichin) which flows into the Rio Negro. This was the course which we took.

The road from San Carlos to San Fernando de Atabapo is far more disagreeable, and is half as long again by the Cassiquiare as by Javita and the Cano Pimichin. In this region I determined, by means of a chronometer by Berthoud, and by the meridional heights of stars, the situation of San Balthasar de Atabapo, Javita, San Carlos del Rio Negro, the rock Culimacavi, and Esmeralda. When no roads exist save tortuous and intertwining rivers, when little villages are hidden amid thick forests, and when, in a country entirely flat, no mountain, no elevated object is visible from two points at once, it is only in the sky that we can read where we are upon the earth.

361

San Fernando de Atabapo stands near the confluence of three great rivers; the Orinoco, the Guaviare, and the Atabapo. Its situation is similar to that of Saint Louis or of New Madrid, at the junction of the Mississippi with the Missouri and the Ohio. In proportion as the activity of commerce increases in these countries traversed by immense rivers, the towns situated at their confluence will necessarily become bustling ports, depots of merchandise, and centre points of civilization. Father Gumilla confesses, that in his time no person had any knowledge of the course of the Orinoco above the mouth of the Guaviare.

D'Anville, in the first edition of his great map of South America, laid down the Rio Negro as an arm of the Orinoco, that branched off from the principal body of the river between the mouths of the Meta and the Vichada, near the cataract of Atures. That great geographer was entirely ignorant of the existence of the Cassiquiare and the Atabapo; and he makes the Orinoco or Rio Paragua, the Japura, and the Putumayo, take their rise from three branchings of the Caqueta. The expedition of the boundaries, commanded by Iturriaga and Solano, corrected these errors. Solano, who was the geographical engineer of this expedition, advanced in 1756 as far as the mouth of the Guaviare, after having passed the Great Cataracts. He found that, to continue to go up the Orinoco, he must direct his course towards the east; and that the river received, at the point of its great inflection, in latitude 4° 4', the waters of the Guaviare, which two miles higher had received those of the Atabapo. Interested in approaching the Portuguese possessions as near as possible, Solano resolved to proceed onward to the south. At the confluence of the Atabapo and the Guaviare he found an Indian settlement of the warlike nation of the Guaypunaves. He gained their favour by presents, and with their aid founded the mission of San Fernando, to which he gave the appellation of villa, or town.

To make known the political importance of this Mission, we must recollect what was at that period the balance of power between the petty Indian tribes of Guiana. The banks of the Lower Orinoco had been long ensanguined by the obstinate struggle between two powerful nations, the Cabres and the Caribs. The latter, whose principal abode since the close of the seventeenth century has been between the sources of the Carony, the Essequibo, the Orinoco, and the Rio Parima, once not only held sway as far as the Great Cataracts, but made incursions also into the Upper Orinoco, employing portages between the Paruspa* and the Caura, the Erevato and the Ventuari, the Conorichite and the Atacavi. None knew better than the Caribs the intertwinings of the rivers, the proximity of the tributary streams, and the roads by which distances might be diminished. The Caribs had vanquished and almost exterminated the Cabres. Having made themselves masters of the Lower Orinoco, they met with resistance from the Guaypunaves, who had founded their dominion on the Upper Orinoco; and who, together with the Cabres, the Manitivitanos, and the Parenis, are the greatest cannibals of these countries. They originally inhabited the banks of the great river Inirida, at its confluence with the Chamochiquini, and the hilly country of Mabicore. About the year 1744, their chief, or as the natives call him, their king (apoto), was named Macapu. He was a man no less distinguished by his intelligence than his valour; had led a part of the nation to the banks of the Atabapo; and when the Jesuit Roman made his memorable expedition from the Orinoco to the Rio Negro, Macapu suffered that missionary to take with him some families of the Guaypunaves to settle them at Uruana, and near the

---

* The Rio Paruspa falls into the Rio Paragua, and the latter into the Rio Carony, which is one of the tributary streams of the Lower Orinoco. There is also an ancient portage of the Caribs between the Paruspa and the Rio Chavaro, which flows into the Rio Caura above the mouth of the Erevato. In going up the Erevato you reach the savannahs that are traversed by the Rio Manipiare above the tributary streams of the Ventuari. The Caribs in their distant excursions sometimes passed from the Rio Caura to the Ventuari, thence to the Padamo, and then by the Upper Orinoco to the Atacavi, which, westward of Manuteso, takes the name of the Atabapo.

cataract of Maypures. This people are connected by their language with the great branch of the Maypure nations. They are more industrious, we might also say more civilized, than the other nations of the Upper Orinoco. The missionaries relate, that the Guaypunaves, at the time of their sway in those countries, were generally clothed, and had considerable villages. After the death of Macapu, the command devolved on another warrior, Cuseru, called by the Spaniards El capitan Cusero. He established lines of defence on the banks of the Inirida, with a kind of little fort, constructed of earth and timber. The piles were more than sixteen feet high, and surrounded both the house of the apoto and a magazine of bows and arrows. These structures, remarkable in a country in other respects so wild, have been described by Father Forneri.

The Marepizanas and the Manitivitanos were the preponderant nations on the banks of the Rio Negro. The former had for its chiefs, about the year 1750, two warriors called Imu and Cajamu. The king of the Manitivitanos was Cocuy, famous for his cruelty. The chiefs of the Guaypunaves and the Manitivitanos fought with small bodies of two or three hundred men; but in their protracted struggles they destroyed the missions, in some of which the poor monks had only fifteen or twenty Spanish soldiers at their disposal. When the expedition of Iturriaga and Solano arrived at the Orinoco, the missions had no longer to fear the incursions of the Caribs. Cuseru, the chief of the Guaypunaves, had fixed his dwelling behind the granitic mountains of Sipapo. He was the friend of the Jesuits; but other nations of the Upper Orinoco and the Rio Negro, led by Imu, Cajamu, and Cocuy, penetrated from time to time to the north of the Great Cataracts. They had other motives for fighting than that of hatred; they hunted men, as was formerly the custom of the Caribs, and is still the practice in Africa. Sometimes they furnished slaves (poitos) to the Dutch (in their language, Paranaquiri—inhabitants of the sea); sometimes they sold them to the Portuguese (Iaranavi—sons of musicians).[*] In America, as in Africa, the cupidity of the Europeans has produced the same evils, by exciting the natives to make war, in order to procure slaves. Everywhere the contact of nations, widely different from each other in the scale of civilization, leads to the abuse of physical strength, and of intellectual preponderance. The Phoenicians and Carthaginians formerly sought slaves in Europe. Europe now presses in her turn both on the countries whence she gathered the first germs of science, and on those where she now almost involuntarily spreads them by carrying thither the produce of her industry.

I have faithfully recorded what I could collect on the state of these countries, where the vanquished nations have become gradually extinct, leaving no other signs of their existence than a few words of their language, mixed with that of the conquerors. In the north, beyond the cataracts, the preponderant nations were at first the Caribs and the Cabres; towards the south, on the Upper Orinoco, the Guaypunaves; and on the Rio Negro, the Marepizanos and the Manitivitanos. The long resistance which the Cabres, united under a valiant chief, had made to the Caribs, became fatal to the latter subsequently to the year 1720. They at first vanquished their enemies near the mouth of the Rio Caura; and a great number of Caribs perished in a precipitate flight, between the rapids of Torno and the Isla del Infierno. The prisoners were devoured; and, by one of those refinements of cunning and cruelty which are common to the savage nations of both North and South America, the Cabres spared the life of one Carib, whom they forced to climb up a tree to witness this barbarous spectacle, and carry back the tidings to the vanquished. The triumph of Tep, the chief of the Cabres, was but of short duration. The Caribs returned in such great numbers that

---

[*] The savage tribes designate every commercial nation of Europe by surnames, the origin of which appears altogether accidental. The Spaniards were called clothed men, Pongheme or Uavemi, by way of distinction.

only a feeble remnant of the Cabres was left on the banks of the Cuchivero.

Cocuy and Cuseru were carrying on a war of extermination on the Upper Orinoco when Solano arrived at the mouth of the Guaviare. The former had embraced the cause of the Portuguese; the latter was a friend of the Jesuits, and gave them warning whenever the Manitivitanos were marching against the christian establishments of Atures and Carichana. Cuseru became a christian only a few days before his death; but in battle he had for some time worn on his left hip a crucifix, given him by the missionaries, and which he believed rendered him invulnerable. We were told an anecdote that paints the violence of his character. He had married the daughter of an Indian chief of the Rio Temi. In a paroxysm of rage against his father-in-law, he declared to his wife that he was going to fight against him. She reminded him of the courage and singular strength of her father; when Cuseru, without uttering a single word, took a poisoned arrow, and plunged it into her bosom. The arrival of a small body of Spaniards in 1756, under the order of Solano, awakened suspicion in this chief of the Guaypunaves. He was on the point of attempting a contest with them, when the Jesuits made him sensible that it would be his interest to remain at peace with the Christians. Whilst dining at the table of the Spanish general, Cuseru was allured by promises, and the prediction of the approaching fall of his enemies. From being a king he became the mayor of a village; and consented to settle with his people at the new mission of San Fernando de Atabapo. Such is most frequently the end of those chiefs whom travellers and missionaries style Indian princes. "In my mission," says the honest father Gili "I had five reyecillos, or petty kings, those of the Tamanacs, the Avarigotes, the Parecas, the Quaquas, and the Maypures. At church I placed them in file on the same bench; but I took care to give the first place to Monaiti, king of the Tamanacs, because he had helped me to found the village; and he seemed quite proud of this precedency."

When Cuseru, the chief of the Guaypunaves, saw the Spanish troops pass the cataracts, he advised Don Jose Solano to wait a whole year before he formed a settlement on the Atabapo; predicting the misfortuncs which were not slow to arrive. "Let me labour with my people in clearing the ground," said Cuseru to the Jesuits; I will plant cassava, and you will find hereafter wherewith to feed all these men." Solano, impatient to advance, refused to listen to the counsel of the Indian chief, and the new inhabitants of San Fernando had to suffer all the evils of scarcity. Canoes were sent at a great expense to New Grenada, by the Meta and the Vichada, in search of flour. The provision arrived too late, and many Spaniards and Indians perished of those diseases which are produced in every climate by want and moral dejection.

Some traces of cultivation are still found at San Fernando. Every Indian has a small plantation of cacao-trees, which produce abundantly in the fifth year; but they cease to bear fruit sooner than in the valleys of Aragua. There are some savannahs and good pasturage round San Fernando, but hardly seven or eight cows are to be found, the remains of a considerable herd which was brought into these countries at the expedition for settling the boundaries. The Indians are a little more civilized here than in the rest of the missions, and we found to our surprise a blacksmith of the native race.

In the mission of San Fernando, a tree which gives a peculiar physiognomy to the landscape, is the piritu or pirijao palm. Its trunk, armed with thorns, is more than sixty feet high; its leaves are pinnated, very thin, undulated, and frizzled towards the points. The fruits of this tree are very extraordinary; every cluster contains from fifty to eighty; they are yellow like apples, grow purple in proportion as they ripen, two or three inches thick, and generally, from abortion, without a kernel. Among the eighty or ninety species of palm-trees peculiar to the New Continent, which I have enumerated in the Nova Genera Plantarum Aequinoctialum, there are none in which the sarcocarp is developed in a manner so extraordinary. The fruit of the pirijao furnishes a farinaceous substance, as yellow as the yolk of an egg, slightly saccharine, and extremely nutritious. It is eaten like plantains or potatoes, boiled or roasted in the ashes, and affords a wholesome and agreeable aliment. The Indians and the missionaries are unwearied in their praises of this noble palm-tree, which might be called the peach-palm. We found it cultivated in abundance at San Fernando, San Balthasar, Santa Barbara, and wherever we advanced towards the south or the east along the banks of the Atabapo and the Upper Orinoco. In those wild regions we are involuntarily reminded of the assertion of Linnaeus, that the country of palm-trees was the first abode of our species, and that man is essentially palmivorous.[*] On examining the provision accumulated in the huts of the Indians, we perceive that their subsistence during several months of the year depends as much on the farinaceous fruit of the pirijao, as on the cassava and plantain. The tree bears fruit but once a year, but to the amount of three clusters, consequently from one hundred and fifty to two hundred fruits.

San Fernando de Atabapo, San Carlos, and San Francisco Solano, are the most considerable settlements among the missions of the Upper Orinoco. At San Fernando, as well as in the neighbouring villages of San Balthasar and Javita, the abodes of the priests are neatly-built houses, covered by lianas, and surrounded by gardens. The tall trunks of the pirijao palms were the most beautiful ornaments of these plantations. In our walks, the president of the mission gave us an animated account of his incursions on the Rio Guaviare. He related to us how much these journeys, undertaken "for the conquest of souls;" are desired by the Indians of the missions. All, even women and old men, take

---

[*] Homo *habitat* intra tropicos, vescitur palmis, lotophagus; *hospitatur* extra tropicos sub novercante Cerere, carnivorus. Man *dwells naturally* within the tropics, and lives on the fruits of the palm-tree; he *exists* in other parts of the world, and there makes shift to feed on corn and flesh. Syst. Nat. volume 1 page 24.

part in them. Under the pretext of recovering neophytes who have deserted the village, children above eight or ten years of age are carried off, and distributed among the Indians of the missions as serfs, or poitos. According to the astronomical observations I took on the banks of the Atabapo, and on the western declivity of the Cordillera of the Andes, near the Paramo de la suma Paz, the distance is one hundred and seven leagues only from San Fernando to the first villages of the provinces of Caguan and San Juan de los Llanos. I was assured also by some Indians, who dwelt formerly to the west of the island of Amanaveni, beyond the confluence of the Rio Supavi, that going in a boat on the Guaviare (in the manner of the savages) beyond the strait (angostura) and the principal cataract, they met, at three days' distance, bearded and clothed men, who came in search of the eggs of the terekay turtle. This meeting alarmed the Indians so much, that they fled precipitately, redescending the Guaviare. It is probable, that these bearded white men came from the villages of Aroma and San Martin, the Rio Guaviare being formed by the union of the rivers Ariari and Guayavero. We must not be surprised that the missionaries of the Orinoco and the Atabapo little suspect how near they live to the missionaries of Mocoa, Rio Fragua, and Caguan. In these desert countries, the real distances can be known only by observations of the longitude. It was in consequence of astronomical data, and the information I gathered in the convents of Popayan and of Pasto, to the west of the Cordillera of the Andes, that I formed an accurate idea of the respective situations of the christian settlements on the Atabapo, the Guayavero, and the Caqueta.[*]

Everything changes on entering the Rio Atabapo; the constitution of the atmosphere, the colour of the waters, and the form of the trees that cover the shore. You no longer suffer during the day the torment of mosquitos; and the long-legged gnats (zancudos) become rare during the night. Beyond the

---

[*] The Caqueta bears, lower down, the name of the Yupura.

mission of San Fernando these nocturnal insects disappear altogether. The water of the Orinoco is turbid, and loaded with earthy matter; and in the coves, from the accumulation of dead crocodiles and other putrescent substances, it diffuses a musky and faint smell. We were sometimes obliged to strain this water through a linen cloth before we drank it. The water of the Atabapo, on the contrary, is pure, agreeable to the taste, without any trace of smell, brownish by reflected, and of a pale yellow by transmitted light. The people call it light, in opposition to the heavy and turbid waters of the Orinoco. Its temperature is generally two degrees, and when you approach the mouth of the Rio Temi, three degrees, cooler than the temperature of the Upper Orinoco. After having been compelled during a whole year to drink water at 27 or 28°, a lowering of a few degrees in the temperature produces a very agreeable sensation. I think this lowering of the temperature may be attributed to the river being less broad, and without the sandy beach, the heat of which, at the Orinoco, is by day more than 50°, and also to the thick shade of the forests which are traversed by the Atabapo, the Temi, the Tuamini, and the Guainia, or Rio Negro.

The extreme purity of the black waters is proved by their limpidity, their transparency, and the clearness with which they reflect the images and colours of surrounding objects. The smallest fish are visible in them at a depth of twenty or thirty feet; and most commonly the bottom of the river may be distinguished, which is not a yellowish or brownish mud, like the colour of the water, but a quartzose and granitic sand of dazzling whiteness. Nothing can be compared to the beauty of the banks of the Atabapo. Loaded with plants, among which rise the palms with feathery leaves; the banks are reflected in the waters, and this reflex verdure seems to have the same vivid hue as that which clothes the real vegetation. The surface of the fluid is homogeneous, smooth, and destitute of that mixture of suspended sand and decomposed organic matter, which roughens and streaks the surface of less limpid rivers.

On quitting the Orinoco, several small rapids must be passed, but without any appearance of danger. Amid these raudalitos, according to the opinion of the missionaries, the Rio Atabapo falls into the Orinoco. I am however disposed to think that the Atabapo falls into the Guaviare. The Rio Guaviare, which is much wider than the Atabapo, has white waters, and in the aspect of its banks, its fishing-birds, its fish, and the great crocodiles which live in it, resembles the Orinoco much more than that part of the Atabapo which comes from the Esmeralda. When a river springs from the junction of two other rivers, nearly alike in size, it is difficult to judge which of the two confluent streams must be regarded as its source. The Indians of San Fernando affirm that the Orinoco rises from two rivers, the Guaviare and the Rio Paragua. They give this latter name to the Upper Orinoco, from San Fernando and Santa Barbara to beyond the Esmeralda, and they say that the Cassiquiare is not an arm of the Orinoco, but of the Rio Paragua. It matters but little whether or not the name of Orinoco be given to the Rio Paragua, provided we trace the course of these rivers as it is in nature, and do not separate by a chain of mountains, (as was done previously to my travels,) rivers that communicate together, and form one system. When we would give the name of a large river to one of the two branches by which it is formed, it should be applied to that branch which furnishes most water. Now, at the two seasons of the year when I saw the Guaviare and the Upper Orinoco or Rio Paragua (between the Esmeralda and San Fernando), it appeared to me that the latter was not so large as the Guaviare. Similar doubts have been entertained by geographers respecting the junction of the Upper Mississippi with the Missouri and the Ohio, the junction of the Maranon with the Guallaga and the Ucayale, and the junction of the Indus with the Chunab (Hydaspes of Cashmere) and the Gurra, or Sutlej.[*] To avoid

---

[*] The Hydaspes is properly a tributary stream of the Chunab or Acesines. The Sutlej or Hysudrus forms, together with the Beyah or Hyphases, the river Gurra. These are the beautiful regions of the Pundjab and Douab, celebrated

embroiling farther a nomenclature of rivers so arbitrarily fixed, I will not propose new denominations. I shall continue, with Father Caulin and the Spanish geographers, to call the river Esmeralda the Orinoco, or Upper Orinoco; but I must observe that if the Orinoco, from San Fernando de Atabapo as far as the delta which it forms opposite the island of Trinidad, were regarded as the continuance of the Rio Guaviare, and if that part of the Upper Orinoco between the Esmeralda and the mission of San Fernando were considered a tributary stream, the Orinoco would preserve, from the savannahs of San Juan de los Llanos and the eastern declivity of the Andes to its mouth, a more uniform and natural direction, that from south-west to north-east.

The Rio Paragua, or that part of the Orinoco east of the mouth of the Guaviare, has clearer, more transparent, and purer water than the part of the Orinoco below San Fernando. The waters of the Guaviare, on the contrary, are white and turbid; they have the same taste, according to the Indians (whose organs of sense are extremely delicate and well practised), as the waters of the Orinoco near the Great Cataracts. "Bring me the waters of three or four great rivers of these countries," an old Indian of the mission of Javita said to us; "on tasting each of them I will tell you, without fear of mistake, whence it was taken; whether it comes from a white or black river; the Orinoco or the Atabapo, the Paragua or the Guaviare." The great crocodiles and porpoises (toninas) which are alike common in the Rio Guaviare and the Lower Orinoco, are entirely wanting, as we were told, in the Rio Paragua (or Upper Orinoco, between San Fernando and the Esmeralda). These are very remarkable differences in the nature of the waters, and the distribution of animals. The Indians do not fail to mention them, when they would prove to travellers that the Upper Orinoco, to the east of San Fernando, is a distinct river which falls into the Orinoco, and that the real origin of the latter must be sought in the sources of the Guaviare.

---

in history from Porus down to Sultan Acbar.

The astronomical observations made in the night of the 25th of April did not give me the latitude with satisfactory precision. The latitude of the mission of San Fernando appeared to me to be 4° 2' 48". In Father Caulin's map, founded on the observations of Solano made in 1756, it is 4° 1'. This agreement proves the justness of a result which, however, I could only deduce from altitudes considerably distant from the meridian. A good observation of the stars at Guapasoso gave me 4° 2' for San Fernando de Atabapo. I was able to fix the longitude with much more precision in my way to the Rio Negro, and in returning from that river. It is 70° 30' 46" (or 4° 0' west of the meridian of Cumana).

On the 26th of April we advanced only two or three leagues, and passed the night on a rock near the Indian plantations or conucos of Guapasoso. The river losing itself by its inundations in the forests, and its real banks being unseen, the traveller can venture to land only where a rock or a small table-land rises above the water. The granite of those countries, owing to the position of the thin laminae of black mica, sometimes resembles graphic granite; but most frequently (and this determines the age of its formation) it passes into a real gneiss. Its beds, very regularly stratified, run from south-west to north-east, as in the Cordillera on the shore of Caracas. The dip of the granite-gneiss is 70° north-west. It is traversed by an infinite number of veins of quartz, which are singularly transparent, and three or four, and sometimes fifteen inches thick. I found no cavity (druse), no crystallized substance, not even rock-crystal; and no trace of pyrites, or any other metallic substance. I enter into these particulars on account of the chimerical ideas that have been spread ever since the sixteenth century, after the voyages of Berreo and Raleigh,[*] "on the immense riches of the great and fine empire of Guiana."

---

[*] Raleigh's work bears the high sounding title of The Discovery of the large, rich, and beautiful Empire of Guiana, London 1596. See also Raleghi

The river Atabapo presents throughout a peculiar aspect; you see nothing of its real banks formed by flat lands eight or ten feet high; they are concealed by a row of palms, and small trees with slender trunks, the roots of which are bathed by the waters. There are many crocodiles from the point where you quit the Orinoco to the mission of San Fernando, and their presence indicates that this part of the river belongs to the Rio Guaviare and not to the Atabapo. In the real bed of the latter river, above the mission of San Fernando, there are no crocodiles: we find there some bavas, a great many fresh-water dolphins, but no manatees. We also seek in vain on these banks for the thick-nosed tapir, the araguato, or great howling monkey, the zamuro, or Vultur aura, and the crested pheasant, known by the name of guacharaca. Enormous water-snakes, in shape resembling the boa, are unfortunately very common, and are dangerous to Indians who bathe. We saw them almost from the first day we embarked, swimming by the side of our canoe; they were at most twelve or fourteen feet long. The jaguars of the banks of the Atabapo and the Temi are large and well fed; they are said, however, to be less daring than the jaguars of the Orinoco.

The night of the 27th was beautiful; dark clouds passed from time to time over the zenith with extreme rapidity. Not a breath of wind was felt in the lower strata of the atmosphere; the breeze was at the height of a thousand toises. I dwell upon this peculiarity; for the movement we saw was not produced by the counter-currents (from west to east) which are sometimes thought to be observed in the torrid zone on the loftiest mountains of the Cordilleras; it was the effect of a real breeze, an east wind. We left the conucos of Guapasoso at two o'clock; and continued to ascend the river toward the south, finding it (or rather that part of its bed which is free from trees) growing more and more narrow. It began to rain toward sunrise. In these forests, which are less inhabited by animals than those of the

---

admiranda Descriptio Regni Guianae, auri abundantissimi, Hondius Noribergae 1599.

Orinoco, we no longer heard the howlings of the monkeys. The dolphins, or toninas, sported by the side of our boat. According to the relation of Mr. Colebrooke, the Delphinus gangeticus, which is the fresh-water porpoise of the Old World, in like manner accompanies the boats that go up towards Benares; but from Benares to the point where the Ganges receives the salt waters is only two hundred leagues, while from the Atabapo to the mouth of the Orinoco is more than three hundred and twenty.

About noon we passed the mouth of the little river Ipurichapano on the east, and afterwards the granitic rock, known by the name of Piedra del Tigre. Between the fourth and fifth degrees of latitude, a little to the south of the mountains of Sipapo, we reach the southern extremity of that chain of cataracts, which I proposed, in a memoir published in 1800, to call the Chain of Parima. At 4° 20' it stretches from the right bank of the Orinoco toward the east and east-south-east. The whole of the land extending from the mountains of the Parima towards the river Amazon, which is traversed by the Atabapo, the Cassiquiare, and the Rio Negro, is an immense plain, covered partly with forests, and partly with grass. Small rocks rise here and there like castles. We regretted that we had not stopped to rest near the Piedra del Tigre; for on going up the Atabapo we had great difficulty to find a spot of dry ground, open and spacious enough to light a fire, and place our instrument and our hammocks.

On the 28th of April, it rained hard after sunset, and we were afraid that our collections would be damaged. The poor missionary had his fit of tertian fever, and besought us to re-embark immediately after midnight. We passed at day-break the Piedra and the Raudalitos[*] of Guarinuma. The rock is on the east bank; it is a shelf of granite, covered with psora, cladonia, and other lichens. I could have fancied myself transported to the north of Europe, to the ridge of the mountains of gneiss and

---

[*] The rock and little cascades.

granite between Freiberg and Marienberg in Saxony. The cladonias appeared to me to be identical with the Lichen rangiferinus, the L. pixidatus, and the L. polymorphus of Linnaeus. After having passed the rapids of Guarinuma, the Indians showed us in the middle of the forest, on our right, the ruins of the mission of Mendaxari, which has been long abandoned. On the east bank of the river, near the little rock of Kemarumo, in the midst of Indian plantations, a gigantic bombax* attracted our curiosity. We landed to measure it; the height was nearly one hundred and twenty feet, and the diameter between fourteen and fifteen. This enormous specimen of vegetation surprised us the more, as we had till then seen on the banks of the Atabapo only small trees with slender trunks, which from afar resembled young cherry-trees. The Indians assured that these small trees do not form a very extensive group. They are checked in their growth by the inundations of the river; while the dry grounds near the Atabapo, the Temi, and the Tuamini, furnish excellent timber for building. These forests do not stretch indefinitely to the east and west, toward the Cassiquiare and the Guaviare; they are bounded by the open savannahs of Manuteso, and the Rio Inirida. We found it difficult in the evening to stem the current, and we passed the night in a wood a little above Mendaxari; which is another granitic rock traversed by a stratum of quartz. We found in it a group of fine crystals of black schorl.

On the 29th, the air was cooler. We had no zancudos, but the sky was constantly clouded, and without stars. I began to regret the Lower Orinoco. We still advanced but slowly from the force of the current, and we stopped a great part of the day to seek for plants. It was night when we arrived at the mission of San Balthasar, or, as the monks style it, the mission of la divina Pastora de Balthasar de Atabapo. We were lodged with a Catalonian missionary, a lively and agreeable man, who displayed in these wild countries the activity that characterises

---

* Bombax ceiba.

his nation. He had planted a garden, where the fig-tree of Europe was found in company with the persea, and the lemon-tree with the mammee. The village was built with that regularity which, in the north of Germany, and in protestant America, we find in the hamlets of the Moravian brethren; and the Indian plantations seemed better cultivated than elsewhere. Here we saw for the first time that white and fungous substance which I have made known by the name of dapicho and zapis.[*] We immediately perceived that it was analogous to india-rubber; but, as the Indians made us understand by signs, that it was found underground, we were inclined to think, till we arrived at the mission of Javita, that the dapicho was a fossil caoutchouc, though different from the elastic bitumen of Derbyshire. A Pomisano Indian, seated by the fire in the hut of the missionary, was employed in reducing the dapicho into black caoutchouc. He had spitted several bits on a slender stick, and was roasting them like meat. The dapicho blackens in proportion as it grows soft, and becomes elastic. The resinous and aromatic smell which filled the hut, seemed to indicate that this coloration is the effect of the decomposition of a carburet of hydrogen, and that the carbon appears in proportion as the hydrogen burns at a low heat. The Indian beat the softened and blackened mass with a piece of brazil-wood, formed at one end like a club; he then kneaded the dapicho into balls of three or four inches in diameter, and let it cool. These balls exactly resemble the caoutchouc of the shops, but their surface remains in general slightly viscous. They are used at San Balthasar in the Indian game of tennis, which is celebrated among the inhabitants of Uruana and Encaramada; they are also cut into cylinders, to be used as corks, and are far preferable to those made of the bark of the cork-tree.

This use of caoutchouc appeared to us the more worthy notice, as we had been often embarrassed by the want of

---

[*] These two words belong to the Poimisano and Paragini tongues.

European corks. The great utility of cork is fully understood in countries where trade has not supplied this bark in plenty. Equinoctial America nowhere produces, not even on the back of the Andes, an oak resembling the Quercus suber; and neither the light wood of the bombax, the ochroma, and other malvaceous plants, nor the rhachis of maize, of which the natives make use, can well supply the place of our corks. The missionary showed us, before the Casa de los Solteros (the house where the young unmarried men reside), a drum, which was a hollow cylinder of wood, two feet long and eighteen inches thick. This drum was beaten with great masses of dapicho, which served as drumsticks; it had openings which could be stopped by the hand at will, to vary the sounds, and was fixed on two light supports. Savage notions love noisy music; the drum and the botuto, or trumpet of baked earth, in which a tube of three or four feet long communicates with several barrels, are indispensable instruments among the Indians for their grand pieces of music.

The night of the 30th of April was sufficiently fine for observing the meridian heights of x of the Southern Cross, and the two large stars in the feet of the Centaur. I found the latitude of San Balthasar 3° 14' 23". Horary angles of the sun gave 70° 14' 21" for the longitude by the chronometer. The dip of the magnetic needle was 27.8° (cent div). We left the mission at a late hour in the morning, and continued to go up the Atabapo for five miles; then, instead of following that river to its source in the east, where it bears the name of Atacavi, we entered the Rio Temi. Before we reached its confluence, a granitic eminence on the western bank, near the mouth of the Guasacavi, fixed our attention: it is called Piedra de la Guahiba (Rock of the Guahiba woman), or the Piedra de la Madre (Mother's Rock.) We inquired the cause of so singular a denomination. Father Zea could not satisfy our curiosity; but some weeks after, another missionary, one of the predecessors of that ecclesiastic, whom we found settled at San Fernando as president of the missions, related to us an event which excited in our minds the most painful feelings. If, in these solitary scenes, man scarcely leaves behind him any trace of his existence, it is doubly humiliating for a European to see perpetuated by so imperishable a monument of nature as a rock, the remembrance of the moral degradation of our species, and the contrast between the virtue of a savage, and the barbarism of civilized man!

In 1797 the missionary of San Fernando had led his Indians to the banks of the Rio Guaviare, on one of those hostile incursions which are prohibited alike by religion and the Spanish laws. They found in an Indian hut a Guahiba woman with her three children (two of whom were still infants), occupied in preparing the flour of cassava. Resistance was impossible; the father was gone to fish, and the mother tried in vain to flee with her children. Scarcely had she reached the savannah when she was seized by the Indians of the mission, who hunt human beings, like the Whites and the Negroes in Africa. The mother and her children were bound, and dragged to the bank of the river. The monk, seated in his boat, waited the issue of an expedition of which he shared not the danger. Had the mother made too violent a resistance the Indians would have killed her, for everything is permitted for the sake of the conquest of souls (la conquista espirituel), and it is particularly desirable to capture children, who may be treated in the Mission as poitos, or slaves of the Christians. The prisoners were carried to San Fernando, in the hope that the mother would be unable to find her way back to her home by land. Separated from her other children who had accompanied their father on the day in which she had been carried off, the unhappy woman showed signs of the deepest despair. She attempted to take back to her home the children who had been seized by the missionary; and she fled with them repeatedly from the village of San Fernando. But the Indians never failed to recapture her; and the missionary, after having caused her to be mercilessly beaten, took the cruel resolution of separating the mother from the two children who had been carried off with her. She was conveyed alone to the missions of the Rio Negro, going up the Atabapo. Slightly bound, she was seated at the bow of the boat, ignorant of the fate that awaited her; but she judged by the direction of the sun, that she was removing farther and farther from her hut and her native country. She succeeded in breaking her bonds, threw herself into the water, and swam to the left bank of the Atabapo. The current carried her to a shelf of rock, which bears her name to this day. She landed and took shelter in the woods, but the president of the missions ordered the Indians to row to the shore, and follow the traces of the Guahiba. In the evening she was brought back. Stretched upon the rock (la Piedra de la Madre) a cruel punishment was inflicted on her with those straps of manatee leather, which serve for whips in that country, and with which the alcaldes are always furnished. This unhappy woman, her hands tied behind her back with strong stalks of mavacure, was then dragged to the mission of Javita.

She was there thrown into one of the caravanserais, called las Casas del Rey. It was the rainy season, and the night was profoundly dark. Forests till then believed to be impenetrable separated the mission of Javita from that of San Fernando, which was twenty-five leagues distant in a straight line. No other route is known than that by the rivers; no man ever attempted to go by land from one village to another. But such difficulties could not deter a mother, separated from her children. The Guahiba was carelessly guarded in the caravanserai. Her arms being wounded, the Indians of Javita had loosened her bonds, unknown to the missionary and the alcaldes. Having succeeded by the help of her teeth in breaking them entirely, she disappeared during the night; and at the fourth sunrise was seen at the mission of San Fernando, hovering around the hut where her children were confined. "What that woman performed," added the missionary, who gave us this sad narrative, "the most robust Indian would not have ventured to undertake!" She traversed the woods at a season when the sky is constantly covered with clouds, and the sun during whole days appears but for a few minutes. Did the course of the waters direct her way? The inundations of the rivers forced her to go far from the banks of the main stream, through the midst of woods where the movement of the water is almost imperceptible. How often must she have been stopped by the thorny lianas, that form a network around the trunks they entwine! How often must she have swum across the rivulets that run into the Atabapo! This unfortunate woman was asked how she had sustained herself during four days. She said that, exhausted with fatigue, she could find no other nourishment than those great black ants called vachacos, which climb the trees in long bands, to suspend on them their resinous nests. We pressed the missionary to tell us whether the Guahiba had peacefully enjoyed the happiness of remaining with her children; and if any repentance had followed this excess of cruelty. He would not satisfy our curiosity; but at our return from the Rio Negro we learned that the Indian mother was again separated from her children, and sent to one of the missions of the Upper Orinoco. There she died, refusing all kind of nourishment, as savages frequently do in great calamities.

Such is the remembrance annexed to this fatal rock, the Piedra de la Madre. In this relation of my travels I feel no desire to dwell on pictures of individual suffering—evils which are frequent wherever there are masters and slaves, civilized Europeans living with people in a state of barbarism, and priests exercising the plenitude of arbitrary power over men ignorant and without defence. In describing the countries through which I passed, I generally confine myself to pointing out what is imperfect, or fatal to humanity, in their civil or religious institutions. If I have dwelt longer on the Rock of the Guahiba, it was to record an affecting instance of maternal tenderness in a race of people so long calumniated; and because I thought some benefit might accrue from publishing a fact, which I had from the monks of San Francisco, and which proves how much the system of the missions calls for the care of the legislator.

Above the mouth of the Guasucavi we entered the Rio Temi, the course of which is from south to north. Had we continued to ascend the Atabapo, we should have turned to east-south-east, going farther from the banks of the Guainia or Rio Negro. The

Temi is only eighty or ninety toises broad, but in any other country than Guiana it would be a considerable river. The country exhibits the uniform aspect of forests covering ground perfectly flat. The fine pirijao palm, with its fruit like peaches, and a new species of bache, or mauritia, its trunk bristled with thorns, rise amid smaller trees, the vegetation of which appears to be retarded by the continuance of the inundations. The Mauritia aculeata is called by the Indians juria or cauvaja; its leaves are in the form of a fan, and they bend towards the ground. At the centre of every leaf, no doubt from the effect of some disease of the parenchyma, concentric circles of alternate blue and yellow appear, the yellow prevailing towards the middle. We were singularly struck by this appearance; the leaves, coloured like the peacock's tail, are supported by short and very thick trunks. The thorns are not slender and long like those of the corozo and other thorny palm-trees; but on the contrary, very woody, short, and broad at the base, like the thorns of the Hura crepitans. On the banks of the Atabapo and the Temi, this palm-tree is distributed in groups of twelve or fifteen stems, close together, and looking as if they rose from the same root. These trees resemble in their appearance, form, and scarcity of leaves, the fan-palms and palmettos of the Old World. We remarked that some plants of the juria were entirely destitute of fruit, and others exhibited a considerable quantity; this circumstance seems to indicate a palm-tree of separate sexes.

Wherever the Rio Temi forms coves, the forest is inundated to the extent of more than half a square league. To avoid the sinuosities of the river and shorten the passage, the navigation is here performed in a very extraordinary manner. The Indians made us leave the bed of the river; and we proceeded southward across the forest, through paths (sendas), that is, through open channels of four or five feet broad. The depth of the water seldom exceeds half a fathom. These sendas are formed in the inundated forest like paths on dry ground. The Indians, in going from one mission to another, pass with their boats as much as possible by the same way; but the communications not being frequent, the force of vegetation sometimes produces unexpected obstacles. An Indian, furnished with a machete (a great knife, the blade of which is fourteen inches long), stood at the head of our boat, employed continually in chopping off the branches that crossed each other from the two sides of the channel. In the thickest part of the forest we were astonished by an extraordinary noise. On beating the bushes, a shoal of toninas (fresh-water dolphins) four feet long, surrounded our boat. These animals had concealed themselves beneath the branches of a fromager, or Bombax ceiba. They fled across the forest, throwing out those spouts of compressed air and water which have given them in every language the name of blowers. How singular was this spectacle in an inland spot, three or four hundred leagues from the mouths of the Orinoco and the Amazon! I am aware that the pleuronectes (dabs) of the Atlantic go up the Loire as far as Orleans; but I am, nevertheless, of opinion that the dolphins of the Temi, like those of the Ganges, and like the skate (raia) of the Orinoco, are of a species essentially different from the dolphins and skates of the ocean. In the immense rivers of South America, and the great lakes of North America, nature seems to repeat several pelagic forms. The Nile has no porpoises:[*] those of the sea go up the Delta no farther than Biana and Metonbis towards Selamoun.

At five in the evening we regained with some difficulty the bed of the river. Our canoe remained fast for some minutes between two trunks of trees; and it was no sooner disengaged than we reached a spot where several paths, or small channels, crossed each other, so that the pilot was puzzled to distinguish the most open path. We navigated through a forest so thick that we could guide ourselves neither by the sun nor by the stars. We

---

[*] Those dolphins that enter the mouth of the Nile, did not escape the observation of the ancients. In a bust in syenite, preserved in the museum at Paris, the sculptor has represented them half concealed in the undulatory beard of the god of the river.

383

were again struck during this day by the want of arborescent ferns in that country; they diminish visibly from the sixth degree of north latitude, while the palm-trees augment prodigiously towards the equator. Fern-trees belong to a climate less hot, and a soil but little mountainous. It is only where there are mountains that these majestic plants descend towards the plains; they seem to avoid perfectly flat grounds, as those through which run the Cassiquiare, the Temi, Inirida, and the Rio Negro. We passed in the night near a rock, called the Piedra de Astor by the missionaries. The ground from the mouth of the Guaviare constantly displays the same geological formation. It is a vast granitic plain, in which from league to league the rock pierces the soil, and forms, not hillocks, but small masses, that resemble pillars or ruined buildings.

On the 1st of May the Indians chose to depart long before sunrise. We were stirring before them, however, because I waited (though vainly) for a star ready to pass the meridian. In those humid regions covered with forests, the nights became more obscure in proportion as we drew nearer to the Rio Negro and the interior of Brazil. We remained in the bed of the river till daybreak, being afraid of losing ourselves among the trees. At sunrise we again entered the inundated forest, to avoid the force of the current. On reaching the junction of the Temi with another little river, the Tuamini, the waters of which are equally black, we proceeded along the latter to the south-west. This direction led us near the mission of Javita, which is founded on the banks of the Tuamini; and at this christian settlement we were to find the aid necessary for transporting our canoe by land to the Rio Negro. We did not arrive at San Antonio de Javita till near eleven in the morning. An accident, unimportant in itself, but which shows the excessive timidity of the little sagoins detained us some time at the mouth of the Tuamini. The noise of the blowers had frightened our monkeys, and one of them fell into the water. Animals of this species, perhaps on account of their extreme meagreness, swim badly; and consequently it was saved with some difficulty.

At Javita we had the pleasure of finding a very intelligent and obliging monk, at whose mission we were forced to remain four or five days, the time required for transporting our boat across the portage of Pimichin. This delay enabled us to visit the surrounding country, as also to relieve ourselves from an annoyance which we had suffered for two days. We felt an extraordinary irritation on the joints of our fingers, and on the backs of our hands. The missionary told us it was caused by the aradores,[*] which get under the skin. We could distinguish with a lens nothing but streaks, or parallel and whitish furrows. It is the form of these furrows, that has obtained for the insect the name of ploughman. A mulatto woman was sent for, who professed to be thoroughly acquainted with all the little insects that burrow in the human skin; the chego, the nuche, the coya, and the arador; she was the curandera, or surgeon of the place. She promised to extirpate, one by one, the insects which caused this smarting irritation. Having heated at a lamp the point a little bit of hard wood, she dug with it into the furrows that marked the skin. After long examination, she announced with the pedantic gravity peculiar to the mulatto race, that an arador was found. I saw a little round bag, which I suspected to be the egg of an acarus. I was to find relief when the mulatto woman had succeeded in taking out three or four of these aradores. Having the skin of both hands filled with acari, I had not the patience to wait the end of an operation, which had already lasted till late at night. The next day an Indian of Javita cured us radically, and with surprising promptitude. He brought us the branch of a shrub, called uzao, with small leaves like those of cassia, very coriaceous and glossy. He made a cold infusion of the bark of this shrub, which had a bluish colour, and the taste of liquorice. When beaten, it yields a great deal of froth. The irritation of the aradores ceased by using simple lotions of this uzao-water. We could not find this shrub in flower, or bearing fruit; it appears to belong to the family of the leguminous plants, the chemical

---

[*] Literally the ploughers.

properties of which are singularly varied. We dreaded so much the sufferings to which we had been exposed, that we constantly kept some branches of the uzao in our boat, till we reached San Carlos. This shrub grows in abundance on the banks of the Pimichin. Why has no remedy been discovered for the irritation produced by the sting of the zancudos, as well as for that occasioned by the aradores or microscopic acari?

In 1755, before the expedition for fixing the boundaries, better known by the name of the expedition of Solano, the whole country between the missions of Javita and San Balthasar was regarded as dependent on Brazil. The Portuguese had advanced from the Rio Negro, by the portage of the Cano Pimichin, as far as the banks of the Temi. An Indian chief of the name of Javita, celebrated for his courage and his spirit of enterprise, was the ally of the Portuguese. He pushed his hostile incursions from the Rio Jupura, or Caqueta, one of the great tributary streams of the Amazon, by the rivers Uaupe and Xie, as far as the black waters of the Temi and the Tuamini, a distance of more than a hundred leagues. He was furnished with letters patent, which authorised him to bring the Indians from the forest, for the conquest of souls. He availed himself amply of this permission; but his incursions had an object which was not altogether spiritual, that of making slaves to sell to the Portuguese. When Solano, the second chief of the expedition of the boundaries, arrived at San Fernando de Atabapo, he had Javita seized, in one of his incursions to the banks of the Temi. He treated him with gentleness, and succeeded in gaining him over to the interests of the Spanish government by promises that were not fulfilled. The Portuguese, who had already formed some stable settlements in these countries, were driven back as far as the lower part of the Rio Negro; and the mission of San Antonio, of which the more usual name is Javita, so called after its Indian founder, was removed farther north of the sources of the Tuamini, to the spot where it is now established. This captain, Javita, was still living, at an advanced age, when we proceeded to the Rio Negro. He was an Indian of great vigour of mind and body. He spoke

Spanish with facility, and preserved a certain influence over the neighbouring nations. As he attended us in all our herborizations, we obtained from his own mouth information so much the more useful, as the missionaries have great confidence in his veracity. He assured us that in his youth he had seen almost all the Indian tribes that inhabit the vast regions between the Upper Orinoco, the Rio Negro, the Inirida, and the Jupura, eat human flesh. The Daricavanas, the Puchirinavis, and the Manitivitanos, appeared to him to be the greatest cannibals among them. He believes that this abominable practice is with them the effect of a system of vengeance; they eat only enemies who are made prisoners in battle. The instances where, by a refinement of cruelty, the Indian eats his nearest relations, his wife, or an unfaithful mistress, are extremely rare. The strange custom of the Scythians and Massagetes, the Capanaguas of the Rio Ucayale, and the ancient inhabitants of the West Indian Islands, of honouring the dead by eating a part of their remains, is unknown on the banks of the Orinoco. In both continents this trait of manners belongs only to nations that hold in horror the flesh of a prisoner. The Indian of Hayti (Saint Domingo) would think himself wanting in regard to the memory of a relation, if he did not throw into his drink a small portion of the body of the deceased, after having dried it like one of the mummies of the Guanches, and reduced it to powder. This gives us just occasion to repeat with an eastern poet, "of all animals man is the most fantastic in his manners, and the most disorderly in his propensities."

The climate of the mission of San Antonio de Javita is extremely rainy. When you have passed the latitude of three degrees north, and approach the equator, you have seldom an opportunity of observing the sun or the stars. It rains almost the whole year, and the sky is constantly cloudy. As the breeze is not felt in these immense forests of Guiana, and the refluent polar currents do not penetrate them, the column of air which reposes on this wooded zone is not renewed by dryer strata. It is saturated with vapours which are condensed into equatorial

387

rains. The missionary assured us that it often rains here four or five months without cessation.

The temperature of Javita is cooler than that of Maypures, but considerably hotter than that of the Guainia or Rio Negro. The centigrade thermometer kept up in the day to twenty-six or twenty-seven degrees; and in the night to twenty-one degrees.

From the 30th of April to the 11th of May, I had not been able to see any star in the meridian so as to determine the latitude of places. I watched whole nights in order to make use of the method of double altitudes; but all my efforts were useless. The fogs of the north of Europe are not more constant than those of the equatorial regions of Guiana. On the 4th of May, I saw the sun for some minutes; and found by the chronometer and the horary angles the longitude of Javita to be 70° 22', or 1° 15' farther west than the longitude of the junction of the Apure with the Orinoco. This result is interesting for laying down on our maps the unknown country lying between the Xie and the sources of the Issana, situated on the same meridian with the mission of Javita.

The Indians of Javita, whose number amounts to one hundred and sixty, now belong for the most part to the nations of the Poimisanos, the Echinavis, and the Paraganis. They are employed in the construction of boats, formed of the trunks of sassafras, a large species of laurel, hollowed by means of fire and the hatchet. These trees are more than one hundred feet high; the wood is yellow, resinous, almost incorruptible in water, and has a very agreeable smell. We saw them at San Fernando, at Javita, and more particularly at Esmeralda, where most of the canoes of the Orinoco are constructed, because the adjacent forests furnish the largest trunks of sassafras.

The forest between Javita and the Cano Pimichin, contains an immense quantity of gigantic trees, ocoteas, and laurels, the Amasonia arborea,[*] the Retiniphyllum secundiflorum, the

---

[*] This is a new species of the genus taligalea of Aublet. On the same spot

curvana, the jacio, the iacifate, of which the wood is red like the brazilletto, the guamufate, with its fine leaves of calophyllum from seven to eight inches long, the Amyris carana, and the mani. All these trees (with the exception of our new genus Retiniphyllum) were more than one hundred or one hundred and ten feet high. As their trunks throw out branches only toward the summit, we had some trouble in procuring both leaves and flowers. The latter were frequently strewed upon the ground at the foot of the trees; but, the plants of different families being grouped together in these forests, and every tree being covered with lianas, we could not, with any degree of confidence, rely on the authority of the natives, when they assured us that a flower belonged to such or such a tree. Amid these riches of nature heborizations caused us more chagrin than satisfaction. What we could gather appeared to us of little interest, compared to what we could not reach. It rained unceasingly during several months, and M. Bonpland lost the greater part of the specimens which he had been compelled to dry by artificial heat. Our Indians distinguished the leaves better than the corollae or the fruit. Occupied in seeking timber for canoes, they are inattentive to flowers. "All those great trees bear neither flowers nor fruits," they repeated unceasingly. Like the botanists of antiquity, they denied what they had not taken the trouble to observe. They were tired with our questions, and exhausted our patience in return.

We have already mentioned that the same chemical properties being sometimes found in the same organs of different families of plants, these families supply each other's places in various climates. Several species of palms[*] furnish the

---

grow the Bignonia magnoliaefolia, B. jasminifolia, Solanum topiro, Justicia pectoralis, Faramea cymosa, Piper javitense, Scleria hirtella, Echites javitensis, Lindsea javitensis, and that curious plant of the family of the verbenaceae, which I have dedicated to the illustrious Leopold von Buch, in whose early labours I participated.

[*] In Africa, the elais or maba; in America the cocoa-tree. In the cocoa-tree

inhabitants of equinoctial America and Africa with the oil which we derive from the olive. What the coniferae are to the temperate zone, the terebinthaceae and the guttiferae are to the torrid. In the forests of those burning climates, (where there is neither pine, thuya, taxodium, nor even a podocarpus,) resins, balsams, and aromatic gums, are furnished by the maronobea, the icica, and the amyris. The collecting of these gummy and resinous substances is a trade in the village of Javita. The most celebrated resin bears the name of mani; and of this we saw masses of several hundred-weight, resembling colophony and mastic. The tree called mani by the Paraginis, which M. Bonpland believes to be the Moronobaea coccinea, furnishes but a small quantity of the substance employed in the trade with Angostura. The greatest part comes from the mararo or caragna, which is an amyris. It is remarkable enough, that the name mani, which Aublet heard among the Galibis[*] of Cayenne, was again heard by us at Javita, three hundred leagues distant from French Guiana. The moronobaea or symphonia of Javita yields a yellow resin; the caragna, a resin strongly odoriferous, and white as snow; the latter becomes yellow where it is adherent to the internal part of old bark.

We went every day to see how our canoe advanced on the portages. Twenty-three Indians were employed in dragging it by land, placing branches of trees to serve as rollers. In this manner

---

it is the perisperm; and in the elais (as in the olive, and the oleineae in general) it is the sarcocarp, or the pulp of the pericarp, that yields oil. This difference, observed in the same family, appears to me very remarkable, though it is in no way contradictory to the results obtained by De Candolle in his ingenious researches on the chemical properties of plants. If our Alfonsia oleifera belong to the genus Elais (as Brown, with great reason believes), it follows, that in the same genus the oil is found in the sarcocarp and in the perisperm.

[*] The Galibis or Caribis (the r has been changed into l, as often happens) are of the great stock of the Carib nations. The products useful in commerce and in domestic life have received the same denomination in every part of America which this warlike and commercial people have overrun.

a small boat proceeds in a day or a day and a half, from the waters of the Tuamini to those of the Cano Pimichin, which flow into the Rio Negro. Our canoe being very large, and having to pass the cataracts a second time, it was necessary to avoid with particular care any friction on the bottom; consequently the passage occupied more than four days. It is only since 1795 that a road has been traced through the forest. By substituting a canal for this portage, as I proposed to the ministry of king Charles IV, the communication between the Rio Negro and Angostura, between the Spanish Orinoco and the Portuguese possessions on the Amazon, would be singularly facilitated.

In this forest we at length obtained precise information respecting the pretended fossil caoutchouc, called dapicho by the Indians. The old chief Javita led us to the brink of a rivulet which runs into the Tuamini; and showed us that, after digging two or three feet deep, in a marshy soil, this substance was found between the roots of two trees known by the name of the jacio and the curvana. The first is the hevea of Aublet, or siphonia of the modern botanists, known to furnish the caoutchouc of commerce in Cayenne and Grand Para; the second has pinnate leaves, and its juice is milky, but very thin, and almost destitute of viscosity. The dapicho appears to be the result of an extravasation of the sap from the roots. This extravasation takes place more especially when the trees have attained a great age, and the interior of the trunk begins to decay. The bark and alburnum crack; and thus is effected naturally, what the art of man performs for the purpose of collecting the milky juices of the hevea, the castilloa, and the caoutchouc fig-tree. Aublet relates, that the Galibis and the Garipons of Cayenne begin by making a deep incision at the foot of the trunk, so as to penetrate into the wood; soon after they join with this horizontal notch others both perpendicular and oblique, reaching from the top of the trunk nearly to the roots. All these incisions conduct the milky juice towards one point, where the vase of clay is placed, in which the caoutchouc is to be deposited. We saw the Indians of Carichana operate nearly in the same manner.

If, as I suppose, the accumulation and overflowing of the milk in the jacio and the curvana be a pathological phenomenon, it must sometimes take place at the extremity of the longest roots, for we found masses of dapicho two feet in diameter and four inches thick, eight feet distant from the trunks. Sometimes the Indians dig in vain at the foot of dead trees; at other times the dapicho is found beneath the hevea or jacio still green. The substance is white, corky, fragile, and resembles by its laminated structure and undulating edge, the Boletus ignarius. The dapicho perhaps takes a long time to form; it is probably a juice thickened by a particular disposition of the vegetable organs, diffused and coagulated in a humid soil secluded from the contact of light; it is caoutchouc in a particular state, I may almost say an etiolated caoutchouc. The humidity of the soil seems to account for the undulating form of the edges of the dapicho, and its division into layers.

I often observed in Peru, that on pouring slowly the milky juice of the hevea, or the sap of the carica, into a large quantity of water, the coagulum forms undulating outlines. The dapicho is certainly not peculiar to the forest that extends from Javita to Pimichin, although that is the only spot where it has hitherto been found. I have no doubt, that on digging in French Guiana beneath the roots and the old trunks of the hevea, those enormous masses of corky caoutchouc,[*] which I have just described, would from time to time be found. As it is observed in Europe, that at the fall of the leaf the sap is conveyed towards the root, it would be curious to examine whether, within the tropics, the milky juices of the urticeae, the euphorbiaceae, and the apocyneae, descend also at certain seasons. Notwithstanding a great equality of temperature, the trees of the torrid zone

---

[*] Thus, at five or six inches depth, between the roots of the Hymenea courbaril, masses of the resin anime (erroneously called copal) are discovered, and are sometimes mistaken for amber in inland places. This phenomenon seems to throw some light on the origin of those large masses of amber which are picked up from time to time on the coast of Prussia.

follow a cycle of vegetation; they undergo changes periodically returning. The existence of the dapicho is more interesting to physiology than to vegetable chemistry. A yellowish-white caoutchouc is now to be found in the shops, which may be easily distinguished from the dapicho, because it is neither dry like cork, nor friable, but extremely elastic, glossy, and soapy. I lately saw considerable quantities of it in London. This caoutchouc, white, and greasy to the touch, is prepared in the East Indies. It exhales that animal and fetid smell which I have attributed in another place to a mixture of caseum and albumen.[*] When we reflect on the immense variety of plants in the equinoctial regions that are capable of furnishing caoutchouc, it is to be regretted that this substance, so eminently useful, is not found among us at a lower price. Without cultivating trees with a milky sap, a sufficient quantity of caoutchouc might be collected in the missions of the Orinoco alone for the consumption of civilized Europe.[†] In the kingdom of New Grenada some successful attempts have been made to make boots and shoes of this substance without a seam. Among the American nations, the Omaguas of the Amazon best understand how to manufacture caoutchouc.

Four days had passed, and our canoe had not yet arrived at the landing-place of the Rio Pimichin. "You want for nothing in my mission," said Father Cereso; "you have plantains and fish; at night you are not stung by mosquitos; and the longer you stay, the better chance you will have of seeing the stars of my country. If your boat be destroyed in the portage, we will give you

---

[*] The pellicles deposited by the milk of hevea, in contact with the atmospheric oxygen, become brown on exposure to the sun. If the dapicho grow black as it is softened before the fire, it is owing to a slight combustion, to a change in the proportion of its elements. I am surprised that some chemists consider the black caoutchouc of commerce, as being mixed with soot, blackened by the smoke to which it has been exposed.
[†] We saw in Guiana, besides the jacio and the curvana, two other trees that yield caoutchouc in abundance; on the banks of the Atabapo the guamaqui with jatropha leaves, and at Maypures the cime.

another; and I shall have had the satisfaction of passing some weeks con gente blanca y de razon." ("With white and rational people." European self-love usually opposes the gente de razon to the gente parda, or coloured people.) Notwithstanding our impatience, we listened with interest to the information given us by the worthy missionary. It confirmed all we had already heard of the moral state of the natives of those countries. They live, distributed in hordes of forty or fifty, under a family government; and they recognise a common chief (apoto, sibierene) only at times when they make war against their neighbours. The mistrust of these hordes towards one another is increased by the circumstance that those who live in the nearest neighbourhood speak languages altogether different. In the open plains, in the countries with savannahs, the tribes are fond of choosing their habitations from an affinity of origin, and a resemblance of manners and idioms. On the table-land of Tartary, as in North America, great families of nations have been seen, formed into several columns, extending their migrations across countries thinly-wooded, and easily traversed. Such were the journeys of the Toltec and Aztec race in the high plains of Mexico, from the sixth to the eleventh century of our era; such probably was also the movement of nations by which the petty tribes of Canada were grouped together. As the immense country between the equator and the eighth degree of north latitude forms one continuous forest, the hordes were there dispersed by following the branchings of the rivers, and the nature of the land compelled them to become more or less agriculturists. Such is the labyrinth of these rivers, that families settled themselves without knowing what race of men lived nearest the spot. In Spanish Guiana a mountain, or a forest half a league broad, sometimes separates hordes who could not meet in less than two days by navigating rivers. In open countries, or in a state of advanced civilization, communication by rivers contributes powerfully to generalize languages, manners, and political institutions; but in the impenetrable forests of the torrid zone, as in the first rude condition of our species, rivers increase the dismemberment of great nations, favour the transition of dialects into languages that appear to us radically distinct, and keep up national hatred and mistrust. Between the banks of the Caura and the Padamo everything bears the stamp of disunion and weakness. Men avoid, because they do not understand, each other; they mutually hate, because they mutually fear.

When we examine attentively this wild part of America, we fancy ourselves transported to those primitive times when the earth was peopled by degrees, and we seem to be present at the birth of human societies. In the old world we see that pastoral life has prepared the hunting nations for agriculture. In the New World we seek in vain these progressive developments of civilization, these intervals of repose, these stages in the life of nations. The luxury of vegetation embarrasses the Indians in the chase; and in their rivers, resembling arms of the sea, the depth of the waters prevents fishing during whole months. Those species of ruminating animals, that constitute the wealth of the nations of the Old World, are wanting in the New. The bison and the musk-ox have never been reduced to a domestic state; the breeding of llamas and guanacos has not created the habits of pastoral life. In the temperate zone, on the banks of the Missouri, as well as on the tableland of New Mexico, the American is a hunter; but in the torrid zone, in the forests of Guiana, he cultivates cassava, plantains, and sometimes maize. Such is the admirable fertility of nature, that the field of the native is a little spot of land, to clear which requires only setting fire to the brambles; and putting a few seeds or slips into the ground is all the husbandry it demands. If we go back in thought to the most remote ages, in these thick forests we must always figure to ourselves nations deriving the greater part of their nourishment from the earth; but, as this earth produces abundance in a small space, and almost without toil, we may also imagine these nations often changing their dwellings along the banks of the same river. Even now the native of the Orinoco travels with his seeds; and transports his farm (conuco) as the Arab transports his tent, and changes his pasturage. The number of cultivated plants found wild amid the woods, proves the nomad habits of an agricultural people. Can we be surprised, that by these habits they lose almost all the advantages that result in the temperate zone from stationary culture, from the growth of corn, which requires extensive lands and the most assiduous labour?

The nations of the Upper Orinoco, the Atabapo, and the Inirida, like the ancient Germans and the Persians, have no other worship than that of the powers of nature. They call the good principle Cachimana; it is the Manitou, the Great Spirit, that regulates the seasons, and favours the harvests. Along with Cachimana there is an evil principle, Iolokiamo, less powerful, but more artful, and in particular more active. The Indians of the forest, when they occasionally visit the missions, conceive with difficulty the idea of a temple or an image. "These good people," said the missionary, "like only processions in the open air. When I last celebrated the festival of San Antonio, the patron of my village, the Indians of Inirida were present at mass. 'Your God,' said they to me, 'keeps himself shut up in a house, as if he were old and infirm; ours is in the forest, in the fields, and on the mountains of Sipapu, whence the rains come.'" Among the more numerous, and on this account less barbarous tribes, religious societies of a singular kind are formed. Some old Indians pretend to be better instructed than others on points regarding divinity; and to them is confided the famous botuto, of which I have spoken, and which is sounded under the palm-trees that they may bear abundance of fruit. On the banks of the Orinoco there exists no idol, as among all the nations who have remained faithful to the first worship of nature, but the botuto, the sacred trumpet, is an object of veneration. To be initiated into the mysteries of the botuto, it is requisite to be of pure morals, and to have lived single. The initiated are subjected to flagellations, fastings, and other painful exercises. There are but a small number of these sacred trumpets. The most anciently celebrated is that upon a hill near the confluence of the Tomo and the Guainia. It is pretended, that it is heard at once on the banks of the Tuamini, and at the mission of San Miguel de Davipe, a distance of ten leagues. Father Cereso assured us, that the Indians speak of the botuto of Tomo as an object of worship common to many surrounding tribes. Fruit and intoxicating liquors arc placed beside the sacred trumpet. Sometimes the Great Spirit himself makes the botuto resound; sometimes he is content to manifest his will through him to whom the keeping of the instrument is entrusted. These juggleries being very ancient (from the fathers of our fathers, say the Indians), we must not be surprised that some unbelievers are already to be found; but they express their disbelief of the mysteries of the botuto only in whispers. Women are not permitted to see this marvellous instrument; and are excluded from all the ceremonies of this worship. If a woman have the misfortune to see the trumpet, she is put to death without mercy. The missionary related to us, that in 1798 he was happy enough to save a young girl, whom a jealous and vindictive lover accused of having followed, from a motive of curiosity, the Indians who sounded the botuto in the plantations. "They would not have murdered her publicly," said father Cesero, "but how was she to be protected from the fanaticism of the natives, in a country where it is so easy to give poison? The young girl told me of her fears, and I sent her to one of the missions of the Lower Orinoco." If the people of Guiana had remained masters of that vast country; if, without having been impeded by Christian settlements, they could follow freely the development of their barbarous institutions; the worship of the botuto would no doubt become of some political importance. That mysterious society of the initiated, those guardians of the sacred trumpet, would be transformed into a ruling caste of priests, and the oracle of Tomo would gradually form a link between the bordering nations.

In the evening of the 4th of May we were informed, that an Indian, who had assisted in dragging our bark over the portage of Pimichin, had been stung by a viper. He was a tall strong man, and was brought to the mission in a very alarming state. He had dropped down senseless; and nausea, vertigo, and congestions in the head, had succeeded the fainting. The liana called vejeco de guaco,[*] which M. Mutis has rendered so celebrated, and which is the most certain remedy for the bite of venomous serpents, is yet unknown in these countries. A number of Indians hastened to the hut of the sick man, and he was cured by an infusion of raiz de mato. We cannot indicate with certainty what plant furnishes this antidote; but I am inclined to think, that the raiz de mato is an apocynea, perhaps the Cerbera thevetia, called by the inhabitants of Cumana lingua de mato or contra-culebra, and which they also use against the bite of serpents. A genus nearly allied to the cerbera[†] is employed in India for the same purpose. It is common enough to find in the same family of plants vegetable poisons, and antidotes against the venom of reptiles. Many tonics and narcotics are antidotes more or less active; and we find these in families very different[‡] from each other, in the aristolochiae, the apocyneae, the gentianae, the polygalae, the solaneae, the compositae, the malvaceae, the drymyrhizeae, and, which is still more surprising, even in the palm-trees.

In the hut of the Indian who had been so dangerously bitten by the viper, we found balls two or three inches in diameter, of an earthy and impure salt called chivi, which is prepared with great care by the natives. At Maypures a conferva is burnt, which is left by the Orinoco on the neighbouring rocks, when, after high swellings, it again enters its bed. At Javita a salt is fabricated by the incineration of the spadix and fruit of the palm-tree seje or chimu. This fine palm-tree, which abounds on the

---

[*] This is a mikania, which was confounded for some time in Europe with the ayapana. De Candolle thinks that the guaco may be the Eupatorium satureiaefolium of Lamarck; but this Eupatorium differs by its lineary leaves, while the Mikania guaco has triangular, oval, and very large leaves.

[†] Ophioxylon serpentinum.

[‡] I shall mention as examples of these nine families; Aristolochia anguicida, Cerbera thevetia, Ophoiorhiza mungos, Polygala senega, Nicotiana tabacum, (One of the remedies most used in Spanish America). Mikanua guaco, Hibiscus abelmoschus (the seeds of which are very active), Lanpujum rumphii, and Kunthia montana (Cana de la Vibora).

banks of the Auvana, near the cataract of Guarinumo, and between Javita and the Cano Pimichin, appears to be a new species of cocoa-tree. It may be recollected, that the fluid contained in the fruit of the common cocoa-tree is often saline, even when the tree grows far from the sea shore. At Madagascar salt is extracted from the sap of a palm-tree called ciro. Besides the spadix and the fruit of the seje palm, the Indians of Javita lixiviate also the ashes of the famous liana called cupana, which is a new species of the genus paullinia, consequently a very different plant from the cupania of Linnaeus. I may here mention, that a missionary seldom travels without being provided with some prepared seeds of the cupana. This preparation requires great care. The Indians scrape the seeds, mix them with flour of cassava, envelope the mass in plantain leaves, and set it to ferment in water, till it acquires a saffron-yellow colour. This yellow paste dried in the sun, and diluted in water, is taken in the morning as a kind of tea. The beverage is bitter and stomachic, but it appeared to me to have a very disagreeable taste.

On the banks of the Niger, and in a great part of the interior of Africa, where salt is extremely rare, it is said of a rich man, "he is so fortunate as to eat salt at his meals." This good fortune is not too common in the interior of Guiana. The whites only, particularly the soldiers of the little fort of San Carlos, know how to procure pure salt, either from the coast of Caracas, or from Chita[*] by the Rio Meta. Here, as throughout America, the Indians eat little meat, and consume scarcely any salt. The chivi of Javita is a mixture of muriate of potash and of soda, of caustic lime, and of several other earthy salts. The Indians dissolve a few particles in water, fill with this solution a leaf of heliconia folded in a conical form, and let drop a little, as from the extremity of a filter, on their food.

---

[*] North of Morocote, at the eastern declivity of the Cordillera of New Grenada. The salt of the coasts, which the Indians call yuquira, costs two piastres the almuda at San Carlos.

On the 5th of May we set off, to follow on foot our canoe, which had at length arrived, by the portage, at the Cano Pimichin. We had to ford a great number of streams; and these passages require some caution on account of the vipers with which the marshes abound. The Indians pointed out to us on the moist clay the traces of the little black bears so common on the banks of the Temi. They differ at least in size from the Ursus americanus. The missionaries call them osso carnicero, to distinguish them from the osso palmero or tamanoir (Myrmecophaga jubata), and from the osso hormigero, or anteater (tamandua). The flesh of these animals is good to eat; the first two defend themselves by rising on their hind feet. The tamanoir of Buffon is called uaraca by the Indians; it is irascible and courageous, which is extraordinary in an animal without teeth. We found, as we advanced, some vistas in the forest, which appeared to us the richer, as it became more accessible. We here gathered some new species of coffee (the American tribe, with flowers in panicles, forms probably a particular genus); the Galega piscatorum, of which the Indians make use, as they do of jacquinia, and of a composite plant of the Rio Temi, as a kind of barbasco, to intoxicate fish; and finally, the liana, known in those countries by the name of vejuco de mavacure, which yields the famous curare poison. It is neither a phyllanthus, nor a coriaria, as M. Willdenouw conjectured, but, as M. Kunth's researches show, very probably a strychnos. We shall have occasion, farther on, to speak of this venomous substance, which is an important object of trade among the savages.

The trees of the forest of Pimichin have the gigantic height of from eighty to a hundred and twenty feet. In these burning climates the laurineae and amyris[*] furnish that fine timber for building, which, on the north-west coast of America, on mountains where the thermometer falls in winter to 20°

---

[*] The great white and red cedars of these countries are not the Cedrela odorata, but the Amyris altissima, which is an icica of Aublet.

centigrade below zero, we find in the family of the coniferae. Such, in every zone, and in all the families of American plants, is the prodigious force of vegetation, that, in the latitude of fifty-seven degrees north, on the same isothermal line with St. Petersburgh and the Orkneys, the Pinus canadensis displays trunks one hundred and fifty feet high, and six feet in diameter.[*] Towards night we arrived at a small farm, in the puerto or landing place of Pimichin. We were shown a cross near the road, which marked the spot where a poor capuchin missionary had been killed by wasps. I state this on the authority of the monks of Javita and the Indians. They talk much in these countries of wasps and venomous ants, but we saw neither one nor the other of these insects. It is well known that in the torrid zone slight stings often cause fits of fever almost as violent as those that with us accompany severe organic injuries. The death of this poor monk was probably the effect of fatigue and damp, rather than of the venom contained in the stings of wasps, which the Indians dread extremely. We must not confound the wasps of Javita with the melipones bees, called by the Spaniards angelitos (little angels) which covered our faces and hands on the summit of the Silla de Caracas.

The landing place of Pimichin is surrounded by a small plantation of cacao-trees; they are very vigorous, and here, as on the banks of the Atabapo and the Guainia, they are loaded with flowers and fruits at all seasons. They begin to bear from the fourth year; on the coast of Caracas they do not bear till the sixth or eighth year. The soil of these countries is sandy, wherever it is not marshy; but the light lands of the Tuamini and Pimichin are

---

[*] Langsdorf informs us that the inhabitants of Norfolk Sound make boats of a single trunk, fifty feet long, four feet and a half broad, and three high at the sides. They contain thirty persons. These boats remind us of the canoes of the Rio Chagres in the isthmus of Panama, in the torrid zone. The Populus balsamifera also attains an immense height, on the mountains that border Norfolk Sound.

extremely productive.* Around the conucos of Pimichin grows, in its wild state, the igua, a tree resembling the Caryocar nuciferum which is cultivated in Dutch and French Guiana, and which, with the almendron of Mariquita (Caryocar amygdaliferum), the juvia of the Esmeralda (Bertholletia excelsa), and the Geoffroea of the Amazon, yields the finest almonds of all South America. No commercial advantage is here made of the igua; but I saw vessels arrive on the coast of Terra Firma, that came from Demerara laden with the fruit of the Caryocar tomentosum, which is the Pekea tuberculosa of Aublet. These trees reach a hundred feet in height, and present, by the beauty of their corolla, and the multitude of their stamens, a magnificent appearance. I should weary the reader by continuing the enumeration of the vegetable wonders which these vast forests contain. Their variety depends on the coexistence of such a great number of families in a small space of ground, on the stimulating power of light and heat, and on the perfect elaboration of the juices that circulate in these gigantic plants.

We passed the night in a hut lately abandoned by an Indian family, who had left behind them their fishing-tackle, pottery, nets made of the petioles of palm-trees; in short, all that composes the household furniture of that careless race of men, little attached to property. A great store of mani (a mixture of the

---

* At Javita, an extent of fifty feet square, planted with Jatropha manihot (yucca) yields in two years, in the worst soil, a harvest of six tortas of cassava: the same extent on a middling soil yields in fourteen months a produce of nine tortas. In an excellent soil, around clumps of mauritia, there is every year from fifty feet square a produce of thirteen or fourteen tortas. A torta weighs three quarters of a pound, and three tortas cost generally in the province of Caracas one silver rial, or one-eighth of a piastre. These statements appear to me to be of some importance, when we wish to compare the nutritive matter which man can obtain from the same extent of soil, by covering it, in different climates, with bread-trees, plantains, jatropha, maize, potatoes, rice, and corn. The tardiness of the harvest of jatropha has, I believe, a beneficial influence on the manners of the natives, by fixing them to the soil, and compelling them to sojourn long on the same spot.

resin of the moronoboea and the Amyris carana) was accumulated round the house. This is used by the Indians here, as at Cayenne, to pitch their canoes, and fix the bony spines of the ray at the points of their arrows. We found in the same place jars filled with a vegetable milk, which serves as a varnish, and is celebrated in the missions by the name of leche para pintar (milk for painting). They coat with this viscous juice those articles of furniture to which they wish to give a fine white colour. It thickens by the contact of the air, without growing yellow, and it appears singularly glossy. We have already mentioned that the caoutchouc is the oily part, the butter of all vegetable milk. It is, no doubt, a particular modification of caoutchouc that forms this coagulum, this white and glossy skin, that seems as if covered with copal varnish. If different colours could be given to this milky varnish, a very expeditious method would be found of painting and varnishing our carriages by one process. The more we study vegetable chemistry in the torrid zone, the more we shall discover, in remote spots, and half-prepared in the organs of plants, products which we believe belong only to the animal kingdom, or which we obtain by processes which are often tedious and difficult. Already we have found the wax that coats the palm-tree of the Andes of Quindiu, the silk of the palm-tree of Mocoa, the nourishing milk of the palo de vaca, the butter-tree of Africa, and the caseous substances obtained from the almost animalized sap of the Carica papaya. These discoveries will be multiplied, when, as the political state of the world seems now to indicate, European civilization shall flow in a great measure toward the equinoctial regions of the New Continent.

The marshy tract between Javita and the embarcadero of Pimichin is infested with great numbers of vipers. Before we took possession of the deserted hut, the Indians killed two great mapanare serpents.[*] These grow to four or five feet long. They

---

[*] This name is given in the Spanish colonies to very different species. The Coluber mapanare of the province of Caracas has one hundred and forty-two

appeared to me to be the same species as those I saw in the Rio Magdalena. This serpent is a beautiful animal, but extremely venomous, white on the belly, and spotted with brown and red on the back. As the inside of the hut was filled with grass, and we were lying on the ground, there being no means of suspending our hammocks, we were not without inquietude during the night. In the morning a large viper was found on lifting the jaguar-skin upon which one of our domestics had slept. The Indians say that these reptiles, slow in their movements when they are not pursued, creep near a man because they are fond of heat. In fact, on the banks of the Magdalena a serpent entered the bed of one of our fellow-travellers, and remained there a part of the night, without injuring him. Without wishing to take up the defence of vipers and rattlesnakes, I believe it may be affirmed that, if these venomous animals had such a disposition for offence as is supposed, the human species would certainly not have withstood their numbers in some parts of America; for instance, on the banks of the Orinoco and the humid mountains of Choco.

We embarked on the 8th of May at sunrise, after having carefully examined the bottom of our canoe. It had become thinner, but had received no crack in the portage. We reckoned that it would still bear the voyage of three hundred leagues, which we had yet to perform, in going down the Rio Negro, ascending the Cassiquiare, and redescending the Orinoco as far as Angostura. The Pimichin, which is called a rivulet (cano) is tolerably broad; but small trees that love the water narrow the bed so much that there remains open a channel of only fifteen or twenty toises. Next to the Rio Chagres this river is one of the most celebrated in America for the number of its windings: it is said to have eighty-five, which greatly lengthen it. They often form right angles, and occur every two or three leagues. To

---

ventral plates, and thirty-eight double caudal scales. The Coluber mapanare of the Rio Magdalena has two hundred and eight ventral plates, and sixty-four double caudal scales.

determine the difference of longitude between the landing-place and the point where we were to enter the Rio Negro, I took by the compass the course of the Cano Pimichin, and noted the time during which we followed the same direction. The velocity of the current was only 2.4 feet in a second; but our canoe made by rowing 4.6 feet. The embarcadero of the Pimichin appeared to me to be eleven thousand toises west of its mouth, and 0° 2' west of the mission of Javita. This Cano is navigable during the whole year, and has but one raudal, which is somewhat difficult to go up; its banks are low, but rocky. After having followed the windings of the Pimichin for four hours and a half we at length entered the Rio Negro.

The morning was cool and beautiful. We had now been confined thirty-six days in a narrow boat, so unsteady that it would have been overset by any person rising imprudently from his seat, without warning the rowers. We had suffered severely from the sting of insects, but we had withstood the insalubrity of the climate; we had passed without accident the great number of waterfalls and bars, which impede the navigation of the rivers, and often render it more dangerous than long voyages by sea. After all we had endured, it may be conceived that we felt no little satisfaction in having reached the tributary streams of the Amazon, having passed the isthmus that separates two great systems of rivers, and in being sure of having fulfilled the most important object of our journey, namely, to determine astronomically the course of that arm of the Orinoco which falls into the Rio Negro, and of which the existence has been alternately proved and denied during half a century. In proportion as we draw near to an object we have long had in view, its interest seems to augment. The uninhabited banks of the Cassiquiare, covered with forests, without memorials of times past, then occupied my imagination, as do now the banks of the Euphrates, or the Oxus, celebrated in the annals of civilized nations. In that interior part of the New Continent one may almost accustom oneself to regard men as not being essential to the order of nature. The earth is loaded with plants,

405

and nothing impedes their free development. An immense layer of mould manifests the uninterrupted action of organic powers. Crocodiles and boas are masters of the river; the jaguar, the peccary, the dante, and the monkeys traverse the forest without fear and without danger; there they dwell as in an ancient inheritance. This aspect of animated nature, in which man is nothing, has something in it strange and sad. To this we reconcile ourselves with difficulty on the ocean, and amid the sands of Africa; though in scenes where nothing recalls to mind our fields, our woods, and our streams, we are less astonished at the vast solitude through which we pass. Here, in a fertile country, adorned with eternal verdure, we seek in vain the traces of the power of man; we seem to be transported into a world different from that which gave us birth. These impressions are the more powerful in proportion as they are of long duration. A soldier, who had spent his whole life in the missions of the Upper Orinoco, slept with us on the bank of the river. He was an intelligent man, who, during a calm and serene night, pressed me with questions on the magnitude of the stars, on the inhabitants of the moon, on a thousand subjects of which I was as ignorant as himself. Being unable by my answers to satisfy his curiosity, he said to me in a firm tone of the most positive conviction: "with respect to men, I believe there are no more up there than you would have found if you had gone by land from Javita to Cassiquiare. I think I see in the stars, as here, a plain covered with grass, and a forest (mucho monte) traversed by a river." In citing these words I paint the impression produced by the monotonous aspect of those solitary regions. May this monotony not be found to extend to the journal of our navigation, and weary the reader accustomed to the description of the scenes and historical memorials of the old continent!

CHAPTER 2.23.

THE RIO NEGRO. BOUNDARIES OF BRAZIL. THE CASSIQUIARE. BIFURCATION OF THE ORINOCO.

The Rio Negro, compared to the Amazon, the Rio de la Plata, or the Orinoco, is but a river of the second order. Its possession has been for ages of great political importance to the Spanish Government, because it is capable of furnishing a rival power, Portugal, with an easy passage into the missions of Guiana, and thereby disturbing the Capitania general of Caracas in its southern limits. Three hundred years have been spent in vain territorial disputes. According to the difference of times, and the degree of civilization among the natives, resource has been had sometimes to the authority of the Pope, and sometimes the support of astronomy; and the disputants being generally more interested in prolonging than in terminating the struggle, the nautical sciences and the geography of the New Continent, have alone gained by this interminable litigation. When the affairs of Paraguay, and the possession of the colony of Del Sacramento, became of great importance to the courts of Madrid and Lisbon, commissioners of the boundaries were sent to the Orinoco, the Amazon, and the Rio Plata.

The little that was known, up to the end of the last century, of the astronomical geography of the interior of the New Continent, was owing to these estimable and laborious men, the French and Spanish academicians, who measured a meridian line at Quito, and to officers who went from Valparaiso to Buenos Ayres to join the expedition of Malaspina. Those persons who know the inaccuracy of the maps of South America, and have seen those uncultivated lands between the Jupura and the Rio Negro, the Madeira and the Ucayale, the Rio Branco and the coasts of Cayenne, which up to our own days have been gravely disputed in Europe, can be not a little surprised at the perseverance with which the possession of a few square leagues is litigated. These disputed grounds are generally separated from the cultivated part of the colonies by deserts, the extent of which is unknown. In the celebrated conferences of Puente de Caya the question was agitated, whether, in fixing the line of demarcation three hundred and seventy Spanish leagues to the west of the Cape Verde Islands, the pope meant that the first meridian should be reckoned from the centre of the island of St. Nicholas, or (as the court of Portugal asserted) from the western extremity of the little island of St. Antonio. In the year 1754, the time of the expedition of Iturriaga and Solano, negociations were entered into respecting the possession of the then desert banks of the Tuamini, and of a marshy tract which we crossed in one evening going from Javita to Cano Pimichin. The Spanish commissioners very recently would have placed the divisional line at the point where the Apoporis falls into the Jupura, while the Portuguese astronomers carried it back as far as Salto Grande.

The Rio Negro and the Jupuro are two tributary streams of the Amazon, and may be compared in length to the Danube. The upper parts belong to the Spaniards, while the lower are occupied by the Portuguese. The Christian settlements are very numerous from Mocoa to the mouth of the Caguan; while on the Lower Jupura the Portuguese have founded only a few villages. On the Rio Negro, on the contrary, the Spaniards have not been able to rival their neighbours. Steppes and forests nearly desert separate, at a distance of one hundred and sixty leagues, the cultivated part of the coast from the four missions of Marsa, Tomo, Davipe, and San Carlos, which are all that the Spanish Franciscans could establish along the Rio Negro. Among the Portuguese of Brazil the military system, that of presides and capitanes pobladores, has prevailed over the government of the missionaries. Grand Para is no doubt far distant from the mouth of the Rio Negro: but the facility of navigation on the Amazon, which runs like an immense canal in one direction from west to east, has enabled the Portuguese population to extend itself rapidly along the river. The banks of the Lower Maranon, from Vistoza as far as Serpa, as well as those of the Rio Negro from Fort da Bara to San Jose da Maravitanos, are embellished by rich cultivation, and by a great number of large villages and towns.

These local considerations are combined with others, suggested by the moral position of nations. The north-west coast of America furnishes to this day no other stable settlements but Russian and Spanish colonies. Before the inhabitants of the United States, in their progressive movement from east to west, could reach the shore between the latitude 41 and 50°, which long separated the Spanish monks and the Siberian hunters,[*] the latter had established themselves south of the Columbia River. Thus in New California the Franciscan missionaries, men estimable for their morals, and their agricultural activity, learnt with astonishment, that Greek priests had arrived in their neighbourhood; and that two nations, who inhabit the eastern and western extremities of Europe, were become neighbours on a coast of America opposite to China. In Guiana circumstances were very different: the Spaniards found on their frontiers those very Portuguese, who, by their language, and their municipal institutions, form with them one of the most noble remains of Roman Europe; but whom mistrust, founded on unequal strength, and too great proximity, has converted into an often hostile, and always rival power.

If two nations adjacent to each other in Europe, the Spaniards and the Portuguese, have alike become neighbours in the New Continent, they are indebted for that circumstance to

---

[*] The hunters connected with military posts, and dependent on the Russian Company, of which the principal shareholders live at Irkutsk. In 1804 the little fortress (krepost) at the bay of Jakutal was still six hundred leagues distant from the most northern Mexican possessions.

the spirit of enterprise and active courage which both displayed at the period of their military glory and political greatness. The Castilian language is now spoken in North and South America throughout an extent of more than one thousand nine hundred leagues in length; if, however, we consider South America apart, we there find the Portuguese language spread over a larger space of ground, and spoken by a smaller number of individuals than the Castilian. It would seem as if the bond that so closely connects the fine languages of Camoens and Lope de Vega, had served only to separate two nations, who have become neighbours against their will. National hatred is not modified solely by a diversity of origin, of manners, and of progress in civilization; whenever it is powerful, it must be considered as the effect of geographical situation, and the conflicting interests thence resulting. Nations detest each other the less, in proportion as they are distant; and when, their languages being radically different, they do not even attempt to combine together. Travellers who have passed through New California, the interior provinces of Mexico, and the northern frontiers of Brazil, have been struck by these shades in the moral dispositions of bordering nations.

When I was in the Spanish Rio Negro, the divergent politics of the courts of Lisbon and Madrid had augmented that system of mistrust which, even in calmer times, the commanders of petty neighbouring forts love to encourage. Boats went up from Barcelos as far as the Spanish missions, but the communications were of rare occurrence. A commandant with sixteen or eighteen soldiers wearied the garrison by measures of safety, which were dictated by the important state of affairs; if he were attacked, he hoped to surround the enemy. When we spoke of the indifference with which the Portuguese government doubtless regarded the four little villages founded by the monks of Saint Francisco, on the Upper Guainia, the inhabitants were hurt by the motives which we alleged with the view to give them confidence. A people who have preserved in vigour, through the revolutions of ages, a national hatred, like occasions of giving it vent. The mind delights in everything impassioned, in the consciousness of an energetic feeling, in the affections, and in rival hatreds that are founded on antiquated prejudices. Whatever constitutes the individuality of nations flows from the mother-country to the most remote colonies; and national antipathies are not effaced where the influence of the same languages ceases. We know, from the interesting narrative of Krusenstern's voyage, that the hatred of two fugitive sailors, one a Frenchman and the other an Englishman, was the cause of a long war between the inhabitants of the Marquesas Islands. On the banks of the Amazon and the Rio Negro, the Indians of the neighbouring Portuguese and Spanish villages detest each other. These poor people speak only the native tongues; they are ignorant of what passes on the other bank of the ocean, beyond the great salt-pool; but the gowns of their missionaries are of a different colour, and this displeases them extremely.

I have stopped to paint the effects of national animosities, which wise statesmen have endeavoured to calm, but have been unable entirely to set at rest. This rivalry has contributed to the imperfection of the geographical knowledge hitherto obtained respecting the tributary rivers of the Amazon. When the communications of the natives are impeded, and one nation is established near the mouth, and another in the upper part of the same river, it is difficult for persons who attempt to construct maps to acquire precise information. The periodical inundations, and still more the portages, by which boats are passed from one stream to another, the sources of which are in the same neighbourhood, have led to erroneous ideas of the bifurcations and branchings of rivers. The Indians of the Portuguese missions, for instance, enter (as I was informed upon the spot) the Spanish Rio Negro on one side by the Rio Guainia and the Rio Tomo; and the Upper Orinoco on the other, by the portages between the Cababuri, the Pacimoni, the Idapa, and the Macava, to gather the aromatic seeds of the puchero laurel beyond the Esmeralda. The Indians, I repeat, are excellent geographers; they outflank the enemy, notwithstanding the limits traced upon the maps, in spite of the forts and the estacamentos; and when the missionaries see them arrive from such distances, and in different seasons, they begin to frame hypotheses of supposed communications of rivers. Each party has an interest in concealing what it knows with certainty; and that love of the mysterious, so general among the ignorant, contributes to perpetuate the doubt. It may also be observed that the various Indian nations, who frequent this labyrinth of rivers, give them names entirely different; and that these names are disguised and lengthened by terminations that signify water, great water, and current. How often have I been perplexed by the necessity of settling the synonyms of rivers, when I have sent for the most intelligent natives, to interrogate them, through an interpreter, respecting the number of tributary streams, the sources of the rivers, and the portages. Three or four languages being spoken in the same mission, it is difficult to make the witnesses agree. Our maps are loaded with names arbitrarily shortened or perverted. To examine how far they may be accurate, we must be guided by the geographical situation of the confluent rivers, I might almost say by a certain etymological tact. The Rio Uaupe, or Uapes of the Portuguese maps, is the Guapue of the Spanish maps, and the Ucayari of the natives. The Anava of the old geographers is the Anauahu of Arrowsmith, and the Uanauhau or Guanauhu of the Indians. The desire of leaving no void in the maps, in order to give them an appearance of accuracy, has caused rivers to be created, to which names have been applied that have not been recognized as synonymous. It is only lately that travellers in America, in Persia, and in the Indies, have felt the importance of being correct in the denomination of places. When we read the travels of Sir Walter Raleigh, it is difficult indeed to recognise in the lake of Mrecabo, the laguna of Maracaybo, and in the Marquis Paraco the name of Pizarro, the destroyer of the empire of the Incas.

The great tributary streams of the Amazon are designated by the missionaries by different names in their upper and lower course. The Iza is called, higher up, Putumayo, the Jupura towards its source bears the name of Caqueta. The researches made in the missions of the Andaquies on the real origin of the Rio Negro have been the more fruitless because the Indian name of the river was unknown. I heard it called Guainia at Javita, Maroa, and San Carlos. Southey, in his history of Brazil, says expressly that the Rio Negro, in the lower part of its course, is called Guiani, or Curana, by the natives; in the upper part, Ueneya. It is the word Gueneya, instead of Guainia; for the Indians of those countries say indifferently Guaranacua or Ouaranacua, Guarapo or Uarapo.

The sources of the Rio Negro have long been an object of contention among geographers. The interest we feel in this question is not merely that which attaches to the origin of all great rivers, but is connected with a crowd of other questions, that comprehend the supposed bifurcations of the Caqueta, the communications between the Rio Negro and the Orinoco, and the local fable of El Dorado, formerly called Enim, or the empire of the Grand Paytiti. When we study with care the ancient maps of these countries, and the history of their geographical errors, we see how by degrees the fable of El Dorado has been transported towards the west with the sources of the Orinoco. It was at first fixed on the eastern declivity of the Andes, to the south-west of the Rio Negro. The valiant Philip de Urre sought for the great city of Manoa by traversing the Guaviare. Even now the Indians of San Jose de Maravitanos relate that, on sailing to the north-east for fifteen days, on the Guape or Uaupe, you reach a famous laguna de oro, surrounded by mountains, and so large that the opposite shore cannot be discerned. A ferocious nation, the Guanes, do not permit the collecting of the gold of a sandy plain that surrounds the lake. Father Acunha places the lake Manoa, or Yenefiti, between the Jupura and the Rio Negro. Some Manoa Indians brought Father Fritz, in 1687, several slips of beaten gold. This nation, the name of which is still known on the banks of the Urarira, between Lamalongo and Moreira, dwelt on the Yurubesh. La Condamine is right in saying that this Mesopotamia, between the Caqueta, the Rio

Negro, the Yurubesh, and the Iquiare, was the first scene of El Dorado. But where shall we find the names of Yurubesh and Iquiare, given by the Fathers Acunha and Fritz? I think I recognise them in the rivers Urubaxi and Iguari,[*] on some manuscript Portuguese maps which I possess. I have long and assiduously studied the geography of South America, north of the Amazon, from ancient maps and unpublished materials. Desirous that my work should preserve the character of a scientific performance, I ought not to hesitate about treating of subjects on which I flatter myself that I can throw some light; namely, on the questions respecting the sources of the Rio Negro and the Orinoco, the communication between these rivers and the Amazon, and the problem of the auriferous soil, which has cost the inhabitants of the New World so much suffering and so much blood.

In the distribution of the waters circulating on the surface of the globe, as well as in the structure of organic bodies, nature has pursued a much less complicated plan than has been believed by those who have suffered themselves to be guided by vague conceptions and a taste for the marvellous. We find, too, that all anomalies, all the exceptions to the laws of hydrography, which the interior of America displays, are merely apparent; that the course of running waters furnishes phenomena equally extraordinary in the old world, but that these phenomena, from their littleness, have less struck the imagination of travellers. When immense rivers may be considered as composed of several parallel furrows of unequal depth; when these rivers are not enclosed in valleys; and when the interior of the great continent is as flat as the shores of the sea with us; the ramifications, the bifurcations, and the interlacings in the form of net-work, must be infinitely multiplied. From what we know of the equilibrium

---

[*] It may be written Urubaji. The j and the x were the same as the German ch to Father Fritz. The Urubaxi, or Hyurubaxi (Yurubesh), falls into the Rio Negro near Santa Isabella; the Iguari (Iquiare?) runs into the Issana, which is also a tributary of the Rio Negro.

of the seas, I cannot think that the New World issued from the waters later than the Old, and that organic life is there younger, or more recent; but without admitting oppositions between the two hemispheres of the same planet, we may conceive that in the hemisphere most abundant in waters the different systems of rivers required more time to separate themselves from one another, and establish their complete independence. The deposits of mud, which are formed wherever the running waters lose somewhat of their swiftness, contribute, no doubt, to raise the beds of the great confluent streams, and augment their inundations; but at length these deposits entirely obstruct the branches of the rivers and the narrow channels that connect the neighbouring streams. The substances washed down by rain-waters form by their accumulation new bars, isthmuses of deposited earth, and points of division that did not before exist. It hence results that these natural channels of communication are by degrees divided into two tributary streams, and from the effect of a transverse rising, acquire two opposite slopes; a part of their waters is turned back towards the principal recipient, and a buttress rises between the two parallel basins, which occasions all traces of their ancient communication to disappear. From this period the bifurcations no longer connect different systems of rivers; and, where they continue to take place at the time of great inundations, we see that the waters diverge from the principal recipient only to enter it again after a longer or shorter circuit. The limits, which at first appeared vague and uncertain, begin to be fixed; and in the lapse of ages, from the action of whatever is moveable on the surface of the globe, from that of the waters, the deposits, and the sands, the basins of rivers separate, as great lakes are subdivided, and as inland seas lose their ancient communications.[*]

---

[*] The geological constitution of the soil seems to indicate that, notwithstanding the actual difference of level in their waters, the Black Sea, the Caspian, and lake Aral, communicated with each other in an era anterior to historic times. The overflowing of the Aral into the Caspian Sea seems

415

The certainty acquired by geographers since the sixteenth century, of the existence of several bifurcations, and the mutual dependence of various systems of rivers in South America, have led them to admit an intimate connection between the five great tributary streams of the Orinoco and the Amazon; the Guaviare, the Inirida, the Rio Negro, the Caqueta or Hyapura, and the Putumayo or Iza.

The Meta, the Guaviare, the Caqueta, and the Putumayo, are the only great rivers that rise immediately from the eastern declivity of the Andes of Santa Fe, Popayan, and Pasto. The Vichada, the Zama, the Inirida, the Rio Negro, the Uaupe, and the Apoporis, which are marked in our maps as extending westward as far as the mountains, take rise at a great distance from them, either in the savannahs between the Meta and the Guaviare, or in the mountainous country which, according to the information given me by the natives, begins at four or five days' journey westward of the missions of Javita and Maroa, and extends through the Sierra Tuhuny, beyond the Xie, towards the banks of the Issana.

It is remarkable that this ridge of the Cordilleras, which contains the sources of so many majestic rivers (the Meta, the Guaviare, the Caqueta, and the Putumayo), is as little covered with snow as the mountains of Abyssinia from which flow the waters of the Blue Nile; but, on the contrary, on going up the tributary streams which furrow the plains, a volcano as found still in activity, before you reach the Cordillera of the Andes. This phenomenon was discovered by the Franciscan monks, who go down from Ceja by the Rio Fragua to Caqueta. A solitary hill, emitting smoke night and day, is found on the north-east of the mission of Santa Rosa, and west of the Puerto del Pescado. This is the effect of a lateral action of the volcanoes of Popayan and Pasto; as Guacamayo and Sangay, situated also at the foot of the

---

even to be partly of a more recent date, and independent of the bifurcation of the Gihon (Oxus), on which one of the most learned geographers of our day, M. Ritter, has thrown new light.

eastern declivity of the Andes, are the effect of a lateral action produced by the system of the volcanoes of Quito. After having closely inspected the banks of the Orinoco and the Rio Negro, where the granite everywhere pierces the soil; when we reflect on the total absence of volcanoes in Brazil, Guiana, on the coast of Venezuela, and perhaps in all that part of the continent lying eastward of the Andes; we contemplate with interest the three burning volcanoes situated near the sources of the Caqueta, the Napo, and the Rio de Macas or Morona.

The little group of mountains with which we became acquainted at the sources of the Guainia, is remarkable from its being isolated in the plain that extends to the south-west of the Orinoco. Its situation with regard to longitude might lead to the belief that it stretches into a ridge, which forms first the strait (angostura) of the Guaviare, and then the great cataracts (saltos, cachoeiras) of the Uaupe and the Jupura. Does this ground, composed probably of primitive rocks, like that which I examined more to the east, contain disseminated gold? Are there any gold-washings more to the south, toward the Uaupe, on the Iquiare (Iguiari, Iguari), and on the Yurubesh (Yurubach, Urubaxi)? It was there that Philip von Huten first sought El Dorado, and with a handful of men fought the battle of Omaguas, so celebrated in the sixteenth century. In separating what is fabulous from the narratives of the Conquistadores, we cannot fail to recognize in the names preserved on the same spots a certain basis of historic truth. We follow the expedition of Huten beyond the Guaviare and the Caqeta; we find in the Guaypes, governed by the cacique of Macatoa, the inhabitants of the river of Uaupe, which also bears the name of Guape, or Guapue; we call to mind, that Father Acunha calls the Iquiari (Quiquiare) a gold river; and that fifty years later Father Fritz, a missionary of great veracity, received, in the mission of Yurimaguas, the Manaos (Manoas), adorned with plates of beaten gold, coming from the country between the Uaupe and the Caqueta, or Jupura. The rivers that rise on the eastern declivity of the Andes (for instance the Napo) carry along with them a great deal of gold, even when their sources are found in trachytic soils. Why may there not be an alluvial auriferous soil to the east of the Cordilleras, as there is to the west, in the Sonoro, at Choco, and at Barbacoas? I am far from wishing to exaggerate the riches of this soil; but I do not think myself authorized to deny the existence of precious metals in the primitive mountains of Guiana, merely because in our journey through that country we saw no metallic veins. It is somewhat remarkable that the natives of the Orinoco have a name in their languages for gold (carucuru in Caribbee, caricuri in Tamanac, cavitta in Maypure), while the word they use to denote silver, prata, is manifestly borrowed from the Spanish.[*] The notions collected by Acunha, Father Fritz, and La Condamine, on the gold-washings south and north of the river Uaupe, agree with what I learnt of the auriferous soil of those countries. However great we may suppose the communications that took place between the nations of the Orinoco before the arrival of Europeans, they certainly did not draw their gold from the eastern declivity of the Cordilleras. This declivity is poor in mines, particularly in mines anciently worked; it is almost entirely composed of volcanic rocks in the provinces of Popayan, Pasto, and Quito. The gold of Guiana probably came from the country east of the Andes. In our days a lump of gold has been found in a ravine near the mission of Encaramada, and we must not be surprised if, since Europeans settled in these wild spots, we hear less of the plates of gold, gold-dust, and amulets of jade-stone, which could heretofore be obtained from the Caribs and other wandering nations by barter. The precious

---

[*] The Parecas say, instead of prata, rata. It is the Castilian word plata ill-pronounced. Near the Yurubesh there is another inconsiderable tributary stream of the Rio Negro, the Curicur-iari. It is easy to recognize in this name the Caribbee word carucur, gold. The Caribs extended their incursions from the mouth of the Orinoco south-west toward the Rio Negro; and it was this restless people who carried the fable of El Dorado, by the same way, but in an opposite direction (from south-west to north-east), from the Mesopotamia between the Rio Negro and the Jupura to the sources of the Rio Branco.

metals, never very abundant on the banks of the Orinoco, the Rio Negro, and the Amazon, disappeared almost entirely when the system of the missions caused the distant communications between the natives to cease.

The banks of the Upper Guainia in general abound much less in fishing-birds than those of Cassiquiare, the Meta, and the Arauca, where ornithologists would find sufficient to enrich immensely the collections of Europe. This scarcity of animals arises, no doubt, from the want of shoals and flat shores, as well as from the quality of the black waters, which (on account of their very purity) furnish less aliment to aquatic insects and fish. However, the Indians of these countries, during two periods of the year, feed on birds of passage, which repose in their long migrations on the waters of the Rio Negro. When the Orinoco begins to swell[*] after the vernal equinox, an innumerable quantity of ducks (patos careteros) remove from the eighth to the third degree of north latitude, to the first and fourth degree of south latitude, towards the south-south-east. These animals then abandon the valley of the Orinoco, no doubt because the increasing depth of waters, and the inundations of the shores, prevent them from catching fish, insects, and aquatic worms. They are killed by thousands in their passage across the Rio Negro. When they go towards the equator they are very fat and savoury; but in the month of September, when the Orinoco decreases and returns into its bed, the ducks, warned either by the voices of the most experienced birds of passage, or by that internal feeling which, not knowing how to define, we call instinct, return from the Amazon and the Rio Branco towards the north. At this period they are too lean to tempt the appetite of the Indians of the Rio Negro, and escape pursuit more easily from being accompanied by a species of herons (gavanes) which are

---

[*] The swellings of the Nile take place much later than those of the Orinoco; after the summer solstice, below Syene; and at Cairo in the beginning of July. The Nile begins to sink near that city generally about the 15th of October, and continues sinking till the 20th of May.

419

excellent eating. Thus the Indians eat ducks in March, and herons in September. We could not learn what becomes of the gavanes during the swellings of the Orinoco, and why they do not accompany the patos careteros in their migration from the Orinoco to the Rio Branco. These regular migrations of birds from one part of the tropics towards another, in a zone which is during the whole year of the same temperature, are very extraordinary phenomena. The southern coasts of the West India Islands receive also every year, at the period of the inundations of the great rivers of Terra Firma, numerous flights of the fishing-birds of the Orinoco, and of its tributary streams. We must presume that the variations of drought and humidity in the equinoctial zone have the same influence as the great changes of temperature in our climates, on the habits of animals. The heat of summer, and the pursuit of insects, call the humming-birds into the northern parts of the United States, and into Canada as far as the parallels of Paris and Berlin: in the same manner a greater facility for fishing draws the web-footed and long-legged birds from the north to the south, from the Orinoco towards the Amazon. Nothing is more marvellous, and nothing is yet known less clearly in a geographical point of view, than the direction, extent, and term of the migrations of birds.

After having entered the Rio Negro by the Pimichin, and passed the small cataract at the confluence of the two rivers, we discovered, at the distance of a quarter of a league, the mission of Maroa. This village, containing one hundred and fifty Indians, presented an appearance of ease and prosperity. We purchased some fine specimens of the toucan alive; a courageous bird, the intelligence of which is developed like that of our domestic ravens. We passed on the right, above Maroa, first the mouth of the Aquio[*], then that of the Tomo.[†] On the banks of the latter river dwell the Cheruvichahenas, some families of whom I have seen at San Francisco Solano. The Tomo lies near the Rio

---

[*] Aqui, Aaqui, Ake, of the most recent maps.
[†] Tomui, Temujo, Tomon.

Guaicia (Xie), and the mission of Tomo receives by that way fugitive Indians from the Lower Guainia. We did not enter the mission, but Father Zea related to us with a smile, that the Indians of Tomo and Maroa had been one day in full insurrection, because an attempt was made to force them to dance the famous dance of the devils. The missionary had taken a fancy to have the ceremonies by which the piaches (who are at once priests, physicians, and conjurors) evoke the evil spirit Iolokiamo, represented in a burlesque manner. He thought that the dance of the devils would be an excellent means of proving to the neophytes that Iolokiamo had no longer any power over them. Some young Indians, confiding in the promises of the missionary, consented to act the devils, and were already decorated with black and yellow plumes, and jaguar-skins with long sweeping tails. The place where the church stands was surrounded by the soldiers who are distributed in the missions, in order to add more effect to the counsels of the monks; and those Indians who were not entirely satisfied with respect to the consequences of the dance, and the impotency of the evil spirit, were brought to the festivity. The oldest and most timid of the Indians, however, imbued all the rest with a superstitious dread; all resolved to flee al monte, and the missionary adjourned his project of turning into derision the demon of the natives. What extravagant ideas may sometimes enter the imagination of an idle monk, who passes his life in the forests, far from everything that can recall human civilization to his mind. The violence with which the attempt was made to execute in public at Tomo the mysterious dance of the devils is the more strange, as all the books written by the missionaries relate the efforts they have used to prevent the funereal dances, the dances of the sacred trumpet, and that ancient dance of serpents, the Queti, in which these wily animals are represented as issuing from the forests, and coming to drink with the men in order to deceive them, and carry off the women.

After two hours' navigation from the mouth of the Tomo we arrived at the little mission of San Miguel de Davipe, founded in

1775, not by monks, but by a lieutenant of militia, Don Francisco Bobadilla. The missionary of the place, Father Morillo, with whom we spent some hours, received us with great hospitality. He even offered us Madeira wine, but, as an object of luxury, we should have preferred wheaten bread. The want of bread becomes more sensibly felt in length of time than that of a strong liquor. The Portuguese of the Amazon carry small quantities of Madeira wine, from time to time, to the Rio Negro; and the word madera, signifying wood in the Castilian language, the monks, who are not much versed in the study of geography, had a scruple of celebrating mass with Madeira wine, which they took for a fermented liquor extracted from the trunk of some tree, like palm-wine; and requested the guardian of the missions to decide, whether the vino de madera were wine from grapes, or the juice of a tree. At the beginning of the conquest, the question was agitated, whether it were allowable for the priests, in celebrating mass, to use any fermented liquor analogous to grape-wine. The question, as might have been foreseen, was decided in the negative.

At Davipe we bought some provisions, among which were fowls and a pig. This purchase greatly interested our Indians, who had been a long while deprived of meat. They pressed us to depart, in order to reach the island of Dapa, where the pig was to be killed and roasted during the night. We had scarcely time to examine in the convent (convento) the great stores of mani resin, and cordage of the chiquichiqui palm, which deserves to be more known in Europe. This cordage is extremely light; it floats upon the water, and is more durable in the navigation of rivers than ropes of hemp. It must be preserved at sea by being often wetted, and little exposed to the heat of the tropical sun. Don Antonio Santos, celebrated in the country for his journey in search of lake Parima, taught the Indians of the Spanish Rio Negro to make use of the petioles of the chiquichiqui, a palm-tree with pinnate leaves, of which we saw neither the flowers nor the fruit. This officer is the only white man who ever came from Angostura to Grand Para, passing by land from the sources of the Rio Carony to those of the Rio Branco. He had studied the mode of fabricating ropes from the chiquichiqui in the Portuguese colonies; and, on his return from the Amazon, he introduced this branch of industry into the missions of Guiana. It were to be wished that extensive rope-walks could be established on the banks of the Rio Negro and the Cassiquiare, in order to make these cables an article of trade with Europe. A small quantity is already exported from Angostura to the West Indies; and it costs from fifty to sixty per cent less than cordage of hemp. Young palm-trees only being employed, they must be planted and carefully cultivated.

A little above the mission of Davipe, the Rio Negro receives a branch of the Cassiquiare, the existence of which is a very remarkable phenomenon in the history of the branchings of rivers. This branch issues from the Cassiquiare, north of Vasiva, bearing the name of the Itinivini; and, after flowing for the length of twenty-five leagues through a flat and almost uninhabited country, it falls into the Rio Negro under the name of the Rio Conorichite. It appeared to me to be more than one hundred and twenty toises broad near its mouth. Although the current of the Conorichite is very rapid, this natural canal abridges by three days the passage from Davipe to Esmeralda. We cannot be surprised at a double communication between the Cassiquiare and the Rio Negro when we recollect that so many of the rivers of America form, as it were, deltas at their confluence with other rivers. Thus the Rio Branco and the Rio Jupura enter by a great number of branches into the Rio Negro and the Amazon. At the confluence of the Jupura there is a much more extraordinary phenomenon. Before this river joins the Amazon, the latter, which is the principal recipient, sends off three branches called Uaranapu, Manhama, and Avateparana, to the Jupura, which is but a tributary stream. The Portuguese astronomer, Ribeiro, has proved this important fact. The Amazon gives waters to the Jupura itself, before it receives that tributary stream.

The Rio Conorichite, or Itinivini, formerly facilitated the trade in slaves carried on by the Portuguese in the Spanish territory. The slave-traders went up by the Cassiquiare and the Cano Mee to Conorichite; and thence dragged their canoes by a portage to the rochelas of Manuteso, in order to enter the Atabapo. This abominable trade lasted till about the year 1756; when the expedition of Solano, and the establishment of the missions on the banks of the Rio Negro, put an end to it. Old laws of Charles V and Philip III[*] had forbidden under the most severe penalties (such as the being rendered incapable of civil employment, and a fine of two thousand piastres), the conversion of the natives to the faith by violent means, and sending armed men against them; but notwithstanding these wise and humane laws, the Rio Negro, in the middle of the last century, was no further interesting in European politics, than as it facilitated the entradas, or hostile incursions, and favoured the purchase of slaves. The Caribs, a trading and warlike people, received from the Portuguese and the Dutch, knives, fish-hooks, small mirrors, and all sorts of glass beads. They excited the Indian chiefs to make war against each other, bought their prisoners, and carried off, themselves, by stratagem or force, all whom they found in their way. These incursions of the Caribs comprehended an immense extent of land; they went from the banks of the Essequibo and the Carony, by the Rupunuri and the Paraguamuzi on one side, directly south towards the Rio Branco; and on the other, to the south-west, following the portages between the Rio Paragua, the Caura, and the Ventuario. The Caribs, when they arrived amid the numerous tribes of the Upper Orinoco, divided themselves into several bands, in order to reach, by the Cassiquiare, the Cababury, the Itinivini, and the Atabapo, on a great many points at once, the banks of the Guiainia or Rio Negro, and carry on the slave-trade with the Portuguese. Thus the unhappy natives, before they came into immediate contact with the Europeans, suffered from their

---

[*] 26 January 1523 and 10 October 1618.

proximity. The same causes produce everywhere the same effects. The barbarous trade which civilized nations have carried on, and still partially continue, on the coast of Africa, extends its fatal influence even to regions where the existence of white men is unknown.

Having quitted the mouth of the Conorichite and the mission of Davipe, we reached at sunset the island of Dapa, lying in the middle of the river, and very picturesquely situated. We were astonished to find on this spot some cultivated ground, and on the top of a small hill an Indian hut. Four natives were seated round a fire of brushwood, and they were eating a sort of white paste with black spots, which much excited our curiosity. These black spots proved to be vachacos, large ants, the hinder parts of which resemble a lump of grease. They had been dried, and blackened by smoke. We saw several bags of them suspended above the fire. These good people paid but little attention to us; yet there were more than fourteen persons in this confined hut, lying naked in hammocks hung one above another. When Father Zea arrived, he was received with great demonstrations of joy. The military are in greater numbers on the banks of the Rio Negro than on those of the Orinoco, owing to the necessity of guarding the frontiers; and wherever soldiers and monks dispute for power over the Indians, the latter are most attached to the monks. Two young women came down from their hammocks, to prepare for us cakes of cassava. In answer to some enquiries which we put to them through an interpreter, they answered that cassava grew poorly on the island, but that it was a good land for ants, and food was not wanting. In fact, these vachacos furnish subsistence to the Indians of the Rio Negro and the Guainia. They do not eat the ants as a luxury, but because, according to the expression of the missionaries, the fat of ants (the white part of the abdomen) is a very substantial food. When the cakes of cassava were prepared, Father Zea, whose fever seemed rather to sharpen than to enfeeble his appetite, ordered a little bag to be brought to him filled with smoked vachacos. He mixed these bruised insects with flour of cassava, which he pressed us to taste. It somewhat resembled rancid butter mixed with crumb of bread. The cassava had not an acid taste, but some remains of European prejudices prevented our joining in the praises bestowed by the good missionary on what he called an excellent ant paste.

The violence of the rain obliged us to sleep in this crowded hut. The Indians slept only from eight till two in the morning; the rest of the time they employed in conversing in their hammocks, and preparing their bitter beverage of cupana. They threw fresh fuel on the fire, and complained of cold, although the temperature of the air was at 21°. This custom of being awake, and even on foot, four or five hours before sunrise, is general among the Indians of Guiana. When, in the entradas, an attempt is made to surprise the natives, the hours chosen are those of the first sleep, from nine till midnight.

We left the island of Dapa long before daybreak; and notwithstanding the rapidity of the current, and the activity of our rowers, our passage to the fort of San Carlos del Rio Negro occupied twelve hours. We passed, on the left, the mouth of the Cassiquiare, and, on the right, the small island of Cumarai. The fort is believed in the country to be on the equatorial line; but, according to the observations which I made at the rocks of Culimacari, it is in 1° 54′ 11″.

We lodged at San Carlos with the commander of the fort, a lieutenant of militia. From a gallery in the upper part of the house we enjoyed a delightful view of three islands of great length, and covered with thick vegetation. The river runs in a straight line from north to south, as if its bed had been dug by the hand of man. The sky being constantly cloudy gives these countries a solemn and gloomy character. We found in the village a few juvia-trees which furnish the triangular nuts called in Europe the almonds of the Amazon, or Brazil-nuts. We have made it known by the name of Bertholletia excelsa. The trees attain after eight years' growth the height of thirty feet.

The military establishment of this frontier consisted of seventeen soldiers, ten of whom were detached for the security of the neighbouring missions. Owing to the extreme humidity of the air there are not four muskets in a condition to be fired. The Portuguese have from twenty-five to thirty men, better clothed and armed, at the little fort of San Jose de Maravitanos. We found in the mission of San Carlos but one garita,[*] a square house, constructed with unbaked bricks, and containing six field-pieces. The little fort, or, as they think proper to call it here, the Castillo de San Felipe, is situated opposite San Carlos, on the western bank of the Rio Negro.

The banks of the Upper Guainia will be more productive when, by the destruction of the forests, the excessive humidity of the air and the soil shall be diminished. In their present state of culture maize scarcely grows, and the tobacco, which is of the finest quality, and much celebrated on the coast of Caracas, is well cultivated only on spots amid old ruins, remains of the huts of the pueblo viejo (old town). Indigo grows wild near the villages of Maroa, Davipe, and Tomo. Under a different system from that which we found existing in these countries, the Rio Negro will produce indigo, coffee, cacao, maize, and rice, in abundance.

The passage from the mouth of the Rio Negro to Grand Para occupying only twenty or twenty-five days, it would not have taken us much more time to have gone down the Amazon as far as the coast of Brazil, than to return by the Cassiquiare and the Orinoco to the northern coast of Caracas. We were informed at San Carlos that, on account of political circumstances, it was difficult at that moment to pass from the Spanish to the Portuguese settlements; but we did not know till after our return to Europe the extent of the danger to which we should have been exposed in proceeding as far as Barcellos. It was known at

---

[*] This word literally signifies a sentry-box; but it is here employed in the sense of store-house or arsenal.

427

Brazil, possibly through the medium of the newspapers, that I was going to visit the missions of the Rio Negro, and examine the natural canal which unites two great systems of rivers. In those desert forests instruments had been seen only in the hands of the commissioners of the boundaries; and at that time the subaltern agents of the Portuguese government could not conceive how a man of sense could expose himself to the fatigues of a long journey, to measure lands that did not belong to him. Orders had been issued to seize my person, my instruments, and, above all, those registers of astronomical observations, so dangerous to the safety of states. We were to be conducted by way of the Amazon to Grand Para, and thence sent back to Lisbon. But fortunately for me, the government at Lisbon, on being informed of the zeal of its subaltern agents, instantly gave orders that I should not be disturbed in my operations; but that on the contrary they should be encouraged, if I traversed any part of the Portuguese possessions.

In going down the Guainia, or Rio Negro, you pass on the right the Cano Maliapo, and on the left the Canos Dariba and Eny. At five leagues distance, nearly in 1° 38' of north latitude, is the island of San Josef. A little below that island, in a spot where there are a great number of orange-trees now growing wild, the traveller is shown a small rock, two hundred feet high, with a cavern called by the missionaries the Glorieta de Cocuy. This summer-house (for such is the signification of the word glorieta in Spanish) recalls remembrances that are not the most agreeable. It was here that Cocuy, the chief of the Manitivitanos,[*] had his harem of women, and where he devoured the finest and fattest. The tradition of the harem and the orgies of Cocuy is more current in the Lower Orinoco than on the banks of the Guainia. At San Carlos the very idea that the chief of the

---

[*] At San Carlos there is still preserved an instrument of music, a kind of large drum, ornamented with very rude Indian paintings, which relate to the exploits of Cocuy.

Manitivitanos could be guilty of cannibalism is indignantly rejected.

The Portuguese government has established many settlements even in this remote part of Brazil. Below the Glorieta, in the Portuguese territory, there are eleven villages in an extent of twenty-five leagues. I know of nineteen more as far as the mouth of the Rio Negro, beside the six towns of Thomare, Moreira (near the Rio Demenene, or Uaraca, where dwelt anciently the Guiana Indians), Barcellos, San Miguel del Rio Branco, near the river of the same name (so well known in the fictions of El Dorado), Moura, and Villa de Rio Negro. The banks of this tributary stream of the Amazon alone are consequently ten times more thickly peopled than all the shores of the Upper and Lower Orinoco, the Cassiquiare, the Atabapo, and the Spanish Rio Negro.

Among the tributary streams which the Rio Negro receives from the north, three are particularly deserving of attention, because on account of their branchings, their portages, and the situation of their sources, they are connected with the often-discussed problem of the origin of the Orinoco. The most southern of these tributary streams are the Rio Branco,[*] which was long believed to issue conjointly with the Orinoco from lake Parime, and the Rio Padaviri, which communicates by a portage with the Mavaca, and consequently with the Upper Orinoco, to the east of the mission of Esmeralda. We shall have occasion to speak of the Rio Branco and the Padaviri, when we arrive in that mission; it suffices here to pause at the third tributary stream of

---

[*] The Portuguese name, Rio Branco, signifies White Water. Rio Parime is a Caribbean name, signifying Great Water. These names having also been applied to different tributary streams, have caused many errors in geography. The great Rio Branco, or Parime, often mentioned in this work, is formed by the Urariquera and the Tacutu, and flows, between Carvoeyro and Villa de Moura, into the Rio Negro. It is the Quecuene of the natives; and forms at its confluence with the Rio Negro a very narrow delta, between the principal trunk and the Amayauhau, which is a little branch more to the west.

the Rio Negro, the Cababury, the interbranchings of which with the Cassiquiare are alike important in their connexion with hydrography, and with the trade in sarsaparilla.

The lofty mountains of the Parime, which border the northern bank of the Orinoco in the upper part of its course above Esmeralda, send off a chain towards the south, of which the Cerro de Unturan forms one of the principal summits. This mountainous country, of small extent but rich in vegetable productions, above all, in the mavacure liana, employed in preparing the wourali poison, in almond-trees (the juvia, or Bertholletia excelsa), in aromatic pucheries, and in wild cacao-trees, forms a point of division between the waters that flow to the Orinoco, the Cassiquiare, and the Rio Negro. The tributary streams on the north, or those of the Orinoco, are the Mavaca and the Daracapo; those on the west, or of the Cassiquiare, are the Idapa and the Pacimoni; and those on the south, or of the Rio Negro, are the Padaviri and the Cababuri. The latter is divided near its source into two branches, the westernmost of which is known by the name of Baria. The Indians of the mission of San Francisco Solano gave us the most minute description of its course. It affords the very rare example of a branch by which an inferior tributary stream, instead of receiving the waters of the superior stream, sends to it a part of its own waters in a direction opposite to that of the principal recipient.

The Cababuri runs into the Rio Negro near the mission of Nossa Senhora das Caldas; but the rivers Ya and Dimity, which are higher tributary streams, communicate also with the Cababuri; so that, from the little fort of San Gabriel de Cachoeiras as far as San Antonio de Castanheira the Indians of the Portuguese possessions can enter the territory of the Spanish missions by the Baria and the Pacimoni.

The chief object of these incursions is the collection of sarsaparilla and the aromatic seeds of the puchery-laurel (Laurus pichurim). The sarsaparilla of these countries is celebrated at Grand Para, Angostura, Cumana, Nueva Barcelona, and in other parts of Terra Firma, by the name of zarza del Rio Negro. It is much preferred to the zarza of the Province of Caracas, or of the mountains of Merida; it is dried with great care, and exposed purposely to smoke, in order that it may become blacker. This liana grows in profusion on the humid declivities of the mountains of Unturan and Achivaquery. Decandolle is right in suspecting that different species of smilax are gathered under the name of sarsaparilla. We found twelve new species, among which the Smilax siphylitica of the Cassiquaire, and the Smilax officinalis of the river Magdalena, are most esteemed on account of their diuretic properties. The quantity of sarsaparilla employed in the Spanish colonies as a domestic medicine is very considerable. We see by the works of Clusius, that at the beginning of the Conquista, Europe obtained this salutary medicament from the Mexican coast of Honduras and the port of Guayaquil. The trade in zarza is now more active in those ports which have interior communications with the Orinoco, the Rio Negro, and the Amazon.

The trials made in several botanical gardens of Europe prove that the Smilax glauca of Virginia, which it is pretended is the S. sarsaparilla of Linnaeus, may be cultivated in the open air, wherever the mean winter temperature rises above six or seven degrees of the centigrade thermometer[*]: but those species that possess the most active virtues belong exclusively to the torrid zone, and require a much higher degree of heat. In reading the works of Clusius, it can scarcely be conceived why our writers on the Materia Medica persist in considering a plant of the

---

[*] The winter temperature at London and Paris is 4.2 and 3.7; at Montpelier, 6.7; at Rome, 7.7°. In that part of Mexico, and the Terra Firma, where we saw the most active species of the sarsaparilla growing (that which supplies the trade of the Spanish and Portuguese colonies), the temperature is from twenty to twenty-six degrees. The roots of another family of monocotyledons (of some cyperaceae) possess also diaphoretic and resolvent properties. The Carex arenaria, the C. hirta, etc. furnish the German sarsaparilla of druggists. According to Clusius, Europe received the first sarsaparilla from Yucatan, and the island of Puna, opposite Guayaquil.

United States as the most ancient type of the officinal species of the genus smilax.

We found in the possession of the Indians of the Rio Negro some of those green stones, known by the name of Amazon stones, because the natives pretend, according to an ancient tradition, that they come from the country of the women without husbands (Cougnantainsecouima), or women living alone (Aikeambenano[*]). We were told at San Carlos, and in the neighbouring villages, that the sources of the Orinoco, which we found east of the Esmeralda, and in the missions of the Carony and at Angostura, that the sources of the Rio Branco are the native spots of the green stones. These statements confirm the report of an old soldier of the garrison of Cayenne (mentioned by La Condamine), who affirmed that those mineral substances were obtained from the country of women, west of the rapids of the Oyapoc. The Indians who inhabit the fort of Topayos on the Amazon five degrees east of the mouth of the Rio Negro, possessed formerly a great number of these stones. Had they received them from the north, that is, from the country pointed out by the Indians of the Rio Negro, which extends from the mountains of Cayenne towards the sources of the Essequibo, the Carony, the Orinoco, the Parime, and the Rio Trombetas? or did they come from the south by the Rio Topayos, which descends from the vast table-land of the Campos Parecis? Superstition attaches great importance to these mineral substances: they are worn suspended from the neck as amulets, because, according to popular belief, they preserve the wearer from nervous complaints, fevers, and the stings of venomous serpents. They have consequently been for ages an article of trade among the natives, both north and south of the Orinoco. The Caribs, who may be considered as the Bucharians of the New World, made them known along the coasts of Guiana; and the same stones, like money in circulation, passed successively from nation to

---

[*] This word is of the Tamanac language; these women are the sole Donne of the Italian missionaries.

nation in opposite directions: their quantity is perhaps not augmented, and the spot which produces them is probably unknown rather than concealed. In the midst of enlightened Europe, on occasion of a warm contest respecting native bark, a few years ago, the green stones of the Orinoco were gravely proposed as a powerful febrifuge. After this appeal to the credulity of Europeans, we cannot be surprised to learn that the Spanish planters share the predilection of the Indians for these amulets, and that they are sold at a very considerable price. The form given to them most frequently is that of the Babylonian cylinders,[*] longitudinally perforated, and loaded with inscriptions and figures. But this is not the work of the Indians of our days, the natives of the Orinoco and the Amazon, whom we find in the last degree of barbarism. The Amazon stones, like the perforated and sculptured emeralds, found in the Cordilleras of New Grenada and Quito, are vestiges of anterior civilization. The present inhabitants of those countries, particularly in the hot region, so little comprehend the possibility of cutting hard stones (the emerald, jade, compact feldspar and rock-crystal), that they imagine the green stone is soft when taken out of the earth, and that it hardens after having been moulded by the hand.

The natural soil of the Amazon-stone is not in the valley of the river Amazon. It does not derive its name from the river, but like the river itself, the stone has been named after a nation of warlike women, whom Father Acunha, and Oviedo, in his letter to cardinal Bembo, compare to the Amazons of the ancient world. What we see in our cabinets under the false denomination of Amazon-stone, is neither jade, nor compact feldspar, but a common feldspar of an apple-green colour, that comes from the Ural mountains and on lake Onega in Russia, but which I never saw in the granitic mountains of Guiana. Sometimes also this very rare and hard Amazon-stone is confounded with the hatchet-nephrite (beilstein[†]) of Werner, which has much less

---

[*] The price of a cylinder two inches long is from twelve to fifteen piastres.

[†] Punamustein (jade axinien). The stone hatchets found in America, for

tenacity. The substance which I obtained from the hands of the Indians, belongs to the saussurite,[*] to the real jade, which resembles compact feldspar, and which forms one of the constituent parts of the verde de Corsica, or gabbro.[†] It takes a fine polish, and passes from apple-green to emerald-green; it is translucent at the edges, extremely tenacious, and in a high degree sonorous. These Amazon stones were formerly cut by the natives into very thin plates, perforated at the centre, and suspended by a thread, and these plates yield an almost metallic sound if struck by another hard body.[‡] This fact confirms the connection which we find, notwithstanding the difference of fracture and of specific gravity between the saussurite and the siliceous basis of the porphyrschiefer, which is the phonolite (klingstein). I have already observed, that, as it is very rare to find in America nephrite, jade, or compact feldspar, in its native place, we may well be astonished at the quantity of hatchets which are everywhere discovered in digging the earth, from the banks of the Ohio as far as Chile. We saw in the mountains of Upper Orinoco, or of Parime, only granular granites containing a little hornblende, granites passing into gneiss, and schistoid hornblendes. Has nature repeated on the east of Esmeralda, between the sources of the Carony, the Essequibo, the Orinoco, and the Rio Branco, the transition-formation of Tucutunemo reposing on mica-schist? Does the Amazon-stone come from the rocks of euphotide, which form the last member of the series of primitive rocks?

We find among the inhabitants of both hemispheres, at the first dawn of civilization, a peculiar predilection for certain

---

instance in Mexico, are not of beilstein, but of compact feldspar.

[*] Jade of Saussure, according to the system of Brongniart; tenacious jade, and compact tenacious feldspar of Hauy; some varieties of the variolithe of Werner.

[†] Euphotide of Hauy, or schillerfels, of Raumer.

[‡] M. Brongniart, to whom I showed these plates on my return to Europe, very justly compared these jades of Parime to the sonorous stones employed by the Chinese in their musical instruments called king.

stones; not only those which, from their hardness, may be useful to man as cutting instruments, but also for mineral substances, which, on account of their colour and their natural form, are believed to bear some relation to the organic functions, and even to the propensities of the soul. This ancient worship of stones, these benign virtues attributed to jade and haematite, belong to the savages of America as well as to the inhabitants of the forests of Thrace. The human race, when in an uncultivated state, believes itself to have sprung from the ground; and feels as if it were enchained to the earth, and the substances contained in her bosom. The powers of nature, and still more those which destroy than those which preserve, are the first objects of its worship. It is not solely in the tempest, in the sound that precedes the earthquake, in the fire that feeds the volcano, that these powers are manifested; the inanimate rock; stones, by their lustre and hardness; mountains, by their mass and their solitude; act upon the untaught mind with a force which, in a state of advanced civilization, can no longer be conceived. This worship of stones, when once established, is preserved amidst more modern forms of worship; and what was at first the object of religious homage, becomes a source of superstitious confidence. Divine stones are transformed into amulets, which are believed to preserve the wearer from every ill, mental and corporeal. Although a distance of five hundred leagues separates the banks of the Amazon and the Orinoco from the Mexican table-land; although history records no fact that connects the savage nations of Guiana with the civilized nations of Anahuac, the monk Bernard de Sahagun, at the beginning of the conquest, found preserved as relics at Cholula, certain green stones which had belonged to Quetzalcohuatl. This mysterious personage is the Mexican Buddha; he appeared in the time of the Toltecs, founded the first religious associations, and established a government similar to that of Meroe and of Japan.

The history of the jade, or the green stones of Guiana, is intimately connected with that of the warlike women whom the travellers of the sixteenth century named the Amazons of the

New World. La Condamine has produced many testimonies in favour of this tradition. Since my return from the Orinoco and the river Amazon, I have often been asked, at Paris, whether I embraced the opinion of that learned man, or believed, like several of his contemporaries, that he undertook the defence of the Cougnantainsecouima (the independent women who received men into their society only in the month of April), merely to fix, in a public sitting of the Academy, the attention of an audience somewhat eager for novelties. I may take this opportunity of expressing my opinion on a tradition which has so romantic an appearance; and I am farther led to do this as La Condamine asserts that the Amazons of the Rio Cayame[*] crossed the Maranon to establish themselves on the Rio Negro. A taste for the marvellous, and a wish to invest the descriptions of the New Continent with some of the colouring of classic antiquity, no doubt contributed to give great importance to the first narratives of Orellana. In perusing the works of Vespucci, Fernando Columbus, Geraldini, Oviedo, and Pietro Martyr, we recognize this tendency of the writers of the sixteenth century to

---

[*] Orellana, arriving at the Maranon by the Rio Coca and the Napo, fought with the Amazons, as it appears, between the mouth of the Rio Negro and that of the Xingu. La Condamine asserts that in the seventeenth century they passed the Maranon between Tefe and the mouth of the Rio Puruz, near the Cano Cuchivara, which is a western branch of the Puruz. These women therefore came from the banks of the Rio Cayame, or Cayambe, consequently from the unknown country which extends south of the Maranon, between the Ucayale and the Madeira. Raleigh also places them on the south of the Maranon, but in the province of Topayos, and on the river of the same name. He says they were rich in golden vessels, which they had acquired in exchange for the famous green stones, or piedras hijadas. (Raleigh means, no doubt, piedros del higado, stones that cure diseases of the liver.) It is remarkable enough that, one hundred and forty-eight years after, La Condamine still found those green stones (divine stones), which differ neither in colour nor in hardness from oriental jade, in greater numbers among the Indians who live near the mouth of the Rio Topayos, than elsewhere. The Indians said that they inherited these stones, which cure the nephritic colic and epilepsy, from their fathers, who received them from the women without husbands.

find among the newly discovered nations all that the Greeks have related to us of the first age of the world, and of the manners of the barbarous Scythians and Africans. But if Oviedo, in addressing his letters to cardinal Bembo, thought fit to flatter the taste of a man so familiar with the study of antiquity, Sir Walter Raleigh had a less poetic aim. He sought to fix the attention of Queen Elizabeth on the great empire of Guiana, the conquest of which he proposed. He gave a description of the rising of that gilded king (el dorado),[*] whose chamberlains, furnished with long tubes, blew powdered gold every morning over his body, after having rubbed it over with aromatic oils: but nothing could be better adapted to strike the imagination of queen Elizabeth, than the warlike republic of women without husbands, who resisted the Castilian heroes. Such were the motives which prompted exaggeration on the part of those writers who have given most reputation to the Amazons of America; but these motives do not, I think, suffice for entirely rejecting a tradition, which is spread among various nations having no communications one with another.

Thirty years after La Condamine visited Quito, a Portuguese astronomer, Ribeiro, who has traversed the Amazon, and the tributary streams which run into that river on the northern side, has confirmed on the spot all that the learned Frenchman had advanced. He found the same traditions among the Indians; and he collected them with the greater impartiality as he did not himself believe that the Amazons formed a separate horde. Not knowing any of the tongues spoken on the Orinoco and the Rio Negro, I could learn nothing certain respecting the popular traditions of the women without husbands, or the origin of the green stones, which are believed to be intimately connected with them. I shall, however, quote a modern testimony of some

---

[*] The term el dorado, which signifies the gilded, was not originally the name of the country. The territory subsequently distinguished by that appellation was at first known as the country of el Rey Dorado, the Gilded King.

weight, that of Father Gili. "Upon inquiring," says this well-informed missionary, "of a Quaqua Indian, what nations inhabited the Rio Cuchivero, he named to me the Achirigotos, the Pajuros, and the Aikeambenanos.* Being well acquainted," pursues he, "with the Tamanac tongue, I instantly comprehended the sense of this last word, which is a compound, and signifies women living alone. The Indian confirmed my observation, and related that the Aikeambenanos were a community of women, who manufactured blow-tubes[†], and other weapons of war. They admit, once a year, the men of the neighbouring nation of Vokearos into their society, and send them back with presents. All the male children born in this horde of women are killed in their infancy." This history seems framed on the traditions which circulate among the Indians of the Maranon, and among the Caribs; yet the Quaqua Indian, of whom Father Gili speaks, was ignorant of the Castilian language; he had never had any communication with white men; and certainly knew not, that south of the Orinoco there existed another river, called the river of the Aikeambenanos, or Amazons.

What must we conclude from this narration of the old missionary of Encaramada? Not that there are Amazons on the banks of the Cuchivero, but that women in different parts of America, wearied of the state of slavery in which they were held by the men, united themselves together; that the desire of preserving their independence rendered them warriors; and that they received visits from a neighbouring and friendly horde. This society of women may have acquired some power in one part of Guiana. The Caribs of the continent held intercourse with those of the islands; and no doubt in this way the traditions of the Maranon and the Orinoco were propagated toward the north. Before the voyage of Orellana, Christopher Columbus imagined he had found the Amazons in the Caribbee Islands. This great

---

* In Italian, Acchirecolti, Pajuri, and Aicheam-benano.
† Long tubes made from a hollow cane, which the natives use to propel their poisoned arrows.

man was told, that the small island of Madanino (Montserrat) was inhabited by warlike women, who lived the greater part of the year separate from men. At other times also, the conquistadores imagined that the women, who defended their huts in the absence of their husbands, were republics of Amazons; and, by an error less excusable, formed a like supposition respecting the religious congregations, the convents of Mexican virgins, who, far from admitting men at any season of the year into their society, lived according to the austere rule of Quetzalcohuatl. Such was the disposition of men's minds, that in the long succession of travellers, who crowded on each other in their discoveries and in narrations of the marvels of the New World, every one readily declared he had seen what his predecessors had announced.

We passed three nights at San Carlos del Rio Negro. I count the nights, because I watched during the greater part of them, in the hope of seizing the moment of the passage of some star over the meridian. That I might have nothing to reproach myself with, I kept the instruments always ready for an observation. I could not even obtain double altitudes, to calculate the latitude by the method of Douwes. What a contrast between two parts of the same zone; between the sky of Cumana, where the air is constantly pure as in Persia and Arabia, and the sky of the Rio Negro, veiled like that of the Feroe islands, without sun, or moon or stars!

On the 10th of May, our canoe being ready before sunrise, we embarked to go up the Rio Negro as far as the mouth of the Cassiquiare, and to devote ourselves to researches on the real course of that river, which unites the Orinoco to the Amazon. The morning was fine; but, in proportion as the heat augmented, the sky became obscured. The air is so saturated by water in these forests, that the vesicular vapours become visible on the least increase of evaporation at the surface of the earth. The breeze being never felt, the humid strata are not displaced and renewed by dryer air. We were every day more grieved at the aspect of the cloudy sky. M. Bonpland was losing by this excessive humidity the plants he had collected; and I, for my part, was afraid lest I should again find the fogs of the Rio Negro in the valley of the Cassiquiare. No one in these missions for half a century past had doubted the existence of communication between two great systems of rivers; the important point of our voyage was confined therefore to fixing by astronomical observations the course of the Cassiquiare, and particularly the point of its entrance into the Rio Negro, and that of the bifurcation of the Orinoco. Without a sight of the sun and the stars this object would be frustrated, and we should have exposed ourselves in vain to long and painful privations. Our fellow travellers would have returned by the shortest way, that of the Pimichin and the small rivers; but M. Bonpland preferred, like me, persisting in the plan of the voyage, which we had traced for ourselves in passing the Great Cataracts. We had already travelled one hundred and eighty leagues in a boat from San Fernando de Apure to San Carlos, on the Rio Apure, the Orinoco, the Atabapo, the Temi, the Tuamini, and the Rio Negro. In again entering the Orinoco by the Cassiquiare we had to navigate three hundred and twenty leagues, from San Carlos to Angostura. By this way we had to struggle against the currents during ten days; the rest was to be performed by going down the stream of the Orinoco. It would have been blamable to have suffered ourselves to be discouraged by the fear of a cloudy sky, and by the mosquitos of the Cassiquiare. Our Indian pilot, who had been recently at Mandavaca, promised us the sun, and those great stars that eat the clouds, as soon as we should have left the black waters of the Guaviare. We therefore carried out our first project of returning to San Fernando de Atabapo by the Cassiquiare; and, fortunately for our researches, the prediction of the Indian was verified. The white waters brought us by degrees a more serene sky, stars, mosquitos, and crocodiles.

We passed between the islands of Zaruma and Mini, or Mibita, covered with thick vegetation; and, after having ascended the rapids of the Piedra de Uinumane, we entered the

Rio Cassiquiare at the distance of eight miles from the small fort of San Carlos. The Piedra, or granitic rock which forms the little cataract, attracted our attention on account of the numerous veins of quartz by which it is traversed. These veins are several inches broad, and their masses proved that their date and formation are very different. I saw distinctly that, wherever they crossed each other, the veins containing mica and black schorl traversed and drove out of their direction those which contained only white quartz and feldspar. According to the theory of Werner, the black veins were consequently of a more recent formation than the white. Being a disciple of the school of Freyberg, I could not but pause with satisfaction at the rock of Uinumane, to observe the same phenomena near the equator, which I had so often seen in the mountains of my own country. I confess that the theory which considers veins as clefts filled from above with various substances, pleases me somewhat less now than it did at that period; but these modes of intersection and driving aside, observed in the stony and metallic veins, do not the less merit the attention of travellers as being one of the most general and constant of geological phenomena. On the east of Javita, all along the Cassiquiare, and particularly in the mountains of Duida, the number of veins in the granite increases. These veins are full of holes and druses; and their frequency seems to indicate that the granite of these countries is not of very ancient formation.

We found some lichens on the rock Uinumane, opposite the island of Chamanare, at the edge of the rapids; and as the Cassiquiare near its mouth turns abruptly from east to south-west, we saw for the first time this majestic branch of the Orinoco in all its breadth. It much resembles the Rio Negro in the general aspect of the landscape. The trees of the forest, as in the basin of the latter river, advance as far as the beach, and there form a thick coppice; but the Cassiquiare has white waters, and more frequently changes its direction. Its breadth, near the rapids of Uinumane, almost surpasses that of the Rio Negro. I found it everywhere from two hundred and fifty to two hundred and eighty toises, as far as above Vasiva. Before we passed the island of Garigave, we perceived to the north-east, almost at the horizon, a little hill with a hemispheric summit; the form which in every zone characterises mountains of granite. Continually surrounded by vast plains, the solitary rocks and hills excite the attention of the traveller. Contiguous mountains are only found more to the east, towards the sources of the Pacimoni, Siapa, and Mavaca. Having arrived on the south of the Raudal of Caravine, we perceived that the Cassiquiare, by the windings of its course, again approached San Carlos. The distance from this fort to the mission of San Francisco Solano, where we slept, is only two leagues and a half by land, but it is reckoned seven or eight by the river. I passed a part of the night in the open air, waiting vainly for stars. The air was misty, notwithstanding the aguas blancas, which were to lead us beneath an ever-starry sky.

The mission of San Francisco Solano, situated on the left bank of the Cassiquiare, was founded, as were most of the Christian settlements south of the Great Cataracts of the Orinoco, not by monks, but by military authority. At the time of the expedition of the boundaries, villages were built in proportion as a subteniente, or a corporal, advanced with his troops. Part of the natives, in order to preserve their independence, retired without a struggle; others, of whom the most powerful chiefs had been gained, joined the missions. Where there was no church, they contented themselves with erecting a great cross of red wood, close to which they constructed a casa fuerte, or block-house, the walls of which were formed of large beams resting horizontally upon each other. This house had two stories; in the upper story two cannon of small calibre were placed; and two soldiers lived on the ground-floor, and were served by an Indian family. Those of the natives with whom they were at peace cultivated spots of land round the casa fuerte. The soldiers called them together by the sound of the horn, or a botuto of baked earth, whenever any hostile attack was dreaded. Such were the pretended nineteen Christian settlements founded by Don Antonio Santos in the way from Esmeralda to the Erevato. Military posts, which had no influence on the civilization of the natives, figured on the maps, and in the works of the missionaries, as villages (pueblos) and reducciones apostolicas.* The preponderance of the military was maintained on the banks of the Orinoco till 1785, when the system of the monks of San Francisco began. The small number of missions founded, or rather re-established, since that period, owe their existence to the Fathers of the Observance; for the soldiers now distributed among the missions are dependent on the missionaries, or at least are reputed to be so, according to the pretensions of the ecclesiastical hierarchy.

The Indians whom we found at San Francisco Solano were of two nations; Pacimonales and Cheruvichahenas. The latter being descended from a considerable tribe settled on the Rio Tomo, near the Manivas of the Upper Guainia, I tried to gather from them some ideas respecting the upper course and the sources of the Rio Negro; but the interpreter whom I employed could not make them comprehend my questions. Their continually-repeated answer was, that the sources of the Rio Negro and the Inirida were as near to each other as "two fingers of the hand." In one of the huts of the Pacimonales we purchased two fine large birds, a toucan (piapoco) and an ana, a species of macaw, seventeen inches long, having the whole body of a purple colour. We had already in our canoe seven parrots, two manakins (pipa), a motmot, two guans, or pavas de monte, two manaviris (cercoleptes or Viverra caudivolvula), and eight monkeys, namely, two ateles,[†] two titis,[‡] one viudita, (Simia lugens.) two douroucoulis or nocturnal monkeys,[§] and a short-tailed cacajao[**]. Father Zea whispered some complaints at the daily augmentation of this ambulatory collection. The toucan

---

[*] Signifying apostolic conquests or conversions.
[†] Marimonda of the Great Cataracts, Simia belzebuth, Brisson.
[‡] Simia sciurea, the saimiri of Buffon.
[§] Cusiensi, or Simia trivirgata.
[**] Simia melanocephala, mono feo. These last three species are new.

resembles the raven in manners and intelligence. It is a courageous animal, but easily tamed. Its long and stout beak serves to defend it at a distance. It makes itself master of the house, steals whatever it can come at, and loves to bathe often and fish on the banks of the river. The toucan we had bought was very young; yet it took delight, during the whole voyage, in teasing the cusicusis, or nocturnal monkeys, which are melancholy and irritable. I did not observe what has been related in some works of natural history, that the toucan is forced, from the structure of its beak, to swallow its food by throwing it up into the air. It raises it indeed with some difficulty from the ground, but, having once seized it with the point of its enormous beak, it has only to lift it up by throwing back its head, and holding it perpendicularly whilst in the act of swallowing. This bird makes extraordinary gestures when preparing to drink. The monks say that it makes the sign of the cross upon the water; and this popular belief has obtained for the toucan, from the creoles, the singular name of diostede.*

Most of our animals were confined in small wicker cages; others ran at full liberty in all parts of the boat. At the approach of rain the macaws sent forth noisy cries, the toucan wanted to reach the shore to fish, and the little monkeys (the titis) went in search of Father Zea, to take shelter in the large sleeves of his Franciscan habit. These incidents sometimes amused us so much that we forgot the torment of the mosquitos. At night we placed a leather case (petaca), containing our provisions, in the centre; then our instruments, and the cages of our animals; our hammocks were suspended around the cages, and beyond were those of the Indians. The exterior circle was formed by the fires which are lighted to keep off the jaguars. Such was the order of our encampment on the banks of the Cassiquiare. The Indians often spoke to us of a little nocturnal animal, with a long nose, which surprises the young parrots in their nests, and in eating

---

* Dios te de, God gives it thee.

makes use of its hands like the monkeys and the maniveris, or kinkajous. They call it the guachi; it is, no doubt, a coati, perhaps the Viverra nasua, which I saw wild in Mexico. The missionaries gravely prohibit the natives from eating the flesh of the guachi, to which, according to far-spread superstitious ideas, they attribute the same stimulating qualities which the people of the East believe to exist in the skink, and the Americans in the flesh of the alligator.

On the 11th of May, we left the mission of San Francisco Solano at a late hour, to make but a short day's journey. The uniform stratum of vapours began to be divided into clouds with distinct outlines: and there was a light east wind in the upper regions of the air. We recognized in these signs an approaching change of the weather; and were unwilling to go far from the mouth of the Cassiquiare, in the hope of observing during the following night the passage of some star over the meridian. We descried the Cano Daquiapo to the south, the Guachaparu to the north, and a few miles further, the rapids of Cananivacari. The velocity of the current being 6.3 feet in a second, we had to struggle against the turbulent waves of the Raudal. We went on shore, and M. Bonpland discovered within a few steps of the beach a majestic almendron, or Bertholletia excelsa. The Indians assured us, that the existence of this valuable plant of the banks of the Cassiquiare was unknown at San Francisco Solano, Vasiva, and Esmeralda. They did not think that the tree we saw, which was more than sixty feet high, had been sown by some passing traveller. Experiments made at San Carlos have shown how rare it is to succeed in causing the bertholletia to germinate, on account of its ligneous pericarp, and the oil contained in its nut which so readily becomes rancid. Perhaps this tree denoted the existence of a forest of bertholletia in the inland country on the east and north-east. We know, at least, with certainty, that this fine tree grows wild in the third degree of latitude, in the Cerro de Guanaya. The plants that live in society have seldom marked limits, and it happens, that before we reach a palmar or a pinar,* we find solitary palm-trees and pines. They are somewhat like colonists that have advanced in the midst of a country peopled with different vegetable productions.

Four miles distant from the rapids of Cunanivacari, rocks of the strangest form rise in the plains. First appears a narrow wall eighty feet high, and perpendicular; and at the southern extremity of this wall are two turrets, the courses of which are of granite, and nearly horizontal. The grouping of the rocks of Guanari is so symmetrical that they might be taken for the ruins of an ancient edifice. Are they the remains of islets in the midst of an inland sea, that covered the flat ground between the Sierra Parime and the Parecis mountains?† or have these walls of rock, these turrets of granite, been upheaved by the elastic forces that still act in the interior of our planet? We may be permitted to meditate a little on the origin of mountains, after having seen the position of the Mexican volcanoes, and of trachyte summits on an elongated crevice; having found in the Andes of South America primitive and volcanic rocks in a straight line in the same chain; and when we recollect the island, three miles in circumference, and of a great height, which in modern times issued from the depths of the ocean near Oonalaska.

The banks of the Cassiquiare are adorned with the chiriva palm-tree with pinnate leaves, silvery on the under part. The rest of the forest furnishes only trees with large, coriaceous, glossy leaves, that have plain edges. This peculiar physiognomy‡ of the

---

* Two Spanish words, which, according to a Latin form, denote a forest of palm-trees, palmetum, and of pines, pinetum.

† The Sierra de la Parime, or of the Upper Orinoco, and the Sierra (or Campos) dos Parecis, are part of the mountains of Matto Grosso, and form the northern back of the Sierra de Chiquitos. I here name the two chains of mountains running from east to west, and bordering the plains or basins of the Cassiquiare, the Rio Negro, and the Amazon, between 5° 30′ north, and 14° south latitude.

‡ This physiognomy struck us forcibly, in the vast forests of Spanish Guiana, only between the second and third degrees of north latitude.

vegetation of the Guainia, the Tuamini, and the Cassiquiare, is owing to the preponderance of the families of the guttiferae, the sapotae, and the laurineae, in the equatorial regions. The serenity of the sky promising us a fine night, we resolved, at five in the evening, to rest near the Piedra de Culimacari, a solitary granite rock, like all those which I have described between the Atabapo and the Cassiquiare. We found by the bearings of the sinuosities of the river, that this rock is nearly in the latitude of the mission of San Francisco Solano. In those desert countries, where man has hitherto left only fugitive traces of his existence, I constantly endeavoured to make my observations near the mouth of a river, or at the foot of a rock distinguishable by its form. Such points only as are immutable by their nature can serve for the basis of geographical maps. I obtained, in the night of the 10th of May, a good observation of latitude by alpha of the Southern Cross; the longitude was determined, but with less precision, by the chronometer, taking the altitudes of the two beautiful stars which shine in the feet of the Centaur. This observation made known to us at the same time, with sufficient precision for the purposes of geography, the positions of the mouth of the Pacimoni, of the fortress of San Carlos, and of the junction of the Cassiquiare with the Rio Negro. The rock of Culimacari is precisely in latitude 2° 0′ 42″, and probably in longitude 69° 33′ 50″.

Satisfied with our observations, we left the rock of Culimacari at half past one on the morning of the 12th. The torment of mosquitos, to which we were exposed, augmented in proportion as we withdrew from the Rio Negro. There are no zancudos in the valley of Cassiquiare, but the simulia, and all the other insects of the tipulary family, are the more numerous and venomous. Having still eight nights to pass in the open air in this damp and unhealthy climate, before we could reach the mission of Esmeralda, our pilot sought to arrange our passage in such a manner as might enable us to enjoy the hospitality of the missionary of Mandavaca, and some shelter in the village of Vasiva. We went up with difficulty against the current, which was nine feet, and in some places (where I measured it with

447

precision) eleven feet eight inches in a second, that is, almost eight miles an hour. Our resting-place was probably not farther than three leagues in a right line from the mission of Mandavaca; yet, though we had no reason to complain of inactivity on the part of our rowers, we were fourteen hours in making this short passage.

Towards sunrise we passed the mouth of the Rio Pacimoni, a river which I mentioned when speaking of the trade in sarsaparilla, and which (by means of the Baria) intertwines in so remarkable a way with the Cababuri. The Pacimoni rises in a hilly ground, from the confluence of three small rivers,[*] not marked on the maps of the missionaries. Its waters are black, but less so than those of the lake of Vasiva, which also communicates with the Cassiquiare. Between those two tributary streams coming from the east, lies the mouth of the Rio Idapa, the waters of which are white. I shall not recur again to the difficulty of explaining this coexistence of rivers differently coloured, within a small extent of territory, but shall merely observe, that at the mouth of the Pacimoni, and on the borders of the lake Vasiva, we were again struck with the purity and extreme transparency of the brown waters. Ancient Arabian travellers have observed, that the Alpine branch of the Nile, which joins the Bahr el Abiad near Halfaja, has green waters, which are so transparent, that the fish may be seen at the bottom of the river.

We passed some turbulent rapids before we reached the mission of Mandavaca. The village, which bears also the name of Quirabuena, contains only sixty natives. The state of the Christian settlements is in general so miserable that, in the whole course of the Cassiquiare, on a length of fifty leagues, not two hundred inhabitants are found. The banks of this river were indeed more peopled before the arrival of the missionaries; the Indians have withdrawn into the woods, toward the east; for the

---

[*] The Rios Guajavaca, Moreje, and Cachevaynery.

western plains are almost deserted. The natives subsist during a part of the year on those large ants of which I have spoken above. These insects are much esteemed here, as spiders are in the southern hcmisphere, where the savages of Australia deem them delicious. We found at Mandavaca the good old missionary, who had already spent twenty years of mosquitos in the bosques del Cassiquiare, and whose legs were so spotted by the stings of insects, that the colour of the skin could scarcely be perceived. He talked to us of his solitude, and of the sad necessity which often compelled him to leave the most atrocious crimes unpunished in the two missions of Mandavaca and Vasiva. In the latter place, an Indian alcalde had, a few years before, eaten one of his wives, after having taken her to his conuco,[*] and fattened her by good feeding. The cannibalism of the nations of Guiana is never caused by the want of subsistence, or by the superstitions of their religion, as in the islands of the South Sea; but is generally the effect of the vengeance of a conqueror, and (as the missionaries say) "of a vitiated appetite." Victory over a hostile tribe is celebrated by a repast, in which some parts of the body of a prisoner are devoured. Sometimes a defenceless family is surprised in the night; or an enemy, who is met with by chance in the woods, is killed by a poisoned arrow. The body is cut to pieces, and carried as a trophy to the hut. It is civilization only, that has made man feel the unity of the human race; which has revealed to him, as we may say, the ties of consanguinity, by which he is linked to beings to whose language and manners he is a stranger. Savages know only their own family; and a tribe appears to them but a more numerous assemblage of relations. When those who inhabit the missions see Indians of the forest, who are unknown to them, arrive, they make use of an expression, which has struck us by its simple candour: they are, no doubt, my relations; I understand them when they speak to me. But these very savages detest all who are

---

[*] A hut surrounded with cultivated ground; a sort of country-house, which the natives prefer to residing in the missions.

not of their family, or their tribe; and hunt the Indians of a neighbouring tribe, who live at war with their own, as we hunt game. They know the duties of family ties and of relationship, but not those of humanity, which require the feeling of a common tie with beings framed like ourselves. No emotion of pity prompts them to spare the wives or children of a hostile race; and the latter are devoured in preference, at the repast given at the conclusion of a battle or warlike incursion.

The hatred which savages for the most part feel for men who speak another idiom, and appear to them to be of an inferior race, is sometimes rekindled in the missions, after having long slumbered. A short time before our arrival at Esmeralda, an Indian, born in the forest* behind the Duida, travelled alone with another Indian, who, after having been made prisoner by the Spaniards on the banks of the Ventuario, lived peaceably in the village, or, as it is expressed here, within the sound of the bell (debaxo de la campana.) The latter could only walk slowly, because he was suffering from one of those fevers to which the natives are subject, when they arrive in the missions, and abruptly change their diet. Wearied by his delay, his fellow-traveller killed him, and hid the body behind a copse of thick trees, near Esmeralda. This crime, like many others among the Indians, would have remained unknown, if the murderer had not made preparations for a feast on the following day. He tried to induce his children, born in the mission and become Christians, to go with him for some parts of the dead body. They had much difficulty in persuading him to desist from his purpose; and the soldier who was posted at Esmeralda, learned from the domestic squabble caused by this event, what the Indians would have concealed from his knowledge.

---

* En el monte. The Indians born in the missions are distinguished from those born in the woods. The word monte signifies more frequently, in the colonies, a forest (bosque) than a mountain, and this circumstance has led to great errors in our maps, on which chains of mountains (sierras) are figured, where there are only thick forests, (monte espeso.)

It is known that cannibalism and the practice of human sacrifices, with which it is often connected, are found to exist in all parts of the globe, and among people of very different races;[*] but what strikes us more in the study of history is to see human sacrifices retained in a state of civilization somewhat advanced; and that the nations who hold it a point of honour to devour their prisoners are not always the rudest and most ferocious. The painful facts have not escaped the observation of those missionaries who are sufficiently enlightened to reflect on the manners of the surrounding tribes. The Cabres, the Guipunaves, and the Caribs, have always been more powerful and more civilized than the other hordes of the Orinoco; and yet the two former are as much addicted to anthropophagy as the latter are repugnant to it. We must carefully distinguish the different branches into which the great family of the Caribbee nations is divided. These branches are as numerous as those of the Mongols, and the western Tartars, or Turcomans. The Caribs of the continent, those who inhabit the plains between the Lower Orinoco, the Rio Branco, the Essequibo, and the sources of the Oyapoc, hold in horror the practice of devouring their enemies. This barbarous custom,[†] at the first discovery of America,

---

[*] Some casual instances of children carried off by the negroes in the island of Cuba have led to the belief, in the Spanish colonies, that there are tribes of cannibals in Africa. This opinion, though supported by some travellers, is not borne out by the researches of Mr. Barrow on the interior of that country. Superstitious practices may have given rise to imputations perhaps as unjust as those of which Jewish families were the victims in the ages of intolerance and persecution.

[†] See Geraldini Itinerarium page 186 and the eloquent tract of cardinal Bembo on the discoveries of Columbus. "Insularum partem homines incolebant feri trucesque, qui puerorum et virorum carnibus, quos aliis in insulus bello aut latrociniis cepissent, vescebantur; a feminis abstinebant; Canibales appellati." "Some of the islands are inhabited by a cruel and savage race, called cannibals, who eat the flesh of men and boys, and captives and slaves of the male sex, abstaining from that of females." Hist. Venet. 1551. The custom of sparing the lives of female prisoners confirms what I have previously said of the language of the women. Does the word cannibal,

existed only among the Caribs of the West Indies. It is they who have rendered the names of cannibals, Caribbees, and anthropophagi, synonymous; it was their cruelties that prompted the law promulgated in 1504, by which the Spaniards were permitted to make a slave of every individual of an American nation which could be proved to be of Caribbee origin. I believe, however, that the anthropophagy of the inhabitants of the West India Islands was much exaggerated by early travellers, whose stories Herrera, a grave and judicious historian, has not disdained to repeat in his Decades historicas. He has even credited that extraordinary event which led the Caribs to renounce this barbarous custom. The natives of a little island devoured a Dominican monk whom they had carried off from the coast of Porto Rico; they all fell sick, and would never again eat monk or layman.

If the Caribs of the Orinoco, since the commencement of the sixteenth century, have differed in their manners from those of the West India Islands; if they are unjustly accused of anthropophagy; it is difficult to attribute this difference to any superiority of their social state. The strangest contrasts are found blended in this mixture of nations, some of whom live only upon fish, monkeys, and ants; while others are more or less cultivators of the ground, more or less occupied in making and painting pottery, or weaving hammocks or cotton cloth. Several of the latter tribes have preserved inhuman customs altogether unknown to the former. "You cannot imagine," said the old missionary of Mandavaca, "the perversity of this Indian race (familia de Indios). You receive men of a new tribe into the village; they appear to be mild, good, and laborious; but suffer them to take part in an incursion (entrada) to bring in the natives, and you can scarcely prevent them from murdering all they meet, and hiding some portions of the dead bodies." In reflecting

applied to the Caribs of the West India Islands, belong to the language of this archipelago (that of Haiti)? or must we seek for it in an idiom of Florida, which some traditions indicate as the first country of the Caribs?

on the manners of these Indians, we are almost horrified at that combination of sentiments which seem to exclude each other; that faculty of nations to become but partially humanized; that prepondcrance of customs, prejudices, and traditions, over the natural affections of the heart. We had a fugitive Indian from the Guaisia in our canoe, who had become sufficiently civilized in a few weeks to be useful to us in placing the instruments necessary for our observations at night. He was no less mild than intelligent, and we had some desire of taking him into our service. What was our horror when, talking to him by means of an interpreter, we learned, that the flesh of the marimonde monkeys, though blacker, appeared to him to have the taste of human flesh. He told us that his relations (that is, the people of his tribe) preferred the inside of the hands in man, as in bears. This assertion was accompanied with gestures of savage gratification. We inquired of this young man, so calm and so affectionate in the little services which he rendered us, whether he still felt sometimes a desire to eat of a Cheruvichahena. He answered, without discomposure, that, living in the mission, he would only eat what he saw was eaten by the Padres. Reproaches addressed to the natives on the abominable practice which we here discuss, produce no effect; it is as if a Brahmin, travelling in Europe, were to reproach us with the habit of feeding on the flesh of animals. In the eyes of the Indian of the Guaisia, the Cheruvichahena was a being entirely different from himself; and one whom he thought it was no more unjust to kill than the jaguars of the forest. It was merely from a sense of propriety that, whilst he remained in the mission, he would only eat the same food as the Fathers. The natives, if they return to their tribe (al monte), or find themselves pressed by hunger, soon resume their old habits of anthropophagy. And why should we be so much astonished at this inconstancy in the tribes of the Orinoco, when we are reminded, by terrible and well-ascertained examples, of what has passed among civilized nations in times of great scarcity? In Egypt, in the thirteenth century, the habit of eating human flesh pervaded all classes of society; extraordinary snares were spread for physicians in particular. They were called to attend persons who pretended to be sick, but who were only hungry; and it was not in order to be consulted, but devoured. An historian of great veracity, Abd-allatif, has related how a practice, which at first inspired dread and horror, soon occasioned not even the slightest surprise.[*]

Although the Indians of the Cassiquiare readily return to their barbarous habits, they evince, whilst in the missions, intelligence, some love of labour, and, in particular, a great facility in learning the Spanish language. The villages being, for the most part, inhabited by three or four tribes, who do not understand each other, a foreign idiom, which is at the same time that of the civil power, the language of the missionary, affords the advantage of more general means of communication. I heard a Poinave Indian conversing in Spanish with a Guahibo, though both had come from their forests within three months. They uttered a phrase every quarter of an hour, prepared with difficulty, and in which the gerund of the verb, no doubt

---

[*] "When the poor began to eat human flesh, the horror and astonishment caused by repasts so dreadful were such that these crimes furnished the never-ceasing subject of every conversation. But at length the people became so accustomed to it, and conceived such a taste for this detestable food, that people of wealth and respectability were found to use it as their ordinary food, to eat it by way of a treat, and even to lay in a stock of it. This flesh was prepared in different ways, and the practice being once introduced, spread into the provinces, so that instances of it were found in every part of Egypt. It then no longer caused any surprise; the horror it had at first inspired vanished; and it was mentioned as an indifferent and ordinary thing. This mania of devouring one another became so common among the poor, that the greater part perished in this manner. These wretches employed all sorts of artifices, to seize men by surprise, or decoy them into their houses under false pretences. This happened to three physicians among those who visited me; and a bookseller who sold me books, an old and very corpulent man, fell into their snares, and escaped with great difficulty. All the facts which we relate as eye-witnesses fell under our observation accidentally, for we generally avoided witnessing spectacles which inspired us with so much horror." Account of Egypt by Abd-allatif, physician of Bagdad, translated into French by De Sacy pages 360 to 374.

according to the grammatical turn of their own languages, was constantly employed. "When I seeing Padre, Padre to me saying;"[*] instead of, "when I saw the missionary, he said to me." I have mentioned in another place, how wise it appeared to me in the Jesuits to generalize one of the languages of civilized America, for instance that of the Peruvians,[†] and instruct the Indians in an idiom which is foreign to them in its roots, but not in its structure and grammatical forms. This was following the system which the Incas, or king-priests of Peru had employed for ages, in order to humanize the barbarous nations of the Upper Maranon, and maintain them under their domination; a system somewhat more reasonable than that of making the natives of America speak Latin, as was gravely proposed in a provincial concilio at Mexico.

We were told that the Indians of the Cassiquiare and the Rio Negro are preferred on the Lower Orinoco, and especially at Angostura, to the inhabitants of the other missions, on account of their intelligence and activity. Those of Mandavaca are celebrated among the tribes of their own race for the preparation of the curare poison, which does not yield in strength to the curare of Esmeralda. Unhappily the natives devote themselves to this employment more than to agriculture. Yet the soil on the banks of the Cassiquiare is excellent. We find there a granitic sand, of a blackish-brown colour, which is covered in the forests with thick layers of rich earth, and on the banks of the river with clay almost impermeable to water. The soil of the Cassiquiare appears more fertile than that of the valley of the Rio Negro, where maize does not prosper. Rice, beans, cotton, sugar, and indigo yield rich harvests, wherever their cultivation has been tried.[‡] We saw wild indigo around the missions of San Miguel

---

[*] "Quando io mirando Padre, Padre me diciendo."

[†] The Quichua or Inca language, Lengua del Inga.

[‡] M. Bonpland found at Mandavaca, in the huts of the natives, a plant with tuberous roots, exactly like cassava (yucca). It is called cumapana, and is cooked by being baked on the ashes. It grows spontaneously on the banks of

de Davipe, San Carlos, and Mandavaca. No doubt can exist that several nations of America, particularly the Mexicans, long before the conquest, employed real indigo in their hieroglyphic paintings; and that small cakes of this substance were sold at the great market of Tenochtitlan. But a colouring matter, chemically identical, may be extracted from plants belonging to neighbouring genera; and I should not at present venture to affirm that the native indigoferae of America do not furnish some generic difference from the Indigofera anil, and the Indigofera argentea of the Old World. In the coffee-trees of both hemispheres this difference has been observed.

Here, as at the Rio Negro, the humidity of the air, and the consequent abundance of insects, are obstacles almost invincible to new cultivation. Everywhere you meet with those large ants that march in close bands, and direct their attacks the more readily on cultivated plants, because they are herbaceous and succulent, whilst the forests of these countries afford only plants with woody stalks. If a missionary wishes to cultivate salad, or any culinary plant of Europe, he is compelled as it were to suspend his garden in the air. He fills an old boat with good mould, and, having sown the seed, suspends it four feet above the ground with cords of the chiquichiqui palm-tree; but most frequently places it on a slight scaffolding. This protects the young plants from weeds, worms, and those ants which pursue their migration in a right line, and, not knowing what vegetates above them, seldom turn from their course to climb up stakes that are stripped of their bark. I mention this circumstance to prove how difficult, within the tropics, on the banks of great rivers, are the first attempts of man to appropriate to himself a little spot of earth in that vast domain of nature, invaded by animals, and covered by spontaneous plants.

During the night of the 13th of May, I obtained some observations of the stars, unfortunately the last at the

---

the Cassiquiare.

Cassiquiare. The latitude of Mandavaca is 2° 4' 7"; its longitude, according to the chronometer, 69° 27'. I found the magnetic dip 25.25° (cent div), showing that it had increased considerably from the fort of San Carlos. Yet the surrounding rocks are of the same granite, mixed with a little hornblende, which we had found at Javita, and which assumes a syenitic aspect. We left Mandavaca at half-past two in the morning. After six hours' voyage, we passed on the east the mouth of the Idapa, or Siapa, which rises on the mountain of Uuturan, and furnishes near its sources a portage to the Rio Mavaca, one of the tributary streams of the Orinoco. This river has white waters, and is not more than half as broad as the Pacimoni, the waters of which are black. Its upper course has been strangely misrepresented on maps. I shall have occasion hereafter to mention the hypotheses that have given rise to these errors, in speaking of the source of the Orinoco.

We stopped near the raudal of Cunuri. The noise of the little cataract augmented sensibly during the night, and our Indians asserted that it was a certain presage of rain. I recollected that the mountaineers of the Alps have great confidence in the same prognostic.[*] It fell before sunrise, and the araguato monkeys had warned us, by their lengthened howlings, of the approaching rain, long before the noise of the cataract increased.

On the 14th, the mosquitos, and especially the ants, drove us from the shore before two in the morning. We had hitherto been

---

[*] "It is going to rain, because we hear the murmur of the torrents nearer," say the mountaineers of the Alps, like those of the Andes. The cause of the phenomenon is a modification of the atmosphere, which has an influence at once on the sonorous and on the luminous undulations. The prognostic drawn from the increase and the intensity of sound is intimately connected with the prognostic drawn from a less extinction of light. The mountaineers predict a change of weather, when, the air being calm, the Alps covered with perpetual snow seem on a sudden to be nearer the observer, and their outlines are marked with great distinctness on the azure sky. What is it that causes the want of homogeneity in the vertical strata of the atmosphere to disappear instantaneously?

of opinion that the ants did not crawl along the cords by which the hammocks are usually suspended: whether we were correct in this supposition, or whether the ants fell on us from the tops of the trees, I cannot say; but certain it is that we had great difficulty to keep ourselves free from these troublesome insects. The river became narrower as we advanced, and the banks were so marshy, that it was not without much labour M. Bonpland could get to a Carolinea princeps loaded with large purple flowers. This tree is the most beautiful ornament of these forests, and of those of the Rio Negro. We examined repeatedly, during this day, the temperature of the Cassiquiare. The water at the surface of the river was only 24° (when the air was at 25.6°.) This is nearly the temperature of the Rio Negro, but four or five degrees below that of the Orinoco. After having passed on the west the mouth of the Cano Caterico, which has black waters of extraordinary transparency, we left the bed of the river, to land at an island on which the mission of Vasiva is established. The lake which surrounds this mission is a league broad, and communicates by three outlets with the Cassiquiare. The surrounding country abounds in marshes which generate fever. The lake, the waters of which appear yellow by transmitted light, is dry in the season of great heat, and the Indians themselves are unable to resist the miasmata rising from the mud. The complete absence of wind contributes to render the climate of this country more pernicious.

From the 14th to the 21st of May we slept constantly in the open air; but I cannot indicate the spots where we halted. These regions are so wild, and so little frequented, that with the exception of a few rivers, the Indians were ignorant of the names of all the objects which I set by the compass. No observation of a star helped me to fix the latitude within the space of a degree. After having passed the point where the Itinivini separates from the Cassiquiare, to take its course to the west towards the granitic hills of Daripabo, we found the marshy banks of the river covered with bamboos. These arborescent gramina rise to the height of twenty feet; their stem is constantly arched towards the summit. It is a new species of Bambusa with very broad leaves. M. Bonpland fortunately found one in flower; a circumstance I mention, because the genera Nastus and Bambusa had before been very imperfectly distinguished, and nothing is more rare in the New World, than to see these gigantic gramina in flower. N. Mutis herborised during twenty years in a country where the Bambusa guadua forms marshy forests several leagues broad, without having ever been able to procure the flowers. We sent that learned naturalist the first ears of Bambusa from the temperate valleys of Popayan. It is strange that the parts of fructification should develop themselves so rarely in a plant which is indigenous, and which vegetates with such extraordinary rigour, from the level of the sea to the height of nine hundred toises, that is, to a subalpine region the climate of which, between the tropics, resembles that of the south of Spain. The Bambusa latifolia seems to be peculiar to the basins of the Upper Orinoco, the Cassiquiare, and the Amazon; it is a social plant, like all the gramina of the family of the nastoides; but in that part of Spanish Guiana which we traversed it does not grow in those large masses which the Spanish Americans call guadales, or forests of bamboos.

Our first resting-place above Vasiva was easily arranged. We found a little nook of dry ground, free from shrubs, to the south of the Cano Curamuni, in a spot where we saw some capuchin monkeys.[*] They were recognizable by their black beards and their gloomy and sullen air, and were walking slowly on the horizontal branches of a genipa. During the five following nights our passage was the more troublesome in proportion as we approached the bifurcation of the Orinoco. The luxuriance of the vegetation increases in a manner of which it is difficult even for those acquainted with the aspect of the forests between the tropics, to form an idea. There is no longer a bank: a palisade of tufted trees forms the margin of the river. You see a canal two

---

[*] Simia chiropotes.

hundred toises broad, bordered by two enormous walls, clothed with lianas and foliage. We often tried to land, but without success. Towards sunset we sailed along for an hour seeking to discover, not an opening (since none exists), but a spot less wooded, where our Indians by means of the hatchet and manual labour, could clear space enough for a resting-place for twelve or thirteen persons. It was impossible to pass the night in the canoe; the mosquitos, which tormented us during the day, accumulated toward evening beneath the toldo covered with palm-leaves, which served to shelter us from the rain. Our hands and faces had never before been so much swelled. Father Zea, who had till then boasted of having in his missions of the cataracts the largest and fiercest (las mas feroces) mosquitos, at length gradually acknowledged that the sting of the insects of the Cassiquiare was the most painful he had ever felt. We experienced great difficulty, amid a thick forest, in finding wood to make a fire, the branches of the trees in those equatorial regions where it always rains, being so full of sap, that they will scarcely burn. There being no bare shore, it is hardly possible to procure old wood, which the Indians call wood baked in the sun. However, fire was necessary to us only as a defence against the beasts of the forest; for we had such a scarcity of provision that we had little need of fuel for the purpose of preparing our food.

On the 18th of May, towards evening, we discovered a spot where wild cacao-trees were growing on the bank of the river. The nut of these cacaos is small and bitter; the Indians of the forest suck the pulp, and throw away the nut, which is picked up by the Indians of the missions, and sold to persons who are not very nice in the preparation of their chocolate. "This is the Puerto del Cacao" (Cacao Port), said the pilot; "it is here our Padres sleep, when they go to Esmeralda to buy sarbacans[*] and juvias (Brazil nuts). Not five boats, however, pass annually by the Cassiquiare; and since we left Maypures (a whole month

---

[*] The bamboo tubes furnished by the Arundinaria, used for projecting the poisoned arrows of the natives. See Views of Nature page 180.

previously), we had not met one living soul on the rivers we navigated, except in the immediate neighbourhood of the missions. To the south of lake Duractumuni we slept in a forest of palm-trees. It rained violently, but the pothoses, arums, and lianas, furnished so thick a natural trellis, that we were sheltered as under a vault of foliage. The Indians whose hammocks were placed on the edge of the river, interwove the heliconias and other musaceae, so as to form a kind of roof over them. Our fires lighted up, to the height of fifty or sixty feet, the palm-trees, the lianas loaded with flowers, and the columns of white smoke, which ascended in a straight line toward the sky. The whole exhibited a magnificent spectacle; but to have enjoyed it fully, we should have breathed an air clear of insects.

The most depressing of all physical sufferings are those which are uniform in their duration, and can be combated only by long patience. It is probable, that in the exhalations of the forests of the Cassiquiare M. Bonpland imbibed the seeds of a severe malady, under which he nearly sunk on our arrival at Angostura. Happily for him and for me, nothing led us to presage the danger with which he was menaced. The view of the river, and the hum of the insects, were a little monotonous; but some remains of our natural cheerfulness enabled us to find sources of relief during our wearisome passage. We discovered, that by eating small portions of dry cacao ground without sugar, and drinking a large quantity of the river water, we succeeded in appeasing our appetite for several hours. The ants and the mosquitos troubled us more than the humidity and the want of food. Notwithstanding the privations to which we were exposed during our excursions in the Cordilleras, the navigation from Mandavaca to Esmeralda has always appeared to us the most painful part of our travels in America. I advise those who are not very desirous of seeing the great bifurcation of the Orinoco, to take the way of the Atabapo in preference to that of the Cassiquiare.

461

Above the Cano Duractumuni, the Cassiquiare pursues a uniform direction from north-east to south-west. We were surprised to see how much the high steep banks of the Cassiquiare had been undermined on each side by the sudden risings of the water. Uprooted trees formed as it were natural rafts; and being half-buried in the mud, they were extremely dangerous for canoes. We passed the night of the 20th of May, the last of our passage on the Cassiquiare, near the point of the bifurcation of the Orinoco. We had some hope of being able to make an astronomical observation, as falling-stars of remarkable magnitude were visible through the vapours that veiled the sky; whence we concluded that the stratum of vapours must be very thin, since meteors of this kind have scarcely ever been seen below a cloud. Those we now beheld shot towards the north, and succeeded each other at almost equal intervals. The Indians, who seldom ennoble by their expressions the wanderings of the imagination, name the falling-stars the urine; and the dew the spittle of the stars. The clouds thickened anew, and we discerned neither the meteors, nor the real stars, for which we had impatiently waited during several days.

We had been told, that we should find the insects at Esmeralda still more cruel and voracious than in the branch of the Orinoco which we were going up; nevertheless we indulged the hope of at length sleeping in a spot that was inhabited, and of taking some exercise in herbalizing. This anticipation was, however, disturbed at our last resting-place on the Cassiquiare. Whilst we were sleeping on the edge of the forest, we were warned by the Indians, in the middle of the night, that they heard very near us the cries of a jaguar. These cries, they alleged, came from the top of some neighbouring trees. Such is the thickness of the forests in these regions, that scarcely any animals are to be found there but such as climb trees; as, for instance, the monkeys, animals of the weasel tribe, jaguars, and other species of the genus Felis.

As our fires burnt brightly, we paid little attention to the cries of the jaguars. They had been attracted by the smell and noise of our dog. This animal (which was of the mastiff breed) began at first to bark; and when the tiger drew nearer, to howl, hiding himself below our hammocks. how great was our grief, when in the morning, at the moment of re-embarking, the Indians informed us that the dog had disappeared! There could be no doubt that it had been carried off by the jaguars.[*] Perhaps, when their cries had ceased, it had wandered from the fires on the side of the beach; and possibly we had not heard its moans, as we were in a profound sleep. We have often heard the inhabitants of the banks of the Orinoco and the Rio Magdalena affirm, that the oldest jaguars will carry off animals from the midst of a halting-place, cunningly grasping them by the neck so as to prevent their cries. We waited part of the morning, in the hope that our dog had only strayed. Three days after we came back to the same place; we heard again the cries of the jaguars, for these animals have a predilection for particular spots; but all our search was vain. The dog, which had accompanied us from Caracas, and had so often in swimming escaped the pursuit of the crocodiles,[†] had been devoured in the forest.

On the 21st May, we again entered the bed of the Orinoco, three leagues below the mission of Esmeralda. It was now a month since we had left that river near the mouth of the Guaviare. We had still to proceed seven hundred and fifty miles[‡] before reaching Angostura, but we should go with the stream; and this consideration lessened our discouragement. In descending great rivers, the rowers take the middle of the current, where there are few mosquitos; but in ascending, they are obliged, in order to avail themselves of the dead waters and counter-currents, to sail near the shore, where the proximity of

---

[*] See Views of Nature page 195.
[†] Ibid page 198.
[‡] Of nine hundred and fifty toises each, or two hundred and fifty nautical leagues.

the forests, and the remains of organic substances accumulated on the beach, harbour the tipulary insects. The point of the celebrated bifurcation of the Orinoco has a very imposing aspect. Lofty granitic mountains rise on the northern bank; and amidst them are discovered at a distance the Maraguaca and the Duida. There are no mountains on the left bank of the Orinoco, west or east of the bifurcation, till opposite the mouth of the Tamatama. On that spot stands the rock Guaraco, which is said to throw out flames from time to time in the rainy season. When the Orinoco is no longer bounded by mountains towards the south, and when it reaches the opening of a valley, or rather a depression of the ground, which terminates at the Rio Negro, it divides itself into two branches. The principal branch (the Rio Paragua of the Indians) continues its course west-north-west, turning round the group of the mountains of Parime; the other branch forming the communication with the Amazon runs into plains, the general slope of which is southward, but of which the partial planes incline, in the Cassiquiare, to south-west, and in the basin of the Rio Negro, south-east. A phenomenon so strange in appearance, which I verified on the spot, merits particular attention; the more especially as it may throw some light on analogous facts, which are supposed to have been observed in the interior of Africa.

The existence of a communication of the Orinoco with the Amazon by the Rio Negro, and a bifurcation of the Caqueta, was believed by Sanson, and rejected by Father Fritz and by Blaeuw: it was marked in the first maps of De l'Isle, but abandoned by that celebrated geographer towards the end of his days. Those who had mistaken the mode of this communication hastened to deny the communication itself. It is in fact well worthy of remark that, at the time when the Portuguese went up most frequently by the Amazon, the Rio Negro, and the Cassiquiare, and when Father Gumilla's letters were carried (by the natural interbranching of the rivers) from the lower Orinoco to Grand Para, that very missionary made every effort to spread the opinion through Europe that the basins of the Orinoco and the

Amazon are perfectly separate. He asserts that, having several times gone up the former of these rivers as far as the Raudal of Tabaje, situate in the latitude of 1° 4', he never saw a river flow in or out that could be taken for the Rio Negro. He adds further, that a great Cordillera, which stretches from east to west, prevents the mingling of the waters, and renders all discussion on the supposed communication of the two rivers useless. The errors of Father Gumilla arose from his firm persuasion that he had reached the parallel of 1° 4' on the Orinoco. He was in error by more than 5° 10' of latitude; for I found, by observation, at the mission of Atures, thirteen leagues south of the rapids of Tabaje, the latitude to be 5° 37' 34". Gumilla having gone but little above the confluence of the Meta, it is not surprising that he had no knowledge of the bifurcation of the Orinoco, which is found by the sinuosities of the river to be one hundred and twenty leagues distant from the Raudal of Tabaje.

La Condamine, during his memorable navigation on the river Amazon in 1743, carefully collected a great number of proofs of this communication of the rivers, denied by the Spanish Jesuit. The most decisive proof then appeared to him to be the unsuspected testimony of a Cauriacani Indian woman with whom he had conversed, and who had come in a boat from the banks of the Orinoco (from the mission of Pararuma) to Grand Para. Before the return of La Condamine to his own country, the voyage of Father Manuel Roman, and the fortuitous meeting of the missionaries of the Orinoco and the Amazon, left no doubt of this fact, the knowledge of which was first obtained by Acunha.

The incursions undertaken from the middle of the seventeenth century, to procure slaves, had gradually led the Portuguese from the Rio Negro, by the Cassiquiare, to the bed of a great river, which they did not know to be the Upper Orinoco. A flying camp, composed of the troop of ransomers,[*] favoured this inhuman commerce. After having excited the natives to

---

[*] Tropa de rescate; from rescatar, to redeem.

make war, they ransomed the prisoners; and, to give an appearance of equity to the traffic, monks accompanied the troop of ransomers to examine whether those who sold the slaves had a right to do so, by having made them prisoners in open war. From the year 1737 these visits of the Portuguese to the Upper Orinoco became very frequent. The desire of exchanging slaves (poitos) for hatchets, fish-hooks, and glass trinkets, induced the Indian tribes to make war upon one another. The Guipunaves, led on by their valiant and cruel chief Macapu, descended from the banks of the Inirida towards the confluence of the Atabapo and the Orinoco. "They sold," says the missionary Gili, "the slaves whom they did not eat."[*] The Jesuits of the Lower Orinoco became uneasy at this state of things, and the superior of the Spanish missions, Father Roman, the intimate friend of Gumilla, took the courageous resolution of crossing the Great Cataracts, and visiting the Guipunaves, without being escorted by Spanish soldiers. He left Carichana the 4th of February, 1744; and having arrived at the confluence of the Guaviare, the Atabapo, and the Orinoco, where the last mentioned river suddenly changes its previous course from east to west, to a direction from south to north, he saw from afar a canoe as large as his own, and filled with men in European dresses. He caused a crucifix to be placed at the bow of his boat in sign of peace, according to the custom of the missionaries when they navigate in a country unknown to them. The whites, who were Portuguese slave-traders of the Rio Negro, recognized with marks of joy the habit of the order of St. Ignatius. They heard with astonishment that the river on which this meeting took place was the Orinoco; and they brought Father Roman by the Cassiquiare to the

---

[*] "I Guipunavi avventizj abitatori dell' Alto Orinoco, recavan de' danni incredibili alle vicine mansuete nazioni; altre mangiondone, altre conducendone schiave ne' Portoghesi dominj." "The Guipunaves, at their first arrival on the Upper Orinoco, inflicted incredible injuries on the other peaceable tribes who dwelt near them, devouring some, and selling others as slaves to the Portuguese." Gili tome 1 page 31.

Brazilian settlements on the Rio Negro. The superior of the Spanish missions was forced to remain near the flying camp of the troop of ransomers till the arrival of the Portuguese Jesuit Avogadri, who had gone upon business to Grand Para. Father Manuel Roman returned with his Salive Indians by the same way, that of the Cassiquiare and the Upper Orinoco, to Pararuma,[*] a little to the north of Carichana, after an absence of seven months. He was the first white man who went from the Rio Negro, consequently from the basin of the Amazon, without passing his boats over any portage, to the basin of the Lower Orinoco.

The tidings of this extraordinary passage spread with such rapidity that La Condamine was able to announce it[†] at a public sitting of the Academy, seven months after the return of Father Roman to Pararuma. "The communication between the Orinoco and the Amazon," said he, "recently averred, may pass so much the more for a discovery in geography, as, although the junction of these two rivers is marked on the old maps (according to the information given by Acunha), it had been suppressed by all the modern geographers in their new maps, as if in concert. This is not the first time that what is positive fact has been thought fabulous, that the spirit of criticism has been pushed too far, and that this communication has been treated as chimerical by those

---

[*] On the 15th of October, 1774. La Condamine quitted the town of Grand Para December the 29th, 1743; it follows, from a comparison of the dates, that the Indian woman of Pararuma, carried off by the Portuguese, and to whom the French traveller had spoken, had not come with Father Roman, as was erroneously affirmed. The appearance of this woman on the banks of the Amazon is interesting with respect to the researches lately made on the mixture of races and languages: it proves the enormous distances through which the individuals of one tribe are compelled to carry on intercourse with those of another.

[†] The intelligence was communicated to him by Father John Ferreyro, rector of the college of Jesuits at Para. Voyage a l'Amazone page 120. Mem. de l'Acad. 1745 page 450. Caulin page 79. See also, in the work of Gili, the fifth chapter of the first book, published in 1780, with the title: Della scoperta delle communicazione dell' Orinoco col Maragnone.

467

who ought to have been better informed." Since the voyage of Father Roman in 1774, no person in Spanish Guiana, or on the coasts of Cumana and Caracas, has admitted a doubt of the existence of the Cassiquiare and the bifurcation of the Orinoco. Father Gumilla himself; whom Bouguer met at Carthagena, confessed that he had been deceived; and he read to Father Gili, a short time before his death, a supplement to his history of the Orinoco, intended for a new edition, in which he recounts pleasantly the manner in which he had been undeceived. The expedition of the boundaries, under Iturriaga and Solano, completed in detail the knowledge of the geography of the Upper Orinoco, and the intertwinings of this river with the Rio Negro. Solano established himself in 1756 at the confluence of the Atabapo; and from that time the Spanish and Portuguese commissioners often passed in their canoes, by the Cassiquiare, from the Lower Orinoco to the Rio Negro, to visit each other at their head-quarters of Cabruta* and Mariva. Since the year 1767, two or three canoes come annually from the fort of San Carlos, by the bifurcation of the Orinoco to Angostura, to fetch salt and the pay of the troops. These passages, from one basin of a river to another, by the natural canal of the Cassiquiare, excite no more attention in the colonists at present than the arrival of boats that descend the Loire by the canal of Orleans, awakens on the banks of the Seine.

Although, since the journey of Father Roman, in 1744, precise notions have been acquired in the Spanish possessions in America, both of the direction of the Upper Orinoco from east to

---

* General Iturriaga, confined by illness, first at Muitaco, or Real Corona, and afterward at Cabruta, received a visit in 1760 from the Portuguese colonel Don Gabriel de Souza y Figueira, who came from Grand Para, having made a voyage of nearly nine hundred leagues in his boat. The Swedish botanist, Loefling, who was chosen to accompany the expedition of the boundaries at the expense of the Spanish government, so greatly multiplied in his ardent imagination the branchings of the great rivers of South America, that he appeared well persuaded of being able to navigate, by the Rio Negro and the Amazon, to the Rio de la Plata. (Iter page 131.)

west, and of the manner of its communication with the Rio Negro, this knowledge did not reach Europe till a much later period. In 1750, La Condamine and D'Anville* were still of opinion that the Orinoco was a branch of the Caqueta coming from the south-east, and that the Rio Negro issued immediately from it. It was only in the second edition of his South America, that D'Anville (without renouncing that intercommunication of the Caqueta, by means of the Iniricha (Inirida), with the Orinoco and the Rio Negro) describes the Orinoco as taking its rise at the east, near the sources of the Rio Branco, and marks the Rio Cassiquiare as bearing the waters of the Upper Orinoco to the Rio Negro. It is probable that this indefatigable and learned writer had obtained information on the manner of the bifurcation from his frequent communications with the missionaries,† who were then the only geographers of the most inland parts of the continents.

Had the nations of the lower region of equinoctial America participated in the civilization spread over the cold and alpine region, that immense Mesopotamia between the Orinoco and the Amazon would have favoured the development of their industry, animated their commerce, and accelerated the progress of social

---

* See the classical memoir of this great geographer in the Journal des Savans, March 1750 page 184. "One fact," says D'Anville, "which cannot be considered as equivocal, after the proofs with which we have been recently furnished, is the communication of the Rio Negro with the Orinoco; but we must not hesitate to admit, that we are not yet sufficiently informed of the manner in which this communication takes place." I was surprised to see in a very rare map, which I found at Rome (Provincia Quitensis Soc. Jesu in America, auctore Carolo Brentano et Nicolao de la Torre; Romae 1745) that seven years after the discovery of Father Roman, the Jesuits of Quito were ignorant of the existence of the Cassiquiare. The Rio Negro is figured in this map as a branch of the Orinoco.

† According to the Annals of Berredo, it would appear, that as early as the year 1739, the military incursions from the Rio Negro to the Cassiquiare had confirmed the Portuguese Jesuits in the opinion that there was a communication between the Amazon and the Orinoco. Southey's Brazils volume 1 page 658.

order. We see everywhere in the old world the influence of locality on the dawning civilization of nations. The island of Meroe between the Astaboras and the Nile, the Punjab of the Indus, the Douab of the Ganges, and the Mesopotamia of the Euphrates, furnish examples that are justly celebrated in the annals of the human race. But the feeble tribes that wander in the savannahs and the woods of eastern America, have profited little by the advantages of their soil, and the interbranchings of their rivers. The distant incursions of the Caribs, who went up the Orinoco, the Cassiquiare, and the Rio Negro, to carry off slaves and exercise pillage, compelled some rude tribes to rouse themselves from their indolence, and form associations for their common defence; the little good, however, which these wars with the Caribs (the Bedouins of the rivers of Guiana) produced, was but slight compensation for the evils that followed in their train, by rendering the tribes more ferocious, and diminishing their population. We cannot doubt, that the physical aspect of Greece, intersected by small chains of mountains, and mediterranean gulfs, contributed, at the dawn of civilization, to the intellectual development of the Greeks. But the operation of this influence of climate, and of the configuration of the soil, is felt in all its force only among a race of men who, endowed with a happy organization of the mental faculties, are susceptible of exterior impulse. In studying the history of our species, we see, at certain distances, these foci of ancient civilization dispersed over the globe like luminous points; and we are struck by the inequality of improvement in nations inhabiting analogous climates, and whose native soil appears equally favoured by the most precious gifts of nature.

Since my departure from the banks of the Orinoco and the Amazon, a new era has unfolded itself in the social state of the nations of the West. The fury of civil discussions has been succeeded by the blessings of peace, and a freer development of the arts of industry. The bifurcations of the Orinoco, the isthmus of Tuamini, so easy to be made passable by an artificial canal, will ere long fix the attention of commercial Europe. The

Cassiquiare, as broad as the Rhine, and the course of which is one hundred and eighty miles in length, will no longer form uselessly a navigable canal between two basins of rivers which have a surface of one hundred and ninety thousand square leagues. The grain of New Grenada will be carried to the banks of the Rio Negro; boats will descend from the sources of the Napo and the Ucuyabe, from the Andes of Quito and of Upper Peru, to the mouths of the Orinoco, a distance which equals that from Timbuctoo to Marseilles. A country nine or ten times larger than Spain, and enriched with the most varied productions, is navigable in every direction by the medium of the natural canal of the Cassiquiare, and the bifurcation of the rivers. This phenomenon, which will one day be so important for the political connections of nations, unquestionably deserves to be carefully examined.

CHAPTER 2.24.

THE UPPER ORINOCO, FROM THE ESMERALDA TO THE CONFLUENCE OF THE GUAVIARE. SECOND PASSAGE ACROSS THE CATARACTS OF ATURES AND MAYPURES. THE LOWER ORINOCO, BETWEEN THE MOUTH OF THE RIO APURE, AND ANGOSTURA THE CAPITAL OF SPANISH GUIANA.

Opposite to the point where the Orinoco forms its bifurcation, the granitic group of Duida rises in an amphitheatre on the right bank of the river. This mountain, which the missionaries call a volcano, is nearly eight thousand feet high. It is perpendicular on the south and west, and has an aspect of solemn grandeur. Its summit is bare and stony, but, wherever its less steep declivities are covered with mould vast forests appear suspended on its flanks. At the foot of Duida is the mission of Esmeralda, a little hamlet with eighty inhabitants, surrounded by a lovely plain, intersected by rills of black but limpid water. This plain is adorned with clumps of the mauritia palm, the sago-tree of America. Nearer the mountain, the distance of which from the cross of the mission I found to be seven thousand three hundred toises, the marshy plain changes to a savannah, and spends itself along the lower region of the Cordillera. Large pine-apples are there found of a delicious flavour; that species of bromelia always grows solitary among the gramina, like our Colchicum autumnale, while the B. karatas, another species of the same genus, is a social plant, like our whortleberries and heaths. The pine-apples of Esmeralda are cultivated throughout Guiana. There are certain spots in America, as in Europe, where different fruits attain their highest perfection. The sapota-plum (achra) should be eaten at the Island of Margareta or at Cumana: the chirimoya (very different from the custard-apple and sweet-sop of the West India Islands) at Loxa in Peru; the grenadilla, or parcha, at Caracas; and the pine-apple at Esmeralda, or in the island of Cuba. The pine-apple forms the ornament of the fields near the Havannah, where it is planted in parallel rows; on the sides of the Duida it embellishes the turf of the savannahs, lifting its yellow fruit, crowned with a tuft of silvery leaves, above the setaria, the paspalum, and a few cyperaceae. This plant, which the Indians of the Orinoco call ana-curua, has been propagated since the sixteenth century in the interior of China,[*] and some English travellers found it recently, together with other plants indubitably American (maize, cassava, tobacco, and pimento), on the banks of the River Congo, in Africa.

There is no missionary at Esmeralda; the monk appointed to celebrate mass in that hamlet is settled at Santa Barbara, more than fifty leagues distant; and he visits this spot but five or six times in a year. We were cordially received by an old officer, who took us for Catalonian shopkeepers, and who supposed that trade had led to the missions. On seeing packages of paper intended for drying our plants, he smiled at our simple ignorance. "You come," said he, "to a country where this kind of merchandise has no sale; we write little here; and the dried leaves of maize, the platano (plantain-tree), and the vijaho (heliconia), serve us, like paper in Europe, to wrap up needles, fish-hooks, and other little articles of which we are careful." This old officer united in his person the civil and ecclesiastical authority. He taught the children, I will not say the Catechism, but the Rosary; he rang the bells to amuse himself; and impelled by ardent zeal for the service of the church, he sometimes used

---

[*] No doubt remains of the American origin of the Bromelia ananas. See Cayley's Life of Raleigh volume 1 page 61. Gili volume 1 pages 210 and 336. Robert Brown, Geogr. Observ. on the Plants of the River Congo 1818 page 50.

his chorister's wand in a manner not very agreeable to the natives.

Notwithstanding the small extent of the mission, three Indian languages are spoken at Esmeralda; the Idapimanare, the Catarapenno, and the Maquiritan. The last of these prevails on the Upper Orinoco, from the confluence of the Ventuari as far as that of the Padamo[*]; the Caribbee prevails on the Lower Orinoco; the Ottomac, near the confluence of the Apure, at the Great Cataracts; and the Maravitan, on the banks of the Rio Negro. These are the five or six languages most generally spoken. We were surprised to find at Esmeralda many zambos, mulattos, and copper-coloured people, who called themselves Spaniards (Espanoles) and who fancy they are white, because they are not so red as the Indians. These people live in the most absolute misery; they have for the most part been sent hither in banishment (desterrados). Solano, in his haste to found colonies in the interior of the country, in order to guard its entrance against the Portuguese, assembled in the Llanos, and as far as the island of Margareta, vagabonds and malefactors, whom justice had vainly pursued, and made them go up the Orinoco to join the unhappy Indians who had been carried off from the woods. A mineralogical error gave celebrity to Esmeralda. The granites of Duida and Maraguaca contain in open veins fine rock-crystals, some of them of great transparency, others coloured by chlorite or blended with actonite; these were mistaken for diamonds and emeralds.

So near the sources of the Orinoco we heard of nothing in these mountains but the proximity of El Dorado, the lake Parima, and the ruins of the great city of Manoa. A man, still known in the country for his credulity and his love of exaggeration, Don Apollinario Diez de la Fuente, assumed the

---

[*] The Arivirianos of the banks of the Ventuari speak a dialect of the language of the Maquiritares. The latter live, jointly with a tribe of the Macos, in the savannahs that are by the Padamo. They are so numerous, that they have even given their name to this tributary stream of the Orinoco.

pompous title of capitan poblador, and cabo militar (military commander) of the fort of Cassiquiare. This fort consisted of a few trunks of trees, joined together by planks; and to complete the deception, a demand was made at Madrid for the privileges of a villa for the mission of Esmeralda, which but a hamlet with twelve or fifteen huts. A colony composed of elements altogether heterogeneous perished by degrees. The vagabonds of the Llanos had as little taste for labour as the natives, who were compelled to live within the sound of the bell. The former found a motive in their pride to justify their indolence. In the missions, every mulatto who is not decidedly black as an African, or copper-coloured as an Indian, calls himself a Spaniard; he belongs to the gente de razon—the race endued with reason; and that reason (sometimes, it must be admitted, arrogant and indolent) persuaded the whites, and those who fancy they are so, that to till the ground is a task fit only for slaves (poitos) and the native neophytes. The colony of Esmeralda had been founded on the principles of that of Australia; but it was far from being governed with the same wisdom. The American colonists, being separated from their native soil, not by seas, but by forests and savannahs, dispersed; some taking the road northward, towards the Caura and the Carony; others proceeding southward to the Portuguese possessions. Thus the celebrity of this villa, and of the emerald-mines of Duida, vanished in a few years; and Esmeralda, on account of the immense number of insects that obscure the air at all seasons of the year, was regarded by the monks as a place of banishment. The superior of the missions, when he would make the lay-brothers mindful of their duty, threatens sometimes to send them to Esmeralda; that is, say the monks, to be condemned to the mosquitos; to be devoured by those buzzing flies (zancudos gritones) which God appears to have created for the torment and chastisement of man.[*] These

---

[*] "Estos mosquitos que llaman zancudos gritones los parece cria la naturaleza para castigo y tormento de los hombres." "Those mosquitos which are called buzzing zancudos, Nature seems to have created for the especial

strange punishments have not always been confined to the lay-brothers. There happened in 1788 one of those monastic revolutions, of which it is difficult to form a conception in Europe, according to the ideas that prevail of the peaceful state of the Christian settlements in the New World. For a long period the Franciscan monks settled in Guiana had been desirous of forming a separate republic, and rendering themselves independent of the college of Piritu at Nueva Barcelona. Discontented with the election of Fray Gutierez de Aguilera, chosen by a general chapter, and confirmed by the king in the important office of president of the missions, five or six monks of the Upper Orinoco, the Cassiquiare, and the Rio Negro, assembled together at San Fernando de Atabapo; chose hastily a new superior from their own body; and caused the old one, who, unfortunately for himself, had come to visit those parts, to be arrested. They put him in irons, threw him into a boat, and conducted him to Esmeralda, as to a place of proscription. This great distance of the coast from the scene of this revolution led the monks to hope that their crime would remain long unknown beyond the Great Cataracts. They wished to gain time to intrigue, to negotiate, to frame acts of accusation, and employ the little artifices by which, in every country, the invalidity of a first election may be proved. Fray Gutierez do Aguilera languished in his prison at Esmeralda, and fell dangerously ill from the double influence of the excessive heat, and the continual irritation of the mosquitos. Happily for the fallen power the monks did not remain united. A missionary of the Cassiquiare conceived serious alarms respecting the issue of this affair; he dreaded being sent a prisoner to Cadiz, or, as they say in the colonies, having his name on the list (baxo partido de registro). Fear overcame his resolution, and he suddenly disappeared. Indians were placed on the watch at the mouth of the Atabapo, at the Great Cataracts, and wherever the fugitive was likely to pass on his way to the Lower Orinoco.

punishment and torture of man." Fray Pedro Simon.

Notwithstanding these precautions, he arrived at Angostura, and then reached the college of the missions of Piritu, denounced his colleagues, and was appointed, in recompense of this information, to arrest those with whom he had conspired against the president of the missions.[*] At Esmeralda, where the political events that have agitated Europe for thirty years past have not yet been heard of, lively interest is still felt in an event which is called the sedition of the monks, (el alboroto de los frailes.) In this country, as in the East, no conception is formed of any other revolutions than those that are made by rulers themselves; and we have just seen that the effects are not very alarming.

If the villa of Esmeralda, with a population of twelve or fifteen families, be at present considered as a frightful place of abode, this must be attributed to the want of cultivation, the distance from every other inhabited country, and the excessive quantity of mosquitos. The site of the mission is highly picturesque; the surrounding country is lovely, and of great fertility. I never saw plantains of so large a size as these: and indigo, sugar, and cacao might be produced in abundance, if any trouble were taken for their cultivation. The Cerro Duida is surrounded with fine pasturage; and if the Observantins of the college of Piritu partook a little of the industry of the Catalonian Capuchins settled on the banks of the Carony, numerous herds would be seen wandering between the Cunucunumo and the Padamo. At present, not a cow or a horse is to be found; and the inhabitants, victims of their own indolence, are often reduced to eat the flesh of alouate monkeys, and flour made from the bones of fish, of which I shall have occasion to speak hereafter. A little

---

[*] Two of the missionaries, considered as the leaders of the insurrection, were embarked at Angostura, in order to be tried in Spain. The vessel in which they were conveyed became leaky, and put into Spanish Harbour in the island of Trinidad. The governor Chacon intereated himself in the fate of the monks; they were pardoned a violent proceeding somewhat inconsistent with monastic discipline, and were again employed in the missions. I was acquainted with them both during my abode in South America.

cassava and a few plantains only are cultivated; and when the fishery is not abundant, the natives of a country so favoured by nature are exposed to the most cruel privations.

The pilots of the small number of boats that go from the Rio Negro to Angostura by the Cassiquiare are afraid to ascend as far as Esmeralda, and therefore that mission would have been much better placed at the point of the bifurcation of the Orinoco. It is probable that this vast country will not always be doomed to the desertion in which it has hitherto been left, owing to the errors of monkish administration and the spirit of monopoly that characterises corporations. We may even predict on what points of the Orinoco industry and commerce will become most active. In every zone, population is concentred at the mouth of tributary streams. The Rio Apure, by which the productions of the provinces of Varinas and Merida are exported, will give great importance to the little town of Cabruta, which will then be in rivalship with San Fernando de Apure, where all commerce has hitherto centred. Higher up, a new settlement will be formed at the confluence of the Meta, which communicates with New Grenada by the Llanos of Casanare. The two missions of the Cataracts will increase, from the activity to which the transport of boats at those points will give rise; for an unhealthy and damp climate, and the swarming of mosquitos, will as little impede the progress of cultivation at the Orinoco as at the Rio Magdalena, whenever a powerful mercantile interest shall call new settlers thither. Habitual evils are those which are least felt; and men born in America do not suffer the same intensity of pain as Europeans recently arrived. Perhaps, also, the destruction of forests round the inhabited places, although slow, will somewhat tend to diminish the torment of the tipulary insects. San Fernando de Atabapo, Javita, San Carlos, and Esmeralda, appear (from their situation at the mouth of the Guaviare, the portage between Tuamini and the Rio Negro, the confluence of the Cassiquiare, and the point of bifurcation of the Upper Orinoco) to promise a considerable increase of population and prosperity. The same improvement will take place in the fertile but uncultivated countries through which flow the Guallaga, the Amazon, and the Orinoco; as well as at the isthmus of Panama, the lake of Nicaragua, and the Rio Huasacualco, which furnish a communication between the two oceans. The imperfection of political institutions may for ages have converted into deserts places where the commerce of the world should be found concentred; but the time approaches when these obstacles shall exist no longer. A vicious administration cannot always struggle against the united interest of men; and civilization will be carried insensibly into those countries, the great destinies of which nature itself proclaims, by the physical configuration of the soil, the immense windings of the rivers, and the proximity of two seas, that bathe the shores of Europe and of India.

Esmeralda is the most celebrated spot on the Orinoco for the preparation of that active poison, which is employed in war, in the chase, and, singularly enough, as a remedy for gastric derangements. The poison of the ticunas of the Amazon, the upas-tieute of Java, and the curare of Guiana, are the most deleterious substances that are known. Raleigh, about the end of the sixteenth century, had heard of urari[*] as being a vegetable substance with which arrows were envenomed; yet no fixed notions of this poison had reached Europe. The missionaries Gumilla and Gili had not been able to penetrate into the country where the curare is manufactured. Gumilla asserts that this preparation was enveloped in great mystery; that its principal ingredient was furnished by a subterranean plant with a tuberous root, which never puts forth leaves, and which is called specially the root (raiz de si misma); that the venomous exhalations which arise from the manufacture are fatal to the lives of the old women who (being otherwise useless) are chosen to watch over this operation; finally, that these vegetable juices are never thought to be sufficiently concentrated till a few drops produce at a distance a repulsive action on the blood. An Indian wounds

---

[*] In Tamanac marana, in Maypure macuri.

himself slightly; and a dart dipped in the liquid curare is held near the wound. If it make the blood return to the vessels without having been brought into contact with them, the poison is judged to be sufficiently concentrated.

When we arrived at Esmeralda, the greater part of the Indians were returning from an excursion which they had made to the east, beyond the Rio Padamo, to gather juvias, or the fruit of the bertholletia, and the liana which yields the curare. Their return was celebrated by a festival, which is called in the mission la fiesta de las juvias, and which resembles our harvest-homes and vintage-feasts. The women had prepared a quantity of fermented liquor; and during two days the Indians were in a state of intoxication. Among nations who attach great importance to the fruit of the palm, and of some other trees useful for the nourishment of man, the period when these fruits are gathered is marked by public rejoicings, and time is divided according to these festivals, which succeed one another in a course invariably regular. We were fortunate enough to find an old Indian more temperate than the rest, who was employed in preparing the curare poison from freshly-gathered plants. He was the chemist of the place. We found at his dwelling large earthen pots for boiling the vegetable juice, shallower vessels to favour the evaporation by a larger surface, and leaves of the plantain-tree rolled up in the shape of our filters, and used to filtrate the liquids, more or less loaded with fibrous matter. The greatest order and neatness prevailed in this hut, which was transformed into a chemical laboratory. The old Indian was known throughout the mission by the name of the poison-master (amo del curare). He had that self-sufficient air and tone of pedantry of which the pharmacopolists of Europe were formerly accused. "I know," said he, "that the whites have the secret of making soap, and manufacturing that black powder which has the defect of making a noise when used in killing animals. The curare, which we prepare from father to son, is superior to anything you can make down yonder (beyond sea). It is the juice of an herb which kills silently, without any one knowing whence the stroke comes."

This chemical operation, to which the old man attached so much importance, appeared to us extremely simple. The liana (bejuco) used at Esmeralda for the preparation of the poison, bears the same name as in the forests of Javita. It is the bejuco de Mavacure, which is gathered in abundance east of the mission, on the left bank of the Orinoco, beyond the Rio Amaguaca, in the mountainous and rocky tracts of Guanaya and Yumariquin. Although the bundles of bejuco which we found in the hut of the Indian were entirely bare of leaves, we had no doubt of their being produced by the same plant of the strychnos family (nearly allied to the rouhamon of Aublet) which we had examined in the forest of Pimichin.[*] The mavacure is employed fresh or dried indifferently during several weeks. The juice of the liana, when it has been recently gathered, is not regarded as poisonous; possibly it is so only when strongly concentrated. It is the bark and a part of the alburnum which contain this terrible poison. Branches of the mavacure four or five lines in diameter are scraped with a knife, and the bark that comes off is bruised, and reduced into very thin filaments on the stone employed for grinding cassava. The venomous juice being yellow, the whole fibrous mass takes that colour. It is thrown into a funnel nine inches high, with an opening four inches wide. This funnel was of all the instruments of the Indian laboratory that of which the poison-master seemed to be most proud. He asked us repeatedly

---

[*] I may here insert the description of the curare or bejuco de Mavacure, taken from a manuscript, yet unpublished, of my learned fellow-labourer M. Kunth, corresponding member of the Institute. "Ramuli lignosi, oppositi, ramulo altero abortivo, teretiusculi, fuscescenti-tomentosi, inter petiolos lineola pilosa notati, gemmula aut processu filiformi (pedunculo?) terminati. *folia* opposita, breviter petiolata, ovato-oblonga, acuminata, intergerrima, reticulato-triplinervia, nervo medio subtus prominente, membranacea, ciliata, utrinque glabra, nervo medio fuscescente-tomentoso, lacte viridia, subtus pallidiora, 1½ to 2½ pollices longa, 8 to 9 lineas lata. *petioli* lineam longi, tomentosi, inarticulati."

481

if, por alla (out yonder, meaning in Europe) we had ever seen anything to be compared to this funnel (embudo). It was a leaf of the plantain-tree rolled up in the form of a cone, and placed within another stronger cone made of the leaves of the palm-tree. The whole of this apparatus was supported by slight frame-work made of the petioles and ribs of palm-leaves. A cold infusion is first prepared by pouring water on the fibrous matter which is the ground bark of the mavacure. A yellowish water filters during several hours, drop by drop, through the leafy funnel. This filtered water is the poisonous liquor, but it acquires strength only when concentrated by evaporation, like molasses, in a large earthen pot. The Indian from time to time invited us to taste the liquid; its taste, more or less bitter, decides when the concentration by fire has been carried sufficiently far. There is no danger in tasting it, the curare being deleterious only when it comes into immediate contact with the blood. The vapours, therefore, which are disengaged from the pans are not hurtful, notwithstanding all that has been asserted on this point by the missionaries of the Orinoco. Fontana, in his experiments on the poison of the ticuna of the Amazon, long since proved that the vapours arising from this poison, when thrown on burning charcoal, may be inhaled without danger and that the statement of La Condamine, that Indian women, when condemned to death, have been killed by the vapours of the poison of the ticuna, is incorrect.

The most concentrated juice of the mavacure is not thick enough to stick to the darts; and therefore, to give a body to the poison, another vegetable juice, extremely glutinous, drawn from a tree with large leaves, called kiracaguero, is poured into the concentrated infusion. As this tree grows at a great distance from Esmeralda, and was at that period as destitute of flowers and fruits as the bejuco de mavacure, we could not determine it botanically. I have several times mentioned that kind of fatality which withholds the most interesting plants from the examination of travellers, while thousands of others, of the chemical properties of which we are ignorant, are found loaded with flowers and fruits. In travelling rapidly, even within the tropics, where the flowering of the ligneous plants is of such long duration, scarcely one-eighth of the trees can be seen furnishing the essential parts of fructification. The chances of being able to determine, I do not say the family, but the genus and species, is consequently as one to eight; and it may be conceived that this unfavourable chance is felt most powerfully when it deprives us of the intimate knowledge of objects which afford a higher interest than that of descriptive botany.

At the instant when the glutinous juice of the kiracaguero-tree is poured into the venomous liquor well concentrated, and kept in a state of ebullition, it blackens, and coagulates into a mass of the consistence of tar, or of a thick syrup. This mass is the curare of commerce. When we hear the Indians say that the kiracaguero is as necessary as the bejuco do mavacure in the manufacture of the poison, we may be led into error by the supposition that the former also contains some deleterious principle, while it only serves (as the algarrobo, or any other gummy substance would do) to give more body to the concentrated juice of the curare. The change of colour which the mixture undergoes is owing to the decomposition of a hydruret of carbon; the hydrogen is burned, and the carbon is set free. The curare is sold in little calabashes; but its preparation being in the hands of a few families, and the quantity of poison attached to each dart being extremely small, the best curare, that of Esmeralda and Mandavaca, is sold at a very high price. This substance, when dried, resembles opium; but it strongly absorbs moisture when exposed to the air. Its taste is an agreeable bitter, and M. Bonpland and myself have often swallowed small portions of it. There is no danger in so doing, if it be certain that neither lips nor gums bleed. In experiments made by Mangili on the venom of the viper, one of his assistants swallowed all the poison that could be extracted from four large vipers of Italy, without being affected by it. The Indians consider the curare, taken internally, as an excellent stomachic. The same poison prepared by the Piraoas and Salives, though it has some celebrity, is not so much esteemed as that of Esmeralda. The process of this preparation appears to be everywhere nearly the same; but there is no proof that the different poisons sold by the same name at the Orinoco and the Amazon are identical, and derived from the same plants. Orfila, therefore, in his excellent work On Poisons, has very judiciously separated the wourali of Dutch Guiana, the curare of the Orinoco, the ticuna of the Amazon, and all those substances which have been too vaguely united under the name of American poisons. Possibly at some future day, one and the same alkaline principle, similar to morphine and strychnia, will be found in poisonous plants belonging to different genera.

At the Orinoco the curare de raiz (of the root) is distinguished from the curare de bejuco (of lianas, or of the bark of branches). We saw only the latter prepared; the former is weaker, and much less esteemed. At the river Amazon we learned to distinguish the poisons of the Ticuna, Yagua, Peva, and Xibaro Indians, which being all obtained from the same plant, perhaps differ only by a more or less careful preparation. The Ticuna poison, to which La Condamine has given so much celebrity in Europe, and which somewhat improperly begins to bear the name of ticuna, is extracted from a liana which grows in the island of Mormorote, on the Upper Maranon. This poison is employed partly by the Ticunas, who remain independent on the Spanish territory near the sources of the Yacarique; and partly by Indians of the same tribe, inhabiting the Portuguese mission of Loreto. The poisons we have just named differ totally from that of La Peca, and from the poison of Lamas and of Moyobamba. I enter into these details because the vestiges of plants which we were able to examine, proved to us (contrary to the common opinion) that the three poisons of the Ticunas, of La Peca, and of Moyobamba are not obtained from the same species, probably not even from congeneric plants. In proportion as the preparation of the curare is simple, that of the poison of Moyobamba is a long and complicated process. With the juice of the bejuco de ambihuasca, which is the principal ingredient, are mixed pimento, tobacco, barbasco (Jacquinia armillaris), sanango (Tabernae montana), and the milk of some other apocyneae. The fresh juice of the ambihuasca has a deleterious action when in contact with the blood; the juice of the mavacure is a mortal poison only when it is concentrated by fire; and ebullition deprives the juice of the root of Jatropha manihot (the manioc) of all its baneful qualities. In rubbing a long time between my fingers the liana which yields the potent poison of La Peca, when the weather was excessively hot, my hands were benumbed; and a person who was employed with me felt the same effects from this rapid absorption by the uninjured integuments.

I shall not here enter into any detail on the physiological properties of those poisons of the New World which kill with the same promptitude as the strychneae of Asia,[*] but without producing vomiting when they are received into the stomach, and without denoting the approach of death by the violent excitement of the spinal marrow. Scarcely a fowl is eaten on the banks of the Orinoco which has not been killed with a poisoned arrow; and the missionaries allege that the flesh of animals is never so good as when this method is employed. Father Zea, who accompanied us, though ill of a tertian fever, every morning had the live fowls allotted for our food brought to his hammock together with an arrow, and he killed them himself; for he would not confide this operation, to which he attached great importance, to any other person. Large birds, a guan (pava de monte) for instance, or a curassao (alector), when wounded in the thigh, die in two or three minutes; but it is often ten or twelve minutes before life is extinct in a pig or a peccary. M. Bonpland found that the same poison, bought in different villages, varied much. We had procured at the river Amazon some real Ticuna poison which was less potent than any of the varieties of the curare of the Orinoco. Travellers, on arriving in the missions,

---

[*] The nux vomica, the upas tieute, and the bean of St. Ignatius, Strychnos Ignatia.

485

frequently testify their apprehension on learning that the fowls, monkeys, guanas, and even the fish which they eat, have been killed with poisoned arrows. But these fears are groundless. Majendie has proved by his ingenious experiments on transfusion, that the blood of animals on which the bitter strychnos of India has produced a deleterious effect, has no fatal action on other animals. A dog received a considerable quantity of poisoned blood into his veins without any trace of irritation being perceived in the spinal marrow.

I placed the most active curare in contact with the crural nerves of a frog, without perceiving any sensible change in measuring the degree of irritability of the organs, by means of an arc formed of heterogeneous metals. Galvanic experiments succeeded upon birds, some minutes after I had killed them with a poisoned arrow. These observations are not uninteresting, when we recollect that a solution of the upas-poison poured upon the sciatic nerve, or insinuated into the texture of the nerve, produces also a sensible effect on the irritability of the organs by immediate contact with the medullary substance. The danger of the curare, as of most of the other strychneae (for we continue to believe that the mavacure belongs to a neighbouring family), results only from the action of the poison on the vascular system. At Maypures, a zambo descended from an Indian and a negro, prepared for M. Bonpland some of those poisoned arrows, that are shot from blowing-tubes to kill small monkeys or birds. He was a man of remarkable muscular strength. Having had the imprudence to rub the curare between his fingers after being slightly wounded, he fell on the ground seized with a vertigo, that lasted nearly half an hour. Happily the poison was of that diluted kind which is used for very small animals, that is, for those which it is believed can be recalled to life by putting muriate of soda into the wound. During our passage in returning from Esmeralda to Atures, I myself narrowly escaped an imminent danger. The curare, having imbibed the humidity of the air, had become fluid, and was spilt from an imperfectly closed jar upon our linen. The person who washed the linen had neglected to examine the inside of a stocking, which was filled with curare; and it was only on touching this glutinous matter with my hand, that I was warned not to draw on the poisoned stocking. The danger was so much the greater, as my feet at that time were bleeding from the wounds made by chegoes (Pulex penetrans), which had not been well extirpated. This circumstance may warn travellers of the caution requisite in the conveyance of poisons.

An interesting chemical and physiological investigation remains to be accomplished in Europe on the poisons of the New World, when, by more frequent communications, the curare de bejuco, the curare de raiz, and the various poisons of the Amazon, Guallaga, and Brazil, can be procured, without being confounded together, from the places where they are prepared. Since the discovery of prussic acid,[*] and many other new substances eminently deleterious, the introduction of poisons prepared by savage nations is less feared in Europe; we cannot however appeal too strongly to the vigilance of those who keep such noxious substances in the midst of populous cities, the centres of civilization, misery, and depravity. Our botanical knowledge of the plants employed in making poison can be but very slowly acquired. Most of the Indians who make poisoned arrows, are totally ignorant of the nature of the venomous substances they use, and which they obtain from other people. A mysterious veil everywhere covers the history of poisons and of their antidotes. Their preparation among savages is the monopoly of the piaches, who are at once priests, jugglers, and physicians; it is only from the natives who are transplanted to the missions, that any certain notions can be acquired on matters so problematical. Ages elapsed before Europeans became

---

[*] First obtained by Scheele in the year 1782. Gay-Lussac (to whom we are indebted for the complete analysis of this acid) observes that it can never become very dangerous to society, because its peculiar smell (that of bitter almonds) betrays its presence, and the facility with which it is decomposed makes it difficult to preserve.

acquainted through the investigation of M. Mutis, with the bejuco del guaco (Mikania guaco), which is the most powerful of all antidotes against the bite of serpents, and of which we were fortunate enough to give the first botanical description.

The opinion is very general in the missions that no cure is possible, if the curare be fresh, well concentrated, and have stayed long in the wound, to have entered freely into the circulation. Among the specifics employed on the banks of the Orinoco, and in the Indian Archipelago, the most celebrated is muriate of soda.[*] The wound is rubbed with this salt, which is also taken internally. I had myself no direct and sufficiently convincing proof of the action of this specific; and the experiments of Delille and Majendie rather tend to disprove its efficacy. On the banks of the Amazon, the preference among the antidotes is given to sugar; and muriate of soda being a substance almost unknown to the Indians of the forests, it is probable that the honey of bees, and that farinaceous sugar which oozes from plantains dried in the sun, were anciently employed throughout Guiana. In vain have ammonia and eau-de-luce been tried against the curare; it is now known that these specifics are uncertain, even when applied to wounds caused by the bite of serpents. Sir Everard Home has shown that a cure is often attributed to a remedy, when it is owing only to the

---

[*] Oviedo, Sommario delle Indie Orientali, recommends sea-water as an antidote against vegetable poisons. The people in the missions never fail to assure European travellers, that they have no more to fear from arrows dipped in curare, if they have a little salt in their mouths, than from the electric shocks of the gymnoti, when chewing tobacco. Raleigh recommends as an antidote to the ourari (curare) the juice of garlick. [But later experiments have completely proved that if the poison has once fairly entered into combination with the blood there is no remedy, either for man or any of the inferior animals. The wourali and other poisons mentioned by Humboldt have, since the publication of this work, been carefully analysed by the first chemists of Europe, and experiments made on their symptoms and supposed remedies. Artificial inflation of the lungs was found the most successful, but in very few instances was any cure effected.]

slightness of the wound, and to a very circumscribed action of the poison. Animals may with impunity be wounded with poisoned arrows, if the wound be well laid open, and the point imbued with poison be withdrawn immediately after the wound is made. If salt or sugar be employed in these cases, people are tempted to regard them as excellent specifics. Indians, who had been wounded in battle by weapons dipped in the curare, described to us the symptoms they experienced, which were entirely similar to those observed in the bite of serpents. The wounded person feels congestion in the head, vertigo, and nausea. He is tormented by a raging thirst, and numbness pervades all the parts that are near the wound.

The old Indian, who was called the poison-master, seemed flattered by the interest we took in his chemical processes. He found us sufficiently intelligent to lead him to the belief that we knew how to make soap, an art which, next to the preparation of curare, appeared to him one of the finest of human inventions. When the liquid poison had been poured into the vessels prepared for their reception, we accompanied the Indian to the festival of the juvias. The harvest of juvias, or fruits of the Bertholletia excelsa,[*] was celebrated by dancing, and by excesses of wild intoxication. The hut where the natives were assembled, displayed during several days a very singular aspect. There was neither table nor bench; but large roasted monkeys, blackened by smoke, were ranged in regular order against the wall. These were the marimondes (Ateles belzebuth), and those bearded monkeys called capuchins, which must not be confounded with the weeper, or sai (Simia capucina of Buffon). The manner of roasting these anthropomorphous animals contributes to render their appearance extremely disagreeable in the eyes of civilized man. A little grating or lattice of very hard wood is formed, and raised one foot from the ground. The monkey is skinned, and bent into a sitting posture; the head

---

[*] The Brazil-nut.

generally resting on the arms, which are meagre and long; but sometimes these are crossed behind the back. When it is tied on the grating, a very clear fire is kindled below. The monkey, enveloped in smoke and flame, is broiled and blackened at the same time. On seeing the natives devour the arm or leg of a roasted monkey, it is difficult not to believe that this habit of eating animals so closely resembling man in their physical organization, has, to a certain degree, contributed to diminish the horror of cannibalism among these people. Roasted monkeys, particularly those which have very round heads, display a hideous resemblance to a child; and consequently Europeans who are obliged to feed on them prefer separating the head and the hands, and serve up only the rest of the animal at their tables. The flesh of monkeys is so lean and dry, that M. Bonpland has preserved in his collections at Paris an arm and hand, which had been broiled over the fire at Esmeralda; and no smell has arisen from them after the lapse of a great number of years.

We saw the Indians dance. The monotony of their dancing is increased by the women not daring to take part in it. The men, young and old, form a circle, holding each others' hands; and turn sometimes to the right, sometimes to the left, for whole hours, with silent gravity. Most frequently the dancers themselves are the musicians. Feeble sounds, drawn from a series of reeds of different lengths, form a slow and plaintive accompaniment. The first dancer, to mark the time, bends both knees in a kind of cadence. Sometimes they all make a pause in their places, and execute little oscillatory movements, bending the body from one side to the other. The reeds ranged in a line, and fastened together, resemble the Pan's pipes, as we find them represented in the bacchanalian processions on Grecian vases. To unite reeds of different lengths, and make them sound in succession by passing them before the lips, is a simple idea, and has naturally presented itself to every nation. We were surprised to see with what promptitude the young Indians constructed and tuned these pipes, when they found reeds on the bank of the river. Uncivilized men, in every zone, make great use of these gramina with high stalks. The Greeks, with truth, said that reeds had contributed to subjugate nations by furnishing arrows, to soften men's manners by the charm of music, and to unfold their understanding by affording the first instruments for tracing letters. These different uses of reeds mark in some sort three different periods in the life of nations. We must admit that the tribes of the Orinoco are in the first stage of dawning civilization. The reed serves them only as an instrument of war and of hunting; and the Pan's pipes, of which we have spoken, have not yet, on those distant shores, yielded sounds capable of awakening mild and humane feelings.

We found in the hut allotted for the festival, several vegetable productions which the Indians had brought from the mountains of Guanaya, and which engaged our attention. I shall only here mention the fruit of the juvia, reeds of a prodigious length, and shirts made of the bark of marima. The almendron, or juvia, one of the most majestic trees of the forests of the New World, was almost unknown before our visit to the Rio Negro. It begins to be found after a journey of four days east of Esmeralda, between the Padamo and Ocamo, at the foot of the Cerro Mapaya, on the right bank of the Orinoco. It is still more abundant on the left bank, at the Cerro Guanaja, between the Rio Amaguaca and the Gehette. The inhabitants of Esmeralda assured us, that in advancing above the Gehette and the Chiguire, the juvia and cacao-trees become so common that the wild Indians (the Guaicas and Guaharibos) do not disturb the Indians of the missions when gathering in their harvests. They do not envy them the productions with which nature has enriched their own soil. Scarcely any attempt has been made to propagate the almendrones in the settlements of the Upper Orinoco. To this the indolence of the inhabitants is a greater obstacle than the rapidity with which the oil becomes rancid in the amygdaliform seeds. We found only three trees of the kind at the mission of San Carlos, and two at Esmeralda. These majestic trees were eight or ten years old, and had not yet borne flowers.

491

As early as the sixteenth century, the seeds with ligneous and triangular teguments (but not the great drupe like a cocoa-nut, which contains the almonds,) were known in Europe. I recognise them in an imperfect engraving of Clusius.* Raleigh, who knew none of the productions of the Upper Orinoco, does not speak of the juvia; but it appears that he first brought to Europe the fruit of the mauritia palm, of which we have so often spoken. (Fructus elegantissimus, squamosus, similis palmae-pini.) This botanist designates them under the name of almendras del Peru. They had no doubt been carried, as a very rare fruit, to the Upper Maranon, and thence, by the Cordilleras, to Quito and Peru. The Novus Orbis of Laet, in which I found the first account of the cow-tree, furnishes also a description and a figure singularly exact of the fruit of the bertholletia. Laet calls the tree totocke, and mentions the drupe of the size of the human head, which contains the almonds. The weight of these fruits, he says, is so enormous, that the savages dare not enter the forests without covering their heads and shoulders with a buckler of very hard wood. These bucklers are unknown to the natives of Esmeralda, but they told us of the danger incurred when the fruit ripens and falls from a height of fifty or sixty feet. The triangular seeds of the juvia are sold in Portugal under the vague appellation of chesnuts (castanas) of the Amazon, and in England under the name of Brazil-nuts; and it was long believed that, like the fruit of the pekea, they grew on separate stalks. They have furnished an article of trade for a century past to the inhabitants of Grand Para, by whom they are sent either directly to Europe, or to Cayenne, where they are called touka. The celebrated botanist, Correa de Serra, told us that this tree abounds in the forests in the neighbourhood of Macapa, at the mouth of the Amazon; that it there bears the name of capucaya, and that the inhabitants

---

* Clusius distinguishes very properly the almendras del Peru, our Bertholletia excelsa, or juvia, (fructus amygdalae-nucleo, triangularis, dorso lato, in bina latera angulosa desinente, rugosus, paululum cuneiformis) from the pekea, or Amygdala guayanica.

gather the almonds, like those of the lecythis, to express the oil. A cargo of almonds of the juvia, bought into Havre, captured by a privateer, in 1807, was employed for the same purpose.

The tree that yields the Brazil-nuts is generally not more than two or three feet in diameter, but attains one hundred or one hundred and twenty feet in height. It does not resemble the mammee-tree, the star-apple, and several other trees of the tropics, the branches of which (as in the laurel-trees of the temperate zone) rise almost straight towards the sky. The branches of the bertholletia are open, very long, almost entirely bare towards the base, and loaded at their summits with tufts of very close foliage. This disposition of the semicoriaceous leaves, which are a little silvery on their under part, and more than two feet long, makes the branches bend down toward the ground, like the fronds of the palm-tree. We did not see this majestic tree in blossom: it is not loaded with flowers[*] till in its fifteenth year, and they appear about the end of March and the beginning of April. The fruits ripen towards the end of May, and some trees retain them till the end of August. These fruits, which are as large as the head of a child, often twelve or thirteen inches in diameter, make a very loud noise in falling from the tops of the trees. Nothing is more fitted to fill the mind with admiration of the force of organic action in the equinoctial zone than the aspect of those great igneous pericarps, for instance, the cocoa-tree (lodoicea) of the Maldives among the monocotyledons, and the bertholletia and the lecythis among the dicotyledons. In our climates only the cucurbitaceae produce in the space of a few months fruits of an extraordinary size; but these fruits are pulpy and succulent. Within the tropics, the bertholletia forms in less than fifty or sixty days a pericarp, the ligneous part of which is

---

[*] According to accounts somewhat vague, they are yellow, very large, and have some similitude to those of the Bombax ceiba. M. Bonpland says, however, in his botanical journal written on the banks of the Rio Negro, flos violaceus. It was thus the Indians of the river had described to him the colour of the corolla.

half an inch thick, and which it is difficult to saw with the sharpest instruments. A great naturalist has observed, that the wood of fruits attains in general a hardness which is scarcely to be found in the wood of the trunks of trees. The pericarp of the bertholletia has traces of four cells, and I have sometimes found even five. The seeds have two very distinct coverings, and this circumstance renders the structure of the fruit more complicated than in the lecythis, the pekea or caryocar, and the saouvari. The first tegument is osseous or ligneous, triangular, tuberculated on its exterior surface, and of the colour of cinnamon. Four or five, and sometimes eight of these triangular nuts, are attached to a central partition. As they are loosened in time, they move freely in the large spherical pericarp. The capuchin monkeys (Simia chiropotes) are singularly fond of the Brazil nuts; and the noise made by the seeds, when the fruit is shaken as it falls from the tree, excites the appetites of these animals in the highest degree. I have most frequently found only from fifteen to twenty-two nuts in each fruit. The second tegument of the almonds is membranaceous, and of a brown-yellow. Their taste is extremely agreeable when they are fresh; but the oil, with which they abound, and which is so useful in the arts, becomes easily rancid. Although at the Upper Orinoco we often ate considerable quantities of these almonds for want of other food, we never felt any bad effects from so doing. The spherical pericarp of the bertholletia, perforated at the summit, is not dehiscent; the upper and swelled part of the columella forms (according to M. Kunth) a sort of inner cover, as in the fruit of the lecythis, but it seldom opens of itself. Many seeds, from the decomposition of the oil contained in the cotyledons, lose the faculty of germination before the rainy season, in which the ligneous integument of the pericarp opens by the effect of putrefaction. A tale is very current on the banks of the Lower Orinoco, that the capuchin and cacajao monkeys (Simia chiropotes, and Simia melanocephala) place themselves in a circle, and, by striking the shell with a stone, succeed in opening it, so as to take out the triangular nuts. This operation must, however, be impossible, on account of the extreme hardness and thickness of the pericarp. Monkeys may have been seen rolling along the fruit of the bertholletia, but though this fruit has a small hole closed by the upper extremity of the columella, nature has not furnished monkeys with the means of opening the ligneous pericarp, as it has of opening the covercle of the lecythis, called in the missions the covercle of the monkeys' cocoa.[*] According to the report of several Indians, only the smaller rodentia, particularly the cavies (the acuri and the lapa), by the structure of their teeth, and the inconceivable perseverance with which they pursue their destructive operations, succeed in perforating the fruit of the juvia. As soon as the triangular nuts are spread on the ground, all the animals of the forest, the monkeys, the manaviris, the squirrels, the cavies, the parrots, and the macaws, hastily assemble to dispute the prey. They have all strength enough to break the ligneous tegument of the seed; they get out the kernel, and carry it to the tops of the trees. "It is their festival also," said the Indians who had returned from the harvest; and on hearing their complaints of the animals, one may perceive that they think themselves alone the lawful masters of the forest.

One of the four canoes, which had taken the Indians to the gathering of the Juvias, was filled in great part with that species of reeds (carices) of which the blow-tubes are made. These reeds were from fifteen to seventeen feet long, yet no trace of a knot for the insertion of leaves and branches was perceived. They were quite straight, smooth externally, and perfectly cylindrical. These carices come from the foot of the mountains of Yumariquin and Guanaja. They are much sought after, even beyond the Orinoco, by the name of reeds of Esmeralda. A hunter preserves the same blow-tube during his whole life, and boasts of its lightness and precision, as we boast of the same qualities in our fire-arms. What is the monocotyledonous plant[†]

---

[*] La tapa del coco de monos.
[†] The smooth surface of these tubes sufficiently proves that they are not furnished by a plant of the family of umbelliferae.

that furnishes these admirable reeds? Did we see in fact the internodes (parts between the knots) of a gramen of the tribe of nastoides? or may this carex be perhaps a cyperaceous plant[*] destitute of knots? I cannot solve this question, or determine to what genus another plant belongs, which furnishes the shirts of marima. We saw on the slope of the Cerra Duida shirt-trees fifty feet high. The Indians cut off cylindrical pieces two feet in diameter, from which they peel the red and fibrous bark, without making any longitudinal incision. This bark affords them a sort of garment, which resembles sacks of a very coarse texture, and without a seam. The upper opening serves for the head; and two lateral holes are cut for the arms to pass through. The natives wear these shirts of marima in the rainy season: they have the form of the ponchos and ruanas of cotton, which are so common in New Grenada, at Quito, and in Peru. In these climates the riches and beneficence of nature being regarded as the primary causes of the indolence of the inhabitants, the missionaries say in showing the shirts of marima, in the forests of the Orinoco garments are found ready-made on the trees. We may also mention the pointed caps, which the spathes of certain palm-trees furnish, and which resemble coarse network.

At the festival of which we were the spectators, the women, who were excluded from the dance, and every sort of public rejoicing, were daily occupied in serving the men with roasted monkey, fermented liquors, and palm-cabbage. This last production has the taste of our cauliflowers, and in no other country had we seen specimens of such an immense size. The leaves that are not unfolded are united with the young stem, and we measured cylinders of six feet long and five inches in diameter. Another substance, which is much more nutritive, is obtained from the animal kingdom: this is fish-flour (manioc de pescado). The Indians throughout the Upper Orinoco fry fish, dry them in the sun, and reduce them to powder without

---

[*] The caricillo del manati, which grows abundantly on the banks of the Orinoco, attains from eight to ten feet in height.

separating the bones. I have seen masses of fifty or sixty pounds of this flour, which resembles that of cassava. When it is wanted for eating, it is mixed with water, and reduced to a paste. In every climate the abundance of fish has led to the invention of the same means of preserving them. Pliny and Diodorus Siculus have described the fish-bread of the ichthyophagous nations, that dwelt on the Persian Gulf and the shores of the Red Sea.[*]

At Esmeralda, as everywhere else throughout the missions, the Indians who will not be baptized, and who are merely aggregated in the community, live in a state of polygamy. The number of wives differs much in different tribes. It is most considerable among the Caribs, and all the nations that have preserved the custom of carrying off young girls from the neighbouring tribes. How can we imagine domestic happiness in so unequal an association? The women live in a sort of slavery, as they do in most nations which are in a state of barbarism. The husbands being in the full enjoyment of absolute power, no complaint is heard in their presence. An apparent tranquillity prevails in the household; the women are eager to anticipate the wishes of an imperious and sullen master; and they attend without distinction to their own children and those of their rivals. The missionaries assert, what may easily be believed, that this domestic peace, the effect of fear, is singularly disturbed when the husband is long absent. The wife who contracted the first ties then applies to the others the names of concubines and servants. The quarrels continue till the return of the master, who knows how to calm their passions by the sound of his voice, by a mere gesticulation, or, if he thinks it necessary, by means a little more violent. A certain inequality in the rights of the women is sanctioned by the language of the Tamanacs. The husband calls

---

[*] These nations, in a still ruder state than the natives of the Orinoco, contented themselves with drying the raw fish in the sun. They made up the fish-paste in the form of bricks, and sometimes mixed with it the aromatic seed of paliurus (rhamnus), as in Germany, and some other countries, cummin and fennel-seed are mixed with wheaten bread.

the second and third wife the companions of the first; and the first treats these companions as rivals and enemies (ipucjatoje), a term which truly expresses their position. The whole weight of labour being supported by these unhappy women, we must not be surprised if, in some nations, their number is extremely small. Where this happens, a kind of polyandry is formed, which we find more fully displayed in Thibet, and on the lofty mountains at the extremity of the Indian peninsula. Among the Avanos and Maypures, brothers have often but one wife. When an Indian, who lives in polygamy, becomes a christian, he is compelled by the missionaries, to choose among his wives her whom he prefers, and to reject the others. At the moment of separation the new convert sometimes discovers the most valuable qualities in the wives he is obliged to abandon. One understands gardening perfectly; another knows how to prepare chiza, an intoxicating beverage extracted from the root of cassava; all appear to him alike clever and useful. Sometimes the desire of preserving his wives overcomes in the Indian his inclination to christianity; but most frequently, in his perplexity, the husband prefers submitting to the choice of the missionary, as to a blind fatality.

The Indians, who from May to August take journeys to the east of Esmeralda, to gather the vegetable productions of the mountains of Yumariquin, gave us precise notions of the course of the Orinoco to the east of the mission. This part of my itinerary may differ entirely from the maps that preceded it. I shall begin the description of this country with the granitic group of Duida, at the foot of which we sojourned. This group is bounded on the west by the Rio Tamatama, and on the east by the Rio Guapo. Between these two tributary streams of the Orinoco, amid the morichales, or clumps of mauritia palm-trees, which surround Esmeralda, the Rio Sodomoni flows, celebrated for the excellence of the pine-apples that grow upon its banks. I measured, on the 22nd of May, in the savannah at the foot of Duida, a base of four hundred and seventy-five metres in length; the angle, under which the summit of the mountain appeared at the distance of thirteen thousand three hundred and twenty-seven metres, was still nine degrees. A trigonometric measurement, made with great care, gave me for Duida (that is, for the most elevated peak, which is south-west of the Cerro Maraguaca) two thousand one hundred and seventy-nine metres, or one thousand one hundred and eighteen toises, above the plain of Esmeralda. The Cerro Duida thus yields but little in height (scarcely eighty or one hundred toises) to the summit of St. Gothard, or the Silla of Caracas on the shore of Venezuela. It is indeed considered as a colossal mountain in those countries; and this celebrity gives a precise idea of the mean height of Parima and of all the mountains of eastern America. To the east of the Sierra Nevada de Merida, as well as to the south-east of the Paramo de las Rosas, none of the chains that extend in the same parallel line reach the height of the central ridge of the Pyrenees.

The granitic summit of Duida is so nearly perpendicular that the Indians have vainly attempted the ascent. It is a well-known fact that mountains not remarkable for elevation are sometimes the most inaccessible. At the beginning and end of the rainy season, small flames, which seem to change their place, are seen on the top of Duida. This phenomenon, the existence of which is borne out by concurrent testimony, has caused this mountain to be improperly called a volcano. As it stands nearly alone, it might be supposed that lightning from time to time sets fire to the brushwood; but this supposition loses its probability when we reflect on the extreme difficulty with which plants are ignited in these damp climates. It must be observed also that these flames are said to appear often where the rock seems scarcely covered with turf, and that the same igneous phenomena are visible, on days entirely exempt from storms, on the summit of Guaraco or Murcielago, a hill opposite the mouth of the Rio Tamatama, on the southern bank of the Orinoco. This hill is scarcely elevated one hundred toises above the neighbouring plains. If the statements of the natives be correct, it is probable that some subterraneous cause produces these flames on the Duida and the Guaraco; for they never appear on the lofty neighbouring mountains of Jao and Maraguaca, so often wrapped in electric storms. The granite of the Cerro Duida is full of veins, partly open, and partly filled with crystals of quartz and pyrites. Gaseous and inflammable emanations, either of hydrogen or of naphtha, may pass through these veins. Of this the mountains of Caramania, of Hindookho, and of Himalaya, furnish frequent examples. We saw the appearance of flames in many parts of eastern America subject to earthquakes, even from secondary rocks, as at Cuchivero, near Cumanacoa. The fire shows itself when the ground, strongly heated by the sun, receives the first rains; or when, after violent showers, the earth begins to dry. The first cause of these igneous phenomena lies at immense depths below the secondary rocks, in the primitive formations: the rains and the decomposition of atmospheric water act only a secondary part. The hottest springs of the globe issue immediately from granite. Petroleum gushes from mica-schist; and frightful detonations are heard at Encaramada, between the rivers Arauca and Cuchivero, in the midst of the granitic soil of the Orinoco and the Sierra Parima. Here, as everywhere else on the globe, the focus of volcanoes is in the most ancient soils; and it appears that an intimate connection exists between the great phenomena that heave up and liquify the crust of our planet, and those igneous meteors which are seen from time to time on its surface, and which from their littleness we are tempted to attribute solely to the influence of the atmosphere.

Duida, though lower than the height assigned to it by popular belief, is however the most prominent point of the whole group of mountains that separate the basin of the Lower Orinoco from that of the Amazon. These mountains lower still more rapidly on the north-east, toward the Purunama, than on the east, toward the Padamo and the Rio Ocamo. In the former direction the most elevated summits next to Duida are Cuneva, at the sources of the Rio Paru (one of the tributary streams of the Ventuari), Sipapo, Calitamini, which forms one group with Cunavami and the peak of Umiana. East of Duida, on the right bank of the Orinoco, Maravaca, or Sierra Maraguaca, is distinguished by its elevation, between the Rio Caurimoni and the Padamo; and on the left bank of the Orinoco rise the mountains of Guanaja and Yumariquin, between the Rios Amaguaca and Gehette. It is almost superfluous to repeat that the line which passes through these lofty summits (like those of the Pyrenees, the Carpathian mountains, and so many other chains of the old continent) is very distinct from the line that marks the partition of the waters. This latter line, which separates the tributary streams of the Lower and Upper Orinoco, intersects the meridian of 64° in latitude 4°. After having separated the sources of the Rio Branco and the Carony, it runs north-west, sending off the waters of the Padamo, the Jao, and the Ventuari towards the south, and the waters of the Arui, the Caura, and the Cuchivero towards the north.

The Orinoco may be ascended without danger from Esmeralda as far as the cataracts occupied by the Guaica Indians, who prevent all farther progress of the Spaniards. This is a voyage of six days and a half. In the first two days you arrive at the mouth of the Rio Padamo, or Patamo, having passed, on the north, the little rivers of Tamatama, Sodomoni, Guapo, Caurimoni, and Simirimoni; and on the south the Cuca, situate between the rock of Guaraco, which is said to throw out flames, and the Cerro Canclilla. Throughout this course the Orinoco continues to be three or four hundred toises broad. The tributary streams are most frequent on the right bank, because on that side the river is bounded by the lofty cloud-capped mountains of Duida and Maraguaca, while the left bank on the contrary is low and contiguous to a plain, the general slope of which inclines to the south-west. The northern Cordilleras are covered with fine timber. The growth of plants is so enormous in this hot and constantly humid climate, that the trunks of the Bombax ceiba are sixteen feet in diameter. From the mouth of the Rio Padamo, which is of considerable breadth, the Indians arrive, in a day and a half, at the Rio Mavaca. The latter takes its rise in the lofty mountains of Unturan, and communicates with a lake, on the banks of which the Portuguese[*] of the Rio Negro gather the aromatic seeds of the Laurus pucheri, known in trade by the names of the pichurim bean, and toda specie. Between the confluence of the Padamo and that of the Mavaca, the Orinoco receives on the north the Ocamo, into which the Rio Matacona falls. At the sources of the latter live the Guainares, who are much less copper-coloured, or tawny, than the other inhabitants of those countries. This is one of the tribes called by the missionaries fair Indians (Indios blancos). Near the mouth of the Ocamo, travellers are shown a rock, which is the wonder of the country. It is a granite passing into gneiss, and remarkable for the peculiar distribution of the black mica, which forms little ramified veins. The Spaniards call this rock Piedra Mapaya (the map-stone). The little fragment which I procured indicated a stratified rock, rich in white feldspar, and containing, together with spangles of mica, grouped in streaks, and variously twisted, some crystals of hornblende. It is not a syenite, but probably a granite of new formation, analogous to those to which the stanniferous granites (hyalomictes) and the pegmatites, or graphic granites, belong.

Beyond the confluence of the Macava, the Orinoco suddenly diminishes in breadth and depth, becoming extremely sinuous, like an Alpine torrent. Its banks are surrounded by mountains, and the number of its tributary streams on the south augments considerably, yet the Cordillera on the north remains the most elevated. It requires two days to go from the mouth of the Macava, to the Rio Gehette, the navigation being very difficult,

---

[*] The pichurim bean is the puchiri of La Condamine, which abounds at the Rio Xingu, a tributary stream of the Amazon, and on the banks of the Hyurubaxy, or Yurubesh, which runs into the Rio Negro. The puchery, or pichurim, which is grated like nutmeg, differs from another aromatic fruit (a laurel?) known in trade at Grand Para by the names of cucheri, cuchiri, or cravo (clavus) do Maranhao, and which, on account of its odour, is compared with cloves.

and the boats, on account of the want of water, being often dragged along the shore. The tributary streams along this distance are, on the south, the Daracapo and the Amaguaca; which skirt on the west and east the mountains of Guanaya and Yumariquin, where the bertholletias are gathered. The Rio Manaviche flows down from the mountains on the north, the elevation of which diminishes progressively from the Cerro Maraguaca. As we advance further up the Orinoco, the whirlpools and little rapids (chorros y remolinos) become more and more frequent; on the north lies the Cano Chiquire, inhabited by the Guaicas, another tribe of white Indians; and two leagues distant is the mouth of the Gehette, where there is a great cataract. A dyke of granitic rocks crosses the Orinoco these rocks are, as it were, the columns of Hercules, beyond which no white man has been able to penetrate. It appears that this point, known by the name of the great Raudal de Guaharibos, is three-quarters of a degree west of Esmeralda, consequently in longitude 67° 38'. A military expedition, undertaken by the commander of the fort of San Carlos, Don Francisco Bovadilla, to discover the sources of the Orinoco, led to some information respecting the cataracts of the Guaharibos. Bovadilla had heard that some fugitive negroes from Dutch Guiana, proceeding towards the west (beyond the isthmus between the sources of the Rio Carony and the Rio Branco) had joined the independent Indians. He attempted an entrada (hostile incursion) without having obtained the permission of the governor; the desire of procuring African slaves, better fitted for labour than the copper-coloured race, was a far more powerful motive than that of zeal for the progress of geography. Bovadilla arrived without difficulty as far as the little Raudal[*] opposite the Gehette; but having advanced to the foot of the rocky dike that forms the great cataract, he was suddenly attacked, while he was breakfasting, by the Guaharibos and Guaycas, two warlike

---

[*] It is called Raudal de abaxo (Low Cataract) in opposition to the great Raudal de Guaharibos, which is situated higher up toward the east.

tribes, celebrated for the virulence of the curare with which their arrows are empoisoned. The Indians occupied the rocks that rise in the middle of the river, and seeing the Spaniards without bows, and having no knowledge of firearms, they provoked the whites, whom they believed to be without defence. Several of the latter were dangerously wounded, and Bovadilla found himself forced to give the signal for battle. A fearful carnage ensued among the natives, but none of the Dutch negroes, who, as was believed, had taken refuge in those parts, were found. Notwithstanding a victory so easily won, the Spaniards did not dare to advance eastward in a mountainous country, and along a river inclosed by very high banks.

These white Guaharibos have constructed a bridge of lianas above the cataract, supported on rocks that rise, as generally happens in the pongos of the Upper Maranon, in the middle of the river. The existence of this bridge, which is known to all the inhabitants of Esmeralda,[*] seems to indicate that the Orinoco must be very narrow at this point. It is generally estimated by the Indians to be only two or three hundred feet broad. They say that the Orinoco, above the Raudal of the Guaharibos, is no longer a river, but a brook (riachuelo); while a well informed ecclesiastic, Fray Juan Gonzales, who had visited those countries, assured me that the Orinoco, in the part where its farther course is no longer known, is two-thirds of the breadth of the Rio Negro near San Carlos. This opinion appears to me hardly probable; but I relate what I have collected, and affirm nothing positively.

In the rocky dike that crosses the Orinoco, forming the Raudal of the Guaharibos, Spanish soldiers pretend to have found the fine kind of saussurite (Amazon-stone), of which we have spoken. This tradition however is very uncertain; and the

---

[*] The Amazon also is crossed twice on bridges of wood near its source in the lake Lauricocha; first north of Chavin, and then below the confluence of the Rio Aguamiras. These, the only two bridges that have been thrown over the largest river we yet know, are called Puente de Quivilla, and Puente de Guancaybamba.

Indians, whom I interrogated on the subject, assured me that the green stones, called piedras de Macagua[*] at Esmeralda, were purchased from the Guaicas and Guaharibos, who traffic with hordes much farther to the east. The same uncertainty prevails respecting these stones, as that which attaches to many other valuable productions of the Indies. On the coast, at the distance of some hundred leagues, the country where they are found is positively named; but when the traveller with difficulty penetrates into that country, he discovers that the natives are ignorant even of the name of the object of his research. It might be supposed that the amulets of saussurite found in the possession of the Indians of the Rio Negro, come from the Lower Maranon, while those that are received by the missions of the Upper Orinoco and the Rio Carony come from a country situated between the sources of the Essequibo and the Rio Branco. The opinion that this stone is taken in a soft state like paste from the little lake Amucu, though very prevalent at Angostura, is wholly without foundation. A curious geognostic discovery remains to be made in the eastern part of America, that of finding in a primitive soil a rock of euphotide containing the piedra de Macagua.

I shall here proceed to give some information respecting the tribes of dwarf and fair Indians, which ancient traditions have placed near the sources of the Orinoco. I had an opportunity of seeing some of these Indians at Esmeralda, and can affirm that the short stature of the Guaicas, and the fair complexion of the Guaharibos, whom Father Caulin calls Guaribos blancos, have been alike exaggerated. The Guaicas, whom I measured, were in general from four feet seven inches to four feet eight inches high (old measure of France).[†] We were assured that the whole tribe

---

[*] The etymology of this name, which is unknown to me, might lead to the knowledge of the spot where these stones are found. I have sought in vain the name of Macagua among the numerous tributary streams of the Tacutu, the Mahu, the Rupunury, and the Rio Trombetas.

[†] About five feet three inches English measure.

were of this diminutive size; but we must not forget that what is called a tribe constitutes, properly speaking, but one family, owing to the exclusion of all foreign connections. The Indians of the lowest stature next to the Guaicas are the Guainares and the Poignaves. It is singular, that all these nations are found in near proximity to the Caribs, who are remarkably tall. They all inhabit the same climate, and subsist on the same aliments. They are varieties in the race, which no doubt existed previously to the settlement of these tribes (tall and short, fair and dark brown) in the same country. The four nations of the Upper Orinoco, which appeared to me to be the fairest, are the Guaharibos of the Rio Gehette, the Guainares of the Ocamo, the Guaicas of Cano Chiguire, and the Maquiritares of the sources of the Padamo, the Jao, and the Ventuari. It being very extraordinary to see natives with a fair skin beneath a burning sky, and amid nations of a very dark hue, the Spaniards have attempted to explain this phenomenon by the following hypotheses. Some assert, that the Dutch of Surinam and the Rio Essequibo may have intermingled with the Guaharibos and the Guainares; others insist, from hatred to the Capuchins of the Carony, and the Observantins of the Orinoco, that the fair Indians are what are called in Dalmatia muso di frate, children whose legitimacy is somewhat doubtful. In either case the Indios blancos would be mestizos, that is to say, children of an Indian woman and a white man. Now, having seen thousands of mestizos, I can assert that this supposition is altogether inaccurate. The individuals of the fair tribes whom we examined, have the features, the stature, and the smooth, straight, black hair which characterises other Indians. It would be impossible to take them for a mixed race, like the descendants of natives and Europeans. Some of these people are very little, others are of the ordinary stature of the copper-coloured Indians. They are neither feeble nor sickly, nor are they albinos; and they differ from the copper-coloured races only by a much less tawny skin. It would be useless, after these considerations, to insist on the distance of the mountains of the Upper Orinoco from the shores inhabited by the Dutch. I will not deny that descendants of fugitive negroes may have been seen among the Caribs, at the sources of the Essequibo; but no white man ever went from the eastern coast to the Rio Gehette and the Ocamo, in the interior of Guiana. It must also be observed, although we may be struck with the singularity of several fair tribes being found at one point to the east of Esmeralda, it is no less certain, that tribes have been found in other parts of America, distinguished from the neighbouring tribes by the less tawny colour of their skin. Such are the Arivirianos and Maquiritares of the Rio Ventuario and the Padamo, the Paudacotos and Paravenas of the Erevato, the Viras and Araguas of the Caura, the Mologagos of Brazil, and the Guayanas of the Uruguay.[*]

These phenomena are so much the more worthy of attention as they are observed in that great branch of the American nations generally ranked in a class totally opposite to that circumpolar

---

[*] The Cumanagotos, the Maypures, the Mapojos, and some hordes of the Tamanacs, are also fair, but in a less degree than the tribes I have just named. We may add to this list (which the researches of Sommering, Blumenbach, and Pritchard, on the varieties of the human species, have rendered so interesting) the Ojes of the Cuchivero, the Boanes (now almost destroyed) of the interior of Brazil, and in the north of America, far from the north-west coast, the Mandans and the Akanas (Walkenaer, Geogr. page 645. Gili volume 2 page 34. Vater, Amerikan. Sprachen page 81. Southey volume 1 page 603.) The most tawny, we might almost say the blackest of the American race, are the Otomacs and the Guamos. These have perhaps given rise to the confused notions of American negroes, spread through Europe in the early times of the conquest. (Herrera Dec 1 lib 3 cap 9, volume 1 page 79. Garcia, Origen de los Americanos page 259.) Who are those Negros de Quereca, placed by Gomara page 277, in that very isthmus of Panama, whence we received the first absurd tales of an albino American people? In reading with attention the authors of the beginning of the 16th century, we see that the discovery of America and of a new race of men, had singularly awakened the interest of travellers respecting the varieties of our species. Now, if a black race had been mingled with copper-colored men, as in the South-sea Islands, the conquistadores would not have failed to speak of it in a precise manner. Besides, the religious traditions of the Americans relate the appearance, in the heroic times, of white and bearded men as priests and legislators; but none of these traditions make mention of a black race.

branch, namely the Tschougaz-Esquimaux,[*] whose children are fair, and who acquire the Mongol or yellowish tint only from the influence of the air and the humidity. In Guiana, the hordes who live in the midst of the thickest forests are generally less tawny than those who inhabit the shores of the Orinoco, and are employed in fishing. But this slight difference, which is alike found in Europe between the artisans of towns and the cultivators of the fields or the fishermen on the coasts, in no way explains the problem of the Indios blancos. They are surrounded by other Indians of the woods (Indios del monte) who are of a reddish-brown, although now exposed to the same physical influences. The causes of these phenomena are very ancient, and we may repeat with Tacitus, "est durans originis vis."

The fair-complexioned tribes, which we had an opportunity of seeing at the mission of Esmeralda, inhabit part of a mountainous country lying between the sources of six tributaries of the Orinoco; that is to say, between the Padamo, the Jao, the Ventuari, the Erevato, the Aruy, and the Paraguay.[†]) The Spanish and Portuguese missionaries are accustomed to designate this country more particularly by the name of Parima.[‡] Here, as in several other countries of Spanish America, the savages have reconquered what had been wrested from them by civilization, or rather by its precursors, the missionaries. The

---

[*] The Chevalier Gieseke has recently confirmed all that Krantz related of the colour of the skin of the Esquimaux. That race (even in the latitude of seventy-five and seventy-six degrees, where the climate is so rigorous) is not in general so diminutive as it was long believed to be. Ross' Voyage to the North.

[†] (They are six tributary streams on the right bank of the Orinoco; the first three run towards the south, or the Upper Orinoco; the three others towards the north, or the Lower Orinoco.

[‡] The name Parima, which signifies water, great water, is applied sometimes, and more especially, to the land washed by the Rio Parima, or Rio Branco (Rio de Aguas Blancas), a stream running into the Rio Negro; sometimes to the mountains (Sierra Parima), which divide the Upper and Lower Orinoco.

expedition of the boundaries under Solano, and the extravagant zeal displayed by a governor of Guiana for the discovery of El Dorado, partially revived in the latter half of the eighteenth century that spirit of enterprise which characterised the Spaniards at the period of the discovery of America. In going along the Rio Padamo, a road was observed across the forests and savannahs (the length of ten days' journey), from Esmeralda to the sources of the Ventuari; and in two days more, from those sources, by the Erevato, the missions on the Rio Caura were reached. Two intelligent and enterprising men, Don Antonio Santos and Captain Bareto, had established, with the aid of the Miquiritares, a chain of military posts on this line from Esmeralda to the Rio Erevato. These posts consisted of block-houses (casas fuertes), mounted with swivels, such as I have already mentioned. The soldiers, left to themselves, exercised all kinds of vexations on the natives (Indians of peace), who had cultivated pieces of ground around the casas fuertes; and the consequence was that, in 1776, several tribes formed a league against the Spaniards. All the military posts were attacked on the same night, on a line of nearly fifty leagues in length. The houses were burnt, and many soldiers massacred; a very small number only owing their preservation to the pity of the Indian women. This nocturnal expedition is still mentioned with horror. It was concerted in the most profound secrecy, and executed with that spirit of unity which the natives of America, skilled in concealing their hostile passions, well know how to practise in whatever concerns their common interests. Since 1776 no attempt has been made to re-establish the road which leads by land from the Upper to the Lower Orinoco, and no white man has been able to pass from Esmeralda to the Erevato. It is certain, however, that in the mountainous lands, between the sources of the Padamo and the Ventuari (near the sites called by the Indians Aurichapa, Ichuana, and Irique) there are many spots where the climate is temperate, and where there are pasturages capable of feeding numerous herds of cattle. The military posts were very useful in preventing the incursions of the Caribs, who, from time to time carried off slaves, though in very small numbers, between the Erevato and the Padamo. They would have resisted the attacks of the natives, if, instead of leaving them isolated and solely to the control of the soldiery, they had been formed into communities, and governed like the villages of neophyte Indians.

We left the mission of Esmeralda on the 23rd of May. Without being positively ill, we felt ourselves in a state of languor and weakness, caused by the torment of insects, bad food, and a long voyage, in narrow and damp boats. We did not go up the Orinoco beyond the mouth of the Rio Guapo, which we should have done, if we could have attempted to reach the sources of the river. There remains a distance of fifteen leagues from the Guapo to the Raudal of the Guaharibos. At this cataract, which is passed on a bridge of lianas, Indians are posted armed with bows and arrows to prevent the whites, or those who come from their territory from advancing westward. How could we hope to pass a point where the commander of the Rio Negro, Don Francisco Bovadilla, was stopped when, accompanied by his soldiers, he tried to penetrate beyond the Gehette?[*] The carnage then made among the natives has rendered them more distrustful, and more averse to the inhabitants of the missions. It must be remembered that the Orinoco had hitherto offered to geographers two distinct problems, alike important, the situation of its sources, and the mode of its communication with the Amazon. The latter problem formed the object of the journey which I have described; with respect to the discovery of its sources, that remains to be done by the Spanish and Portuguese governments.

Our canoe was not ready to receive us till near three o'clock in the afternoon. It had been filled with innumerable swarms of ants during the navigation of the Cassiquiare; and the toldo, or roof of palm-leaves, beneath which we were again doomed to

---

[*] See above.

remain stretched out during twenty-two days, was with difficulty cleared of these insects. We employed part of the morning in repeating to the inhabitants of Esmeralda the questions we had already put to them, respecting the existence of a lake towards the east. We showed copies of the maps of Surville and La Cruz to old soldiers, who had been posted in the mission ever since its first establishment. They laughed at the supposed communication of the Orinoco with the Rio Idapa, and at the White Sea, which the former river was represented to cross. What we politely call geographical fictions they termed lies of the old world (mentiras de por alla). These good people could not comprehend how men, in making the map of a country which they had never visited, could pretend to know things in minute detail, of which persons who lived on the spot were ignorant. The lake Parima, the Sierra Mey, and the springs which separate at the point where they issue from the earth, were entirely unknown at Esmeralda. We were repeatedly assured that no one had ever been to the east of the Raudal of the Guaharibos; and that beyond that point, according to the opinion of some of the natives, the Orinoco descends like a small torrent from a group of mountains, inhabited by the Coroto Indians. Father Gili, who was living on the banks of the Orinoco when the expedition of the boundaries arrived, says expressly that Don Apollinario Diez was sent in 1765 to attempt the discovery of the source of the Orinoco; that he found the river, east of Esmeralda, full of shoals; that he returned for want of provision; and that he learned nothing, absolutely nothing, of the existence of a lake. This statement perfectly accords with what I heard myself thirty-five years later at Esmeralda. The probability of a fact is powerfully shaken when it can be proved to be totally unknown on the very spot where it ought to be known best; and when those by whom the existence of the lake is affirmed contradict each other, not in the least essential circumstances, but in all that are the most important.

When travellers judge only by their own sensations they differ from each other respecting the abundance of the mosquitos as they do respecting the progressive increase or diminution of the temperature. The state of our organs, the motion of the air, its degree of humidity or dryness, its electric intensity, a thousand circumstances contribute at once to make us suffer more or less from the heat and the insects. My fellow travellers were unanimously of opinion that Esmeralda was more tormented by mosquitos than the banks of the Cassiquiare, and even more than the two missions of the Great Cataracts; whilst I, less sensible than they of the high temperature of the air, thought that the irritation produced by the insects was somewhat less at Esmeralda than at the entrance of the Upper Orinoco. On hearing the complaints that are made of these tormenting insects in hot countries it is difficult to believe that their absence, or rather their sudden disappearance, could become a subject of inquietude; yet such is the fact. The inhabitants of Esmeralda related to us, that in the year 1795, an hour before sunset, when the mosquitos usually form a very thick cloud, the air was observed to be suddenly free from them. During the space of twenty minutes, not one insect was perceived, although the sky was cloudless, and no wind announced rain. It is necessary to have lived in those countries to comprehend the degree of surprise which the sudden disappearance of the insects must have produced. The inhabitants congratulated each other, and inquired whether this state of happiness, this relief from pain (feicidad y alivio), could be of any duration. But soon, instead of enjoying the present, they yielded to chimerical fears, and imagined that the order of nature was perverted. Some old Indians, the sages of the place, asserted that the disappearance of the insects must be the precursor of a great earthquake. Warm discussions arose; the least noise amid the foliage of the trees was listened to with an attentive ear; and when the air was again filled with mosquitos they were almost hailed with pleasure. We could not guess what modification of the atmosphere had caused this phenomenon, which must not be confounded with the periodical replacing of one species of insects by another.

512

After four hours' navigation down the Orinoco we arrived at the point of the bifurcation. Our resting place was on the same beach of the Cassiquiare, where a few days previously our great dog had, as we believe, been carried off by the jaguars. All the endeavours of the Indians to discover any traces of the animal were fruitless. The cries of the jaguars were heard during the whole night.[*] These animals are very frequent in the tracts situated between the Cerro Maraguaca, the Unturan, and the banks of the Pamoni. There also is found that black species of tiger[†] of which I saw some fine skins at Esmeralda. This animal is celebrated for its strength and ferocity; it appears to be still larger than the common jaguar. The black spots are scarcely visible on the dark-brown ground of its skin. The Indians assert, that these tigers are very rare, that they never mingle with the common jaguars, and that they form another race. I believe that Prince Maximilian of Neuwied, who has enriched American zoology by so many important observations, acquired the same information farther to the south, in the hot part of Brazil. Albino varieties of the jaguar have been seen in Paraguay: for the spots of these animals, which may be called the beautiful panthers of America, are sometimes so pale as to be scarcely distinguishable on a very white ground. In the black jaguars, on the contrary, it is the colour of the ground which renders the spots indistinct. It requires to reside long in those countries, and to accompany the Indians of Esmeralda in the perilous chase of the tiger, to decide with certainty upon the varieties and the species. In all the

---

[*] This frequency of large jaguars is somewhat remarkable in a country destitute of cattle. The tigers of the Upper Orinoco are far less bountifully supplied with prey than those of the Pampas of Buenos Ayres and the Llanos of Caracas, which are covered with herds of cattle. More than four thousand jaguars are killed annually in the Spanish colonies, several of them equalling the mean size of the royal tiger of Asia. Two thousand skins of jaguars were formerly exported annually from Buenos Ayres alone.

[†] Gmelin, in his Synonyma, seems to confound this animal, under the name of Felis discolor, with the great American lion (Felis concolor) which is very different from the puma of the Andes of Quito.

mammiferae, and particularly in the numerous family of the apes, we ought, I believe, to fix our attention less on the transition from one colour to another in individuals, than on their habit of separating themselves, and forming distinct bands.

We left our resting place before sunrise on the 24th of May. In a rocky cove, which had been the dwelling of some Durimundi Indians, the aromatic odour of the plants was so powerful, that although sleeping in the open air, and the irritability of our nervous system being allayed by the habits of a life of fatigue, we were nevertheless incommoded by it. We could not ascertain the flowers which diffused this perfume. The forest was impenetrable; but M. Bonpland believed that large clumps of pancratium and other liliaceous plants were concealed in the neighbouring marshes. Descending the Orinoco by favour of the current, we passed first the mouth of the Rio Cunucunumo, and then the Guanami and the Puriname. The two banks of the principal river are entirely desert; lofty mountains rise on the north, and on the south a vast plain extends far as the eye can reach beyond the sources of the Atacavi, which lower down takes the name of the Atabapo. There is something gloomy and desolate in this aspect of a river, on which not even a fisherman's canoe is seen. Some independent tribes, the Abirianos and the Maquiritares, dwell in the mountainous country; but in the neighbouring savannahs,[*] bounded by the Cassiquiare, the Atabapo, the Orinoco, and the Rio Negro, there is now scarcely any trace of a human habitation. I say now; for here, as in other parts of Guiana, rude figures representing the sun, the moon, and different animals, traced on the hardest rocks of granite, attest the anterior existence of a people, very different from those who became known to us on the banks of the Orinoco. According to the accounts of the natives, and of the

---

[*] They form a quadrilateral plot of a thousand square leagues, the opposite sides of which have contrary slopes, the Cassiquiare flowing towards the south, the Atabapo towards the north, the Orinoco towards the north-west, and the Rio Negro towards the south-east.

most intelligent missionaries, these symbolic signs resemble perfectly the characters we saw a hundred leagues more to the north, near Caycara, opposite the mouth of the Rio Apure. (See Chapter 2.18 above.)

In advancing from the plains of the Cassiquiare and the Conorichite, one hundred and forty leagues further eastward, between the sources of the Rio Blanco and the Rio Essequibo, we also meet with rocks and symbolical figures. I have lately verified this curious fact, which is recorded in the journal of the traveller Hortsman, who went up the Rupunuvini, one of the tributary streams of the Essequibo. Where this river, full of small cascades, winds between the mountains of Macarana, he found, before he reached lake Amucu, rocks covered with figures, or (as he says in Portuguese) with varias letras. We must not take this word letters in its real signification. We were also shewn, near the rock Culimacari, on the banks of the Cassiquiare, and at the port of Caycara in the Lower Orinoco, traces which were believed to be regular characters. They were however only misshapen figures, representing the heavenly bodies, together with tigers, crocodiles, boas, and instruments used for making the flour of cassava. It was impossible to recognize in these painted rocks[*] (the name by which the natives denote those masses loaded with figures) any symmetrical arrangement, or characters with regular spaces. The Spanish Americans also call the rock covered with sculptured figures, piedras pintadas; those for instance, which are found on the summit of the Paramo of Guanacas, in New Grenada, and which recall to mind the tepumereme of the Orinoco, the Cassiquiare, and the Rupunuvini.) The traces discovered in the mountains of Uruana, by the missionary Fray Ramon Bueno, approach nearer to alphabetical writing; but are nevertheless very doubtful.

---

[*] In Tamanac tepumereme. (Tepu, a stone, rock; as in Mexican, tetl, a stone, and tepetl, a mountain; in Turco-Tatarian, tepe.

Whatever may be the meaning of these figures, and with whatever view they were traced upon granite, they merit the examination of those who direct their attention to the philosophic history of our species. In travelling from the coast of Caracas towards the equator, we are at first led to believe that monuments of this kind are peculiar to the mountain-chain of Encaramada; they are found at the port of Sedeno, near Caycara,[*] at San Rafael del Capuchino, opposite Cabruta, and in almost every place where the granitic rock pierces the soil, in the savannah which extends from the Cerro Curiquima towards the banks of the Caura. The nations of the Tamanac race, the ancient inhabitants of those countries, have a local mythology, and traditions connected with these sculptured rocks. Amalivaca, the father of the Tamanacs, that is, the creator of the human race (for every nation regards itself as the root of all other nations), arrived in a bark, at the time of the great inundation, which is called the age of water,[†] when the billows of the ocean broke against the mountains of Encaramada in the interior of the land. All mankind, or, to speak more correctly, all the Tamanacs, were drowned, with the exception of one man and one woman, who saved themselves on a mountain near the banks of the Asiveru, called Cuchivero by the Spaniards. This mountain is the Ararat of the Aramean or Semitic nations, and the Tlaloc or Colhuacan of the Mexicans. Amalivaca, sailing in his bark, engraved the figures of the moon and the sun on the Painted Rock (Tepumereme) of Encaramada. Some blocks of granite piled upon one another, and forming a kind of cavern, are still called the house or dwelling of the great forefather of the Tamanacs. The natives show also a large stone near this cavern, in the plains of Maita, which they say was an instrument of music, the drum of Amalivaca. We must here observe, that this heroic personage had a brother, Vochi, who helped him to give the

---

[*] In the Mountains of the Tyrant, Cerros del Tirano.
[†] The Atonatiuh of the Mexicans, the fourth age, the fourth regeneration of the world.

surface of the earth its present form. The Tamanacs relate that the two brothers, in their system of perfectibility, sought, at first, to arrange the Orinoco in such a manner, that the current of the water could always be followed either going down or going up the river. They hoped by this means to spare men trouble in navigating rivers; but, however great the power of these regenerators of the world, they could never contrive to give a double slope to the Orinoco, and were compelled to relinquish this singular plan. Amalivaca had daughters, who had a decided taste for travelling. The tradition states, doubtless with a figurative meaning, that he broke their legs, to render them sedentary, and force them to people the land of the Tamanacs. After having regulated everything in America, on that side of the great water, Amalivaca again embarked, and returned to the other shore, to the same place from whence he came. Since the natives have seen the missionaries arrive, they imagine that Europe is this other shore; and one of them inquired with great simplicity of Father Gili, whether he had there seen the great Amalivaca, the father of the Tamanacs, who had covered the rocks with symbolic figures.

These notions of a great convulsion of nature; of two human beings saved on the summit of a mountain, and casting behind them the fruits of the mauritia palm-tree, to repeople the earth; of that national divinity, Amalivaca, who arrived by water from a distant land, who prescribed laws to nature, and forced the nations to renounce their migrations; these various features of a very ancient system of belief, are well worthy of attention. What the Tamanacs, and the tribes whose languages are analogous to the Tamanac tongue, now relate to us, they have no doubt learned from other people, who inhabited before them the same regions. The name of Amalivaca is spread over a region of more than five thousand square leagues; he is found designated as the father of mankind, or our great grandfather, as far as to the Caribbee nations, whose idiom approaches the Tamanac only in the same degree as the German approaches the Greek, the Persian, and the Sanscrit. Amalivaca is not originally the Great

Spirit, the Aged of Heaven, the invisible being, whose worship springs from that of the powers of nature, when nations rise insensibly to the consciousness of the unity of these powers; he is rather a personage of the heroic times, a man, who, coming from afar, lived in the land of the Tamanacs and the Caribs, sculptured symbolic figures upon the rocks, and disappeared by going back to the country he had previously inhabited beyond the ocean. The anthropomorphism of the divinity has two sources diametrically opposite; and this opposition seems to arise less from the various degrees of intellectual culture, than from the different dispositions of nations, some of which are more inclined to mysticism, and others more governed by the senses, and by external impressions. Sometimes man makes the divinities descend upon earth, charging them with the care of ruling nations, and giving them laws, as in the fables of the East; sometimes, as among the Greeks and other nations of the West, they are the first monarchs, priest-kings, who are stripped of what is human in their nature, to be raised to the rank of national divinities. Amalivaca was a stranger, like Manco-Capac, Bochica, and Quetzalcohuatl; those extraordinary men, who, in the alpine or civilized part of America, on the tablelands of Peru, New Grenada, and Anahuac, organized civil society, regulated the order of sacrifices, and founded religious congregations. The Mexican Quetzalcohuatl, whose descendants Montezuma[*] thought he recognized in the companions of Cortez, displays an additional resemblance to Amalivaca, the mythologic personage of savage America or the plains of the torrid zone. When advanced in age, the high-priest of Tula left the country of Anahuac, which he had filled with his miracles, to return to an unknown region, called Tlalpallan. When the monk Bernard de Sahagun arrived in Mexico, the same questions were put to him, as those which were addressed to Father Gili two hundred years later, in the forests of the Orinoco; he was asked whether he

---

[*] The second king of this name, of the race of Acamapitzin, properly called Montezuma-Ilhuicamina.

came from the other shore (del otro lado), from the countries to which Quetzalcohuatl had retired.

The region of sculptured rocks, or of painted stones, extends far beyond the Lower Orinoco, beyond the country (latitude 7° 5' to 7° 40', longitude 68° 50' to 69° 45') to which belongs what may be called the local fables of the Tamanacs. We again find these same sculptured rocks between the Cassiquiare and the Atabapo (latitude 2° 5' to 3° 20'; longitude 69 to 70°); and between the sources of the Essequibo and the Rio Branco (latitude 3° 50'; longitude 62° 32'). I do not assert that these figures prove the knowledge of the use of iron, or that they denote a very advanced degree of culture; but even on the supposition that, instead of being symbolical, they are the fruits of the idleness of hunting nations, we must still admit an anterior race of men, very different from those who now inhabit the banks of the Orinoco and the Rupunuri. The more a country is destitute of remembrances of generations that are extinct, the more important it becomes to follow the least traces of what appears to be monumental. The eastern plains of North America display only those extraordinary circumvallations that remind us of the fortified camps (the pretended cities of vast extent) of the ancient and modern nomad tribes of Asia. In the oriental plains of South America, the force of vegetation, the heat of the climate, and the too lavish gifts of nature, have opposed obstacles still more powerful to the progress of human civilization. Between the Orinoco and the Amazon I heard no mention of any wall of earth, vestige of a dyke, or sepulchral tumulus; the rocks alone show us (and this through a great extent of country), rude sketches which the hand of man has traced in times unknown, and which are connected with religious traditions.

Before I quitted the wildest part of the Upper Orinoco, I thought it desirable to mention facts which are important only when they are considered in their connection with each other. All I could relate of our navigation from Esmeralda to the mouth of the Atabapo would be merely an enumeration of rivers and uninhabited places. From the 24th to the 27th of May, we slept but twice on land; our first resting-place was at the confluence of the Rio Jao, and our second below the mission of Santa Barbara, in the island of Minisi. The Orinoco being free from shoals, the Indian pilot pursued his course all night, abandoning the boat to the current of the river. Setting apart the time which we spent on the shore in preparing the rice and plantains that served us for food, we took but thirty-five hours in going from Esmeralda to Santa Barbara. The chronometer gave me for the longitude of the latter mission 70° 3′; we had therefore made near four miles an hour, a velocity which was partly owing to the current, and partly to the action of the oars. The Indians assert that the crocodiles do not go up the Orinoco above the mouth of the Rio Jao, and that the manatees are not even found above the cataract of Maypures.

The mission of Santa Barbara is situated a little to the west of the mouth of the Rio Ventuari, or Venituari, examined in 1800 by Father Francisco Valor. We found in this small village of one hundred and twenty inhabitants some traces of industry; but the produce of this industry is of little profit to the natives; it is reserved for the monks, or, as they say in these countries, for the church and the convent. We were assured that a great lamp of massive silver, purchased at the expense of the neophytes, is expected from Madrid. Let us hope that, after the arrival of this treasure, they will think also of clothing the Indians, of procuring for them some instruments of agriculture, and assembling their children in a school. Although there are a few oxen in the savannahs round the mission, they are rarely employed in turning the mill (trapiche), to express the juice of the sugar-cane; this is the occupation of the Indians, who work without pay here as they do everywhere when they are understood to work for the church. The pasturages at the foot of the mountains round Santa Barbara are not so rich as at Esmeralda, but superior to those at San Fernando de Atabapo. The grass is short and thick, yet the upper stratum of earth furnishes only a dry and parched granitic sand. The savannahs (far from fertile) of the banks of the Guaviare, the Meta, and the Upper Orinoco, are equally destitute of the mould which abounds in the surrounding forests, and of the thick stratum of clay, which covers the sandstone of the Llanos, or steppes of Venezuela. The small herbaceous mimosas contribute in this zone to fatten the cattle, but are very rare between the Rio Jao and the mouth of the Guaviare.

During the few hours of our stay at the mission of Santa Barbara, we obtained pretty accurate ideas respecting the Rio Ventuari, which, next to the Guaviare, appeared to me to be the most considerable tributary of the Orinoco. Its banks, heretofore occupied by the Maypures, are still peopled by a great number of independent nations. On going up by the mouth of the Ventuari, which forms a delta covered with palm-trees, you find in the east, after three days' journey, the Cumaruita and the Paru, two streams that rise at the foot of the lofty mountains of Cuneva. Higher up, on the west, lie the Mariata and the Manipiare, inhabited by the Macos and Curacicanas. The latter nation is remarkable for their active cultivation of cotton. In a hostile incursion (entrada) a large house was found containing more than thirty or forty hammocks of a very fine texture of spun cotton, cordage, and fishing implements. The natives had fled; and Father Valor informed us, that the Indians of the mission who accompanied him had set fire to the house before he could save these productions of the industry of the Curacicanas. The neophytes of Santa Barbara, who think themselves very superior to these supposed savages, appeared to me far less industrious. The Rio Manipiare, one of the principal branches of the Ventuari, approaches near its source those lofty mountains, the northern ridge of which gives birth to the Cuchivero. It is a prolongation of the chain of Baraguan; and there Father Gili places the table-land of Siamacu, of which he vaunts the temperate climate. The upper course of the Rio Ventuari, beyond the confluence of the Asisi, and the Great Raudales, is almost unknown. I was informed only that the Upper Ventuari bends so much towards the east that the ancient road from Esmeralda to the Rio Caura crosses the bed of the river. The proximity of the tributary streams of the Carony, the Caura, and the Ventuari, has facilitated for ages the access of the Caribs to the banks of the Upper Orinoco. Bands of this warlike and trading people went up from the Rio Carony, by the Paragua, to the sources of the Paruspa. A portage conducted them to the Chavarro, an eastern tributary stream of the Rio Caura; they descended with their canoes first this stream, and then the Caura itself as far as the mouth of the Erevato. After having gone up this last river south-west, and traversed vast savannahs for three days, they entered by the Manipiare into the great Rio Ventuari. I trace this road with precision not only because it was that by which the traffic of native slaves was carried on, but also to call the attention of those, who at some future day may rule the destiny of Guiana, to the high importance of this labyrinth of rivers.

It is by the four largest tributary streams, which the majestic river of the Orinoco receives on the right (the Carony, the Caura, the Padamo, and the Ventuari), that European civilization will one day penetrate into this region of forests and mountains, which has a surface of ten thousand six hundred square leagues, and which is bounded by the Orinoco on the north, the west, and the south. The Capuchins of Catalonia and the Observantins of Andalusia and Valencia, have already made settlements in the valleys of the Carony and the Caura. The tributary streams of the Lower Orinoco, being the nearest to the coast and to the cultivated region of Venezuela, were naturally the first to receive missionaries, and with them some germs of social life. Corresponding to the Carony and the Caura, which flow toward the north, are two great tributary streams of the Upper Orinoco, that send their waters toward the south; these are the Padamo and the Ventuari. No village has hitherto risen on their banks, though they offer advantages for agriculture and pasturage, which would be sought in vain in the valley of the immense river to which they are tributary. In the centre of these wild countries, where there will long be no other road than the rivers, every project of civilization should be founded on an intimate knowledge of the hydraulic features of the country, and the relative importance of the tributary streams.

In the morning of the 26th of May we left the little village of Santa Barbara, where we found several Indians of Esmeralda, who had come reluctantly, by order of the missionary, to construct for him a house of two stories. During the whole day we enjoyed the view of the fine mountains of Sipapo, which rise at a distance of more than eighteen leagues in the direction of north-north-west. The vegetation of the banks of the Orinoco is singularly varied in this part of the country; the aborescent ferns[*] descend from the mountains, and mingle with the palm-trees of the plain. We rested that night on the island of Minisi; and, after having passed the mouths of the little rivers Quejanuma, Ubua, and Masao, we arrived, on the 27th of May, at San Fernando de Atabapo. We lodged in the same house which we had occupied a month previously, when going up the Rio Negro. We then directed our course towards the south, by the Atabapo and the Temi; we were now returning from the west, having made a long circuit by the Cassiquiare and the Upper Orinoco.

We remained only one day at San Fernando de Atabapo, although that village, adorned as it was by the pirijao palm-tree, with fruit like peaches, appeared to us a delicious abode. Tame pauxis[†] surrounded the Indian huts; in one of which we saw a very rare monkey, which inhabits the banks of the Guaviare. This monkey is the caparro, which I have made known in my Observations on Zoology and comparative Anatomy; it forms, as Geoffroy believes, a new genus (Lagothrix) between the ateles and the alouates. The hair of this monkey is grey, like that of the marten, and extremely soft to the touch. The caparro is distinguished by a round head, and a mild and agreeable

---

[*] The geographical distribution of these plants is extremely singular. Scarcely any are found on the eastern coast of Brazil. See the interesting work of Prince Maximilian of Neuwied, Reise nach Brasilien volume 1 page 274.

[†] Not the ourax of Cuvier, Crax pauxi Linn., but the Crax alector.

expression of countenance. I believe the missionary Gili is the only author who has made mention before me of this curious animal, around which zoologists begin to group other monkeys of Brazil. Having quitted San Fernando on the 27th of May, we arrived, by help of the rapid current of the Orinoco, in seven hours, at the mouth of the Rio Mataveni. We passed the night in the open air, under the granitic rock El Castillito, which rises in the middle of the river, and the form of which reminded us of the ruin called the Mouse-tower (Mausethurm), on the Rhine, opposite Bingen. Here, as on the banks of the Atabapo, we were struck by the sight of a small species of drosera, having exactly the appearance of the drosera of Europe.

The Orinoco had sensibly swelled during the night; and the current, strongly accelerated, bore us, in ten hours, from the mouth of the Mataveni to the Upper Great Cataract, that of Maypures, or Quituna. The distance which we passed over was thirteen leagues. We recalled to mind, with much satisfaction, the scenes where we had reposed in going up the river. We again found the Indians who had accompanied us in our herborizations; and we visited anew the fine spring that issues from a rock of stratified granite behind the house of the missionary: its temperature was not changed more than 0.3°. From the mouth of the Atabapo as far as that of the Apure we seemed to be travelling as through a country which we had long inhabited. We were reduced to the same abstinence; we were stung by the same mosquitos; but the certainty of reaching in a few weeks the term of our physical sufferings kept up our spirits.

The passage of the canoe through the Great Cataract obliged us to stop two days at Maypures. Father Bernardo Zea, missionary at the Raudales, who had accompanied us to the Rio Negro, though ill, insisted on conducting us with his Indians as far as Atures. One of these Indians, Zerepe, the interpreter, who had been so unmercifully punished at the beach of Pararuma, rivetted our attention by his appearance of deep sorrow. We learned that his grief was caused by the loss of a young girl to whom he was engaged, and that he had lost her in consequence of false intelligence which had been spread respecting the direction of our journey. Zerepe, who was a native of Maypures, had been brought up in the woods by his parents, who were of the tribe of the Macos. He had brought with him to the mission a girl of twelve years of age, whom he intended to marry at our return from the Cataracts. The Indian girl was little pleased with the life of the missions, and she was told that the whites would go to the country of the Portuguese (Brazil), and would take Zerepe with them. Disappointed in her hopes, she seized a boat, and with another girl of her own age, crossed the Great Cataract, and fled al monte. The recital of this courageous adventure was the great news of the place. The affliction of Zerepe, however, was not of long duration. Born among the Christians, having travelled as far as the foot of the Rio Negro, understanding Spanish and the language of the Macos, he thought himself superior to the people of his tribe, and he no doubt soon forgot his forest love.

On the 31st of May we passed the rapids of Guahibos and Garcita. The islands which rise in the middle of the waters of the river were overspread with the purest verdure. The rains of winter had unfolded the spathes of the vadgiai palm-tree, the leaves of which rise straight toward the sky. The eye is never wearied of the view of those scenes, where the trees and rocks give the landscape that grand and severe character which we admire in the background of the pictures of Salvator Rosa. We landed before sunset on the eastern bank of the Orinoco, at the Puerto de la Expedicion, in order to visit the cavern of Ataruipe, which is the place of sepulchre of a whole nation destroyed. I shall attempt to describe this cavern, so celebrated among the natives.

We climbed with difficulty, and not without some danger, a steep rock of granite, entirely bare. It would have been almost impossible to fix the foot on its smooth and sloping surface, if large crystals of feldspar, resisting decomposition, did not stand out from the rock, and furnish points of support. Scarcely had we attained the summit of the mountain when we beheld with astonishment the singular aspect of the surrounding country. The foamy bed of the waters is filled with an archipelago of islands covered with palm-trees. Westward, on the left bank of the Orinoco, the wide-stretching savannahs of the Meta and the Casanare resembled a sea of verdure. The setting sun seemed like a globe of fire suspended over the plain, and the solitary Peak of Uniana, which appeared more lofty from being wrapped in vapours which softened its outline, all contributed to augment the majesty of the scene. Immediately below us lay a deep valley, enclosed on every side. Birds of prey and goatsuckers winged their lonely flight in this inaccessible circus. We found a pleasure in following with the eye their fleeting shadows, as they glided slowly over the flanks of the rock.

A narrow ridge led us to a neighbouring mountain, the rounded summit of which supported immense blocks of granite. These masses are more than forty or fifty feet in diameter; and their form is so perfectly spherical, that, as they appear to touch the soil only by a small number of points, it might be supposed, at the least shock of an earthquake, they would roll into the abyss. I do not remember to have seen anywhere else a similar phenomenon, amid the decompositions of granitic soils. If the balls rested on a rock of a different nature, as in the blocks of Jura, we might suppose that they had been rounded by the action of water, or thrown out by the force of an elastic fluid; but their position on the summit of a hill alike granitic, makes it more probable that they owe their origin to the progressive decomposition of the rock.

The most remote part of the valley is covered by a thick forest. In this shady and solitary spot, on the declivity of a steep mountain, the cavern of Ataruipe opens to the view. It is less a cavern than a jutting rock in which the waters have scooped a vast hollow when, in the ancient revolutions of our planet, they attained that height.[*] In this tomb of a whole extinct tribe we soon counted nearly six hundred skeletons well preserved, and regularly placed. Every skeleton reposes in a sort of basket made of the petioles of the palm-tree. These baskets, which the natives call mapires, have the form of a square bag. Their size is proportioned to the age of the dead; there are some for infants cut off at the moment of their birth. We saw them from ten inches to three feet four inches long, the skeletons in them being bent together. They are all ranged near each other, and are so entire that not a rib or a phalanx is wanting. The bones have been prepared in three manners, either whitened in the air and the sun, dyed red with anoto, or, like mummies, varnished with odoriferous resins, and enveloped in leaves of the heliconia or of the plantain-tree. The Indians informed us that the fresh corpse is placed in damp ground, that the flesh may be consumed by degrees; some months afterwards it is taken out, and the flesh remaining on the bones is scraped off with sharp stones. Several hordes in Guiana still observe this custom. Earthen vases half-baked are found near the mapires or baskets. They appear to contain the bones of the same family. The largest of these vases, or funeral urns, are five feet high, and three feet three inches long. Their colour is greenish-grey, and their oval form is pleasing to the eye. The handles are made in the shape of crocodiles or serpents; the edges are bordered with painted meanders, labyrinths, and grecques, in rows variously combined. Such designs are found in every zone among nations the farthest removed from each other, either with respect to their respective positions on the globe, or to the degree of civilization which they have attained. They still adorn the common pottery made by the inhabitants of the little mission of Maypures; they ornament the

---

[*] I saw no vein, no hole (four) filled with crystals. The decomposition of granitic rocks, and their separation into large masses, dispersed in the plains and valleys in the form of blocks and balls with concentric layers, appear to favour the enlarging of these natural excavations, which resemble real caverns.

bucklers of the Otaheitans, the fishing-implements of the Esquimaux, the walls of the Mexican palace of Mitla, and the vases of ancient Greece.

We could not acquire any precise idea of the period to which the origin of the mapires and the painted vases, contained in the bone-cavern of Ataruipe, can be traced. The greater part seemed not to be more than a century old; but it may be supposed that, sheltered from all humidity under the influence of a uniform temperature, the preservation of these articles would be no less perfect if their origin dated from a period far more remote. A tradition circulates among the Guahibos, that the warlike Atures, pursued by the Caribs, escaped to the rocks that rise in the middle of the Great Cataracts; and there that nation, heretofore so numerous, became gradually extinct, as well as its language. The last families of the Atures still existed in 1767, in the time of the missionary Gili. At the period of our voyage an old parrot was shown at Maypures, of which the inhabitants said, and the fact is worthy of observation, that they did not understand what it said, because it spoke the language of the Atures.

We opened, to the great concern of our guides, several mapires, for the purpose of examining attentively the form of the skulls. They were all marked by the characteristics of the American race, with the exception of two or three, which approached indubitably to the Caucasian. In the middle of the Cataracts, in the most inaccessible spots, cases are found strengthened with iron bands, and filled with European tools, vestiges of clothes, and glass trinkets. These articles, which have given rise to the most absurd reports of treasures hidden by the Jesuits, probably belonged to Portuguese traders who had penetrated into these savage countries. May we suppose that the skulls of European race, which we saw mingled with the skeletons of the natives, and preserved with the same care, were the remains of some Portuguese travellers who had died of sickness, or had been killed in battle? The aversion evinced by the natives for whatever is not of their own race renders this hypothesis little probable. Perhaps fugitive mestizos of the missions of the Meta and Apure may have come and settled near the Cataracts, marrying women of the tribe of the Atures. Such mixed marriages sometimes take place in this zone, though they are more rare than in Canada, and in the whole of North America, where hunters of European origin unite themselves with savages, assume their habits, and sometimes acquire great political influence.

We took several skulls, the skeleton of a child of six or seven years old, and two of full-grown men of the nation of the Atures, from the cavern of Ataruipe. All these bones, partly painted red, partly varnished with odoriferous resins, were placed in the baskets (mapires or canastos) which we have just described. They made almost the whole load of a mule; and as we knew the superstitious feelings of the Indians in reference to the remains of the dead after burial, we carefully enveloped the canastos in mats recently woven. Unfortunately for us, the penetration of the Indians, and the extreme quickness of their sense of smelling, rendered all our precautions useless. Wherever we stopped, in the missions of the Caribbees, amid the Llanos, between Angostura and Nueva Barcelona, the natives assembled round our mules to admire the monkeys which we had purchased at the Orinoco. These good people had scarcely touched our baggage, when they announced the approaching death of the beast of burden that carried the dead. In vain we told them that they were deceived in their conjectures; and that the baskets contained the bones of crocodiles and manatees; they persisted in repeating that they smelt the resin that surrounded the skeletons, and that they were their old relations. We were obliged to request that the monks would interpose their authority, to overcome the aversion of the natives, and procure for us a change of mules.

One of the skulls, which we took from the cavern of Ataruipe, has appeared in the fine work published by my old master, Blumenbach, on the varieties of the human species. The skeletons of the Indians were lost on the coast of Africa, together with a considerable part of our collections, in a shipwreck, in which perished our friend and fellow-traveller, Fray Juan Gonzales, the young monk of the order of Saint Francis.

We withdrew in silence from the cavern of Ataruipe. It was one of those calm and serene nights which are so common in the torrid zone. The stars shone with a mild and planetary light. Their scintillation was scarcely sensible at the horizon, which seemed illumined by the great nebulae of the southern hemisphere. An innumerable multitude of insects spread a reddish light upon the ground, loaded with plants, and resplendent with these living and moving fires, as if the stars of the firmament had sunk down on the savannah. On quitting the cavern we stopped several times to admire the beauty of this singular scene. The odoriferous vanilla and festoons of bignonia decorated the entrance; and above, on the summit of the hill, the arrowy branches of the palm-trees waved murmuring in the air. We descended towards the river, to take the road to the mission, where we arrived late in the night. Our imagination was struck by all we had just seen. Occupied continually by the present, in a country where the traveller is tempted to regard human society as a new institution, he is more powerfully interested by remembrances of times past. These remembrances were not indeed of a distant date; but in all that is monumental antiquity is a relative idea, and we easily confound what is ancient with what is obscure and problematic. The Egyptians considered the historical remembrances of the Greeks as very recent. If the Chinese, or, as they prefer calling themselves, the inhabitants of the Celestial Empire, could have communicated with the priests of Heliopolis, they would have smiled at those pretensions of the Egyptians to antiquity. Contrasts not less striking are found in the north of Europe and of Asia, in the New World, and in every region where the human race has not preserved a long consciousness of itself. The migration of the Toltecs, the most ancient historical event on the tableland of Mexico, dates only in the sixth century of our era. The introduction of a good system of intercalation, and the reform of the calendars, the indispensable basis of an accurate chronology, took place in the year 1091. These epochs, which to us appear so modern, fall on fabulous times, when we reflect on the history of our species between the banks of the Orinoco and the Amazon. We there see symbolic figures sculptured on the rocks, but no tradition throws light upon their origin. In the hot part of Guiana we can go back only to the period when the Castilian and Portuguese conquerors, and more recently peaceful monks, penetrated amid so many barbarous nations.

It appears that to the north of the Cataracts, in the strait of Baraguan, there are caverns filled with bones, similar to those I have just described: but I was informed of this fact only after my return; our Indian pilots did not mention it when we landed at the strait. These tombs no doubt have given rise to a fable of the Ottomacs, according to which the granitic and solitary rocks of Baraguan, the forms of which are very singular, are regarded as the grandfathers, the ancient chiefs of the tribe. The custom of separating the flesh from the bones, very anciently practised by the Massagetes, is still known among several hordes of the Orinoco. It is even asserted, and with some probability, that the Guaraons plunge their dead bodies under water enveloped in nets; and that the small caribe-fishes, of which we saw everywhere an innumerable quantity, devour in a few days the muscular flesh, and thus prepare the skeleton. It may be supposed that this operation can be practised only in places where crocodiles are not common. Some tribes, for instance the Tamanacs, are accustomed to lay waste the fields of a deceased relative, and cut down the trees which he has planted. They say that the sight of objects which belonged to their relation makes them melancholy. They like better to efface than to preserve remembrances. These effects of Indian sensibility are very detrimental to agriculture, and the monks oppose with energy these superstitious practices, to which the natives converted to Christianity still adhere in the missions.

The tombs of the Indians of the Orinoco have not been very closely examined, because they do not contain valuable articles like those of Peru; and even on the spot no faith is now lent to the chimerical ideas, which were heretofore formed of the wealth of the ancient inhabitants of El Dorado. The thirst of gold everywhere precedes the desire of instruction, and a taste for researches into antiquity; in all the mountainous part of South America, from Merida and Santa Martha to the table-lands of Quito and Upper Peru, the labours of absolute mining have been undertaken to discover tombs, or, as the Creoles say, employing a word altered from the Inca language, guacas. When in Peru, at Mancichi, I went into the guaca from which, in the sixteenth century, masses of gold of great value were extracted. No trace of the precious metals has been found in the caverns which have served the natives of Guiana for ages as sepulchres. This circumstance proves that even at the period when the Caribs, and other travelling nations, made incursions to the south-west, gold had flowed in very small quantities from the mountains of Peru towards the eastern plains.

Wherever the granitic rocks do not present any of those large cavities caused by their decomposition, or by an accumulation of their blocks, the Indians deposit their dead in the earth. The hammock (chinchorro), a kind of net in which the deceased had reposed during his life, serves for a coffin. This net is fastened tight round the body, a hole is dug in the hut, and there the body is laid. This is the most usual method, according to the account of the missionary Gili, and it accords with what I myself learned from Father Zea. I do not believe that there exists one tumulus in Guiana, not even in the plains of the Cassiquiare and the Essequibo. Some, however, are to be met with in the savannahs of Varinas, as in Canada, to the west of the Alleghenies.[*] It

---

[*] Mummies and skeletons contained in baskets were recently discovered in a cavern in the United States. It is believed they belong to a race of men analogous to that of the Sandwich Islands. The description of these tombs has some similitude with that of the tombs of Ataruipe.

seems remarkable enough that, notwithstanding the extreme abundance of wood in those countries, the natives of the Orinoco were as little accustomed as the ancient Scythians to burn the dead. Sometimes they formed funeral piles for that purpose; but only after a battle, when the number of the dead was considerable. In 1748, the Parecas burned not only the bodies of their enemies, the Tamanacs, but also those of their own people who fell on the field of battle. The Indians of South America, like all nations in a state of nature, are strongly attached to the spots where the bones of their fathers repose. This feeling, which a great writer has beautifully painted in the episode of Atala, is cherished in all its primitive ardour by the Chinese. These people among whom everything is the produce of art, or rather of the most ancient civilization, do not change their dwelling without carrying along with them the bones of their ancestors. Coffins are seen deposited on the banks of great rivers, to be transported, with the furniture of the family, to a remote province. These removals of bones, heretofore more common among the savages of North America, are not practised among the tribes of Guiana; but these are not nomad, like nations who live exclusively by hunting.

We stayed at the mission of Atures only during the time necessary for passing the canoe through the Great Cataract. The bottom of our frail bark had become so thin that it required great care to prevent it from splitting. We took leave of the missionary, Bernardo Zea, who remained at Atures, after having accompanied us during two months, and shared all our sufferings. This poor monk still continued to have fits of tertian ague; they had become to him an habitual evil, to which he paid little attention. Other fevers of a more fatal kind prevailed at Atures on our second visit. The greater part of the Indians could not leave their hammocks, and we were obliged to send in search of cassava-bread, the most indispensable food of the country, to the independent but neighbouring tribe of the Piraoas. We had hitherto escaped these malignant fevers, which I believe to be always contagious.

We ventured to pass in our canoe through the latter half of the Raudal of Atures. We landed here and there, to climb upon the rocks, which like narrow dikes joined the islands to one another. Sometimes the waters force their way over the dikes, sometimes they fall within them with a hollow noise. A considerable portion of the Orinoco was dry, because the river had found an issue by subterraneous caverns. In these solitary haunts the rock-manakin with gilded plumage (Pipra rupicola), one of the most beautiful birds of the tropics, builds its nest. The Raudalito of Carucari is caused by an accumulation of enormous blocks of granite, several of which are spheroids of five or six feet in diameter, and they are piled together in such a manner, as to form spacious caverns. We entered one of these caverns to gather the confervas that were spread over the clefts and humid sides of the rock. This spot displayed one of the most extraordinary scenes of nature that we had contemplated on the banks of the Orinoco. The river rolled its waters turbulently over our heads. It seemed like the sea dashing against reefs of rocks; but at the entrance of the cavern we could remain dry beneath a large sheet of water that precipitated itself in an arch from above the barrier. In other cavities, deeper, but less spacious, the rock was pierced by the effect of successive filtrations. We saw columns of water, eight or nine inches broad, descending from the top of the vault, and finding an issue by clefts, that seemed to communicate at great distances with each other.

The cascades of Europe, forming only one fall, or several falls close to each other, can never produce such variety in the shifting landscape. This variety is peculiar to rapids, to a succession of small cataracts several miles in length, to rivers that force their way across rocky dikes and accumulated blocks of granite. We had the opportunity of viewing this extraordinary sight longer than we wished. Our boat was to coast the eastern bank of a narrow island, and to take us in again after a long circuit. We passed an hour and a half in vain expectation of it. Night approached, and with it a tremendous storm. It rained with violence. We began to fear that our frail bark had been wrecked against the rocks, and that the Indians, conformably to their habitual indifference for the evils of others, had returned tranquilly to the mission. There were only three of us: we were completely wet, and uneasy respecting the fate of our boat: it appeared far from agreeable to pass, without sleep, a long night of the torrid zone amid the noise of the Raudales. M. Bonpland proposed to leave me in the island with Don Nicolas Soto, and to swim across the branches of the river that are separated by the granitic dikes. He hoped to reach the forest, and seek assistance at Atures from Father Zea. We dissuaded him with difficulty from undertaking this hazardous enterprise. He knew little of the labyrinth of small channels, into which the Orinoco is divided. Most of them have strong whirlpools, and what passed before our eyes while we were deliberating on our situation, proved sufficiently that the natives had deceived us respecting the absence of crocodiles in the cataracts. The little monkeys which we had carried along with us for months were deposited on the point of our island. Wet by the rains and sensible of the least lowering of the temperature, these delicate animals sent forth plaintive cries, and attracted to the spot two crocodiles, the size and leaden colour of which denoted their great age. Their unexpected appearance made us reflect on the danger we had incurred by bathing, at our first passing by the mission of Atures, in the middle of the Raudal. After long waiting, the Indians at length arrived at the close of day. The natural coffer-dam by which they had endeavoured to descend in order to make the circuit of the island, had become impassable owing to the shallowness of the water. The pilot sought long for a more accessible passage in this labyrinth of rocks and islands. Happily our canoe was not damaged and in less than half an hour our instruments, provision, and animals, were embarked.

We pursued our course during a part of the night, to pitch our tent again in the island of Panumana. We recognized with pleasure the spots where we had botanized when going up the Orinoco. We examined once more on the beach of Guachaco that small formation of sandstone, which reposes directly on granite. Its position is the same as that of the sandstone which Burckhardt observed at the entrance of Nubia, superimposed on the granite of Syene. We passed, without visiting it, the new mission of San Borga, where (as we learned with regret a few days after) the little colony of Guahibos had fled al monte, from the chimerical fear that we should carry them off; to sell them as poitos, or slaves. After having passed the rapids of Tabaje, and the Raudal of Cariven, near the mouth of the great Rio Meta, we arrived without accident at Carichana. The missionary received us with that kind hospitality which he extended to us on our first passage. The sky was unfavourable for astronomical observations; we had obtained some new ones in the two Great Cataracts; but thence, as far as the mouth of the Apure, we were obliged to renounce the attempt. M. Bonpland had the satisfaction at Carichana of dissecting a manatee more than nine feet long. It was a female, and the flesh appeared to us not unsavoury. I have spoken in another place of the manner of catching this herbivorous cetacea. The Piraoas, some families of whom inhabit the mission of Carichana, detest this animal to such a degree, that they hid themselves, to avoid being obliged to touch it, whilst it was being conveyed to our hut. They said that the people of their tribe die infallibly when they eat of it. This prejudice is the more singular, as the neighbours of the Piraoas, the Guamos and the Ottomacs, are very fond of the flesh of the manatee. The flesh of the crocodile is also an object of horror to some tribes, and of predilection to others.

The island of Cuba furnishes a fact little known in the history of the manatee. South of the port of Xagua, several miles from the coast, there are springs of fresh water in the middle of the sea. They are supposed to be owing to a hydrostatic pressure existing in subterraneous channels, communicating with the lofty mountains of Trinidad. Small vessels sometimes take in water there; and, what is well worthy of observation, large manatees remain habitually in those spots. I have already called the attention of naturalists to the crocodiles which advance from the mouth of rivers far into the sea. Analogous circumstances may have caused, in the ancient catastrophes of our planet, that singular mixture of pelagian and fluviatile bones and petrifactions, which is observed in some rocks of recent formation.

Our stay at Carichana was very useful in recruiting our strength after our fatigues. M. Bonpland bore with him the germs of a cruel malady; he needed repose; but as the delta of the tributary streams included between the Horeda and Paruasi is covered with a rich vegetation, he made long herbalizations, and was wet through several times in a day. We found, fortunately, in the house of the missionary, the most attentive care; we were supplied with bread made of maize flour, and even with milk. The cows yield milk plentifully enough in the lower regions of the torrid zone, wherever good pasturage is found. I call attention to this fact, because local circumstances have spread through the Indian Archipelago the prejudice of considering hot climates as repugnant to the secretion of milk. We may conceive the indifference of the inhabitants of the New World for a milk diet, the country having been originally destitute of animals capable of furnishing it*; but how can we avoid being astonished at this indifference in the immense Chinese population, living in great part beyond the tropics, and in the same latitude with the nomad and pastoral tribes of central Asia? If the Chinese have ever been a pastoral people, how have they lost the tastes and habits so intimately connected with that state, which precedes agricultural institutions? These questions are interesting with

---

* The reindeer are not domesticated in Greenland as they are in Lapland; and the Esquimaux care little for their milk. The bisons taken very young accustom themselves, on the west of the Alleghenies, to graze with herds of European cows. The females in some districts of India yield a little milk, but the natives have never thought of milking them. What is the origin of that fabulous story related by Gomara (chapter 43 page 36) according to which the first Spanish navigators saw, on the coast of South Carolina, stags led to the savannahs by herdsmen? The female bisons, according to Mr. Buchanan and the philosophical historian of the Indian Archipelago, Mr. Crawford, yield more milk than common cows.

respect both to the history of the nations of oriental Asia, and to the ancient communications that are supposed to have existed between that part of the world and the north of Mexico.

We went down the Orinoco in two days, from Carichana to the mission of Uruana, after having again passed the celebrated strait of Baraguan. We stopped several times to determine the velocity of the river, and its temperature at the surface, which was 27.4°. The velocity was found to be two feet in a second (sixty-two toises in 3' 6") in places where the bed of the Orinoco was more than twelve thousand feet broad, and from ten to twelve fathoms deep. The slope of the river is in fact extremely gentle from the Great Cataracts to Angostura; and, if a barometric measurement were wanting, the difference of height might be determined by approximation, by measuring from time to time the velocity of the stream, and the extent of the section in breadth and depth. We had some observations of the stars at Uruana. I found the latitude of the mission to be 7° 8'; but the results from different stars left a doubt of more than 1'. The stratum of mosquitos, which hovered over the ground, was so thick that I could not succeed in rectifying properly the artificial horizon. I tormented myself in vain; and regretted that I was not provided with a mercurial horizon. On the 7th of June, good absolute altitudes of the sun gave me 69° 40' for the longitude. We had advanced from Esmeralda 1° 17' toward the west, and this chronometric determination merits entire confidence on account of the double observations, made in going and returning, at the Great Cataracts, and at the confluence of the Atabapo and of the Apure.

The situation of the mission of Uruana is extremely picturesque. The little Indian village stands at the foot of a lofty granitic mountain. Rocks everywhere appear in the form of pillars above the forest, rising higher than the tops of the tallest trees. The aspect of the Orinoco is nowhere more majestic than when viewed from the hut of the missionary, Fray Ramon Bueno. It is more than two thousand six hundred toises broad, and it runs without any winding, like a vast canal, straight toward the east. Two long and narrow islands (Isla de Uruana and Isla vieja de la Manteca) contribute to give extent to the bed of the river; the two banks are parallel, and we cannot call it divided into different branches. The mission is inhabited by the Ottomacs, a tribe in the rudest state, and presenting one of the most extraordinary physiological phenomena. They eat earth; that is, they swallow every day, during several months, very considerable quantities, to appease hunger, and this practice does not appear to have any injurious effect on their health. Though we could stay only one day at Uruana, this short space of time sufficed to make us acquainted with the preparation of the poya, or balls of earth. I also found some traces of this vitiated appetite among the Guamos; and between the confluence of the Meta and the Apure, where everybody speaks of dirt-eating as of a thing anciently known. I shall here confine myself to an account of what we ourselves saw or heard from the missionary, who had been doomed to live for twelve years among the savage and turbulent tribe of the Ottomacs.

The inhabitants of Uruana belong to those nations of the savannahs called wandering Indians (Indios andantes) who, more difficult to civilize than the nations of the forest (Indios del monte), have a decided aversion to cultivate the land, and live almost exclusively by hunting and fishing. They are men of very robust constitution; but ill-looking, savage, vindictive, and passionately fond of fermented liquors. They are omnivorous animals in the highest degree; and therefore the other Indians, who consider them as barbarians, have a common saying, nothing is so loathsome but that an Ottomac will eat it. While the waters of the Orinoco and its tributary streams are low, the Ottomacs subsist on fish and turtles. The former they kill with surprising dexterity, by shooting them with an arrow when they appear at the surface of the water. When the rivers swell fishing almost entirely ceases.[*] It is then very difficult to procure fish, which often fails the poor missionaries, on fast-days as well as flesh-days, though all the young Indians are under the obligation of fishing for the convent. During the period of these inundations, which last two or three months, the Ottomacs swallow a prodigious quantity of earth. We found heaps of earth-balls in their huts, piled up in pyramids three or four feet high. These balls were five or six inches in diameter. The earth which the Ottomacs eat is a very fine and unctuous clay of a yellowish grey colour; and, when being slightly baked at the fire, the hardened crust has a tint inclining to red, owing to the oxide of iron which is mingled with it. We brought away some of this earth, which we took from the winter-provision of the Indians; and it is a mistake to suppose that it is steatitic, and that it contains magnesia. Vauquelin did not discover any traces of that substance in it but he found that it contained more silex than alumina, and three or four per cent of lime.

The Ottomacs do not eat every kind of clay indifferently; they choose the alluvial beds or strata, which contain the most unctuous earth, and the smoothest to the touch. I inquired of the missionary whether the moistened clay were made to undergo that peculiar decomposition which is indicated by a disengagement of carbonic acid and sulphuretted hydrogen, and which is designated in every language by the term of putrefaction; but he assured us that the natives neither cause the clay to rot, nor do they mingle it with flour of maize, oil of turtle's eggs, or fat of the crocodile. We ourselves examined, both at the Orinoco and after our return to Paris, the balls of earth which we brought away with us, and found no trace of the mixture of any organic substance, whether oily or farinaceous. The savage regards every thing as nourishing that appeases hunger: when, therefore, you inquire of an Ottomac on what he

---

[*] In South America, as in Egypt and Nubia, the swelling of the rivers, which occurs periodically in every part of the torrid zone, is erroneously attributed to the melting of the snows.

subsists during the two months when the river is at its highest flood he shows you his balls of clayey earth. This he calls his principal food at the period when he can seldom procure a lizard, a root of fern, or a dead fish swimming at the surface of the water. If necessity force the Indians to eat earth during two months (and from three quarters to five quarters of a pound in twenty-four hours), he eats it from choice during the rest of the year. Every day in the season of drought, when fishing is most abundant, he scrapes his balls of poya, and mingles a little clay with his other aliment. It is most surprising that the Ottomacs do not become lean by swallowing such quantities of earth: they are, on the contrary, extremely robust. The missionary Fray Ramon Bueno asserts that he never remarked any alteration in the health of the natives at the period of the great risings of the Orinoco.

The Ottomacs during some months eat daily three-quarters of a pound of clay slightly hardened by fire, but which they moisten before swallowing it. It has not been possible to verify hitherto with precision how much nutritious vegetable or animal matter they take in a week at the same time; but they attribute the sensation of satiety which they feel to the clay, and not to the wretched aliments which they take with it occasionally.

No physiological phenomenon being entirely insulated, it may be interesting to examine several analogous phenomena, which I have been able to collect. I observed everywhere within the torrid zone, in a great number of individuals, children, women, and sometimes even full-grown men, an inordinate and almost irresistible desire of swallowing earth; not an alkaline or calcareous earth to neutralize (as it is said) acid juices, but a fat clay, unctuous, and exhaling a strong smell. It is often found necessary to tie the children's hands or to confine them to prevent them eating earth when the rain ceases to fall. At the village of Banco, on the bank of the river Magdalena, I saw the Indian women who make pottery continually swallowing great pieces of clay. These women were not in a state of pregnancy;

541

and they affirmed that earth is an aliment which they do not find hurtful. In other American tribes, people soon fall sick, and waste away, when they yield too much to this mania of eating earth. We found at the mission of San Borja an Indian child of the Guahiba nation, who was as thin as a skeleton. The mother informed us that the little girl was reduced to this lamentable state of atrophy in consequence of a disordered appetite, she having refused during four months to take almost any other food than clay. Yet San Borja is only twenty-five leagues distant from the mission of Uruana, inhabited by that tribe of the Ottomacs, who, from the effect no doubt of a habit progressively acquired, swallow the poya without experiencing any pernicious effects. Father Gumilla asserts that the Ottomacs take as an aperient, oil, or rather the melted fat of the crocodile, when they feel any gastric obstructions; but the missionary whom we found among them was little disposed to confirm this assertion. It may be asked, why the mania of eating earth is much more rare in the frigid and temperate than in the torrid zones; and why in Europe it is found only among women in a state of pregnancy, and sickly children. This difference between hot and temperate climates arises perhaps only from the inert state of the functions of the stomach caused by strong cutaneous perspiration. It has been supposed to be observed that the inordinate taste for eating earth augments among the African slaves, and becomes more pernicious when they are restricted to a regimen purely vegetable and deprived of spirituous liquors.

The negroes on the coast of Guinea delight in eating a yellowish earth, which they call caouac. The slaves who are taken to America endeavour to indulge in this habit; but it proves detrimental to their health. They say that the earth of the West Indies is not so easy of digestion as that of their country. Thibaut de Chanvalon, in his Voyage to Martinico, expresses himself very judiciously on this pathological phenomenon. "Another cause," he says, "of this pain in the stomach is that several of the negroes, who come from the coast of Guinea, eat earth; not from a depraved taste, or in consequence of disease, but from a habit contracted at home in Africa, where they eat, they say, a particular earth, the taste of which they find agreeable, without suffering any inconvenience. They seek in our islands for the earth most similar to this, and prefer a yellowish red volcanic tufa. It is sold secretly in our public markets; but this is an abuse which the police ought to correct. The negroes who have this habit are so fond of caouac, that no chastisement will prevent their eating it."

In the Indian Archipelago, at the island of Java, Labillardiere saw, between Surabaya and Samarang, little square and reddish cakes exposed for sale. These cakes called tanaampo, were cakes of clay, slightly baked, which the natives eat with relish. The attention of physiologists, since my return from the Orinoco, having been powerfully directed to these phenomena of geophagy, M. Leschenault (one of the naturalists of the expedition to the Antarctic regions under the command of captain Baudin) has published some curious details on the tanaampo, or ampo, of the Javanese. "The reddish and somewhat ferruginous clay," he says "which the inhabitants of Java are fond of eating occasionally, is spread on a plate of iron, and baked, after having been rolled into little cylinders in the form of the bark of cinnamon. In this state it takes the name of ampo, and is sold in the public markets. This clay has a peculiar taste, which is owing to the baking: it is very absorbent, and adheres to the tongue, which it dries. In general it is only the Javanese women who eat the ampo, either in the time of pregnancy, or in order to grow thin; the absence of plumpness being there regarded as a kind of beauty. The use of this earth is fatal to health; the women lose their appetite imperceptibly, and take only with relish a very small quantity of food; but the desire of becoming thin, and of preserving a slender shape, induces them to brave these dangers, and maintains the credit of the ampo." The savage inhabitants of New Caledonia also, to appease their hunger in times of scarcity, eat great pieces of a friable Lapis ollaris. Vauquelin analysed this stone, and found in it, beside magnesia and silex in equal portions, a small quantity of oxide of copper. M. Goldberry had seen the negroes in Africa, in the islands of Bunck and Los Idolos, eat an earth of which he had himself eaten, without being incommoded by it, and which also was a white and friable steatite. These examples of earth-eating in the torrid zone appear very strange. We are struck by the anomaly of finding a taste, which might seem to belong only to the inhabitants of the most sterile regions, prevailing among races of rude and indolent men, who live in the finest and most fertile countries on the globe. We saw at Popayan, and in several mountainous parts of Peru, lime reduced to a very fine powder, sold in the public markets to the natives among other articles of food. This powder, when eaten, is mingled with coca, that is, with the leaves of the Erythroxylon peruvianum. It is well known that Indian messengers take no other aliment for whole days than lime and coca: both excite the secretion of saliva, and of the gastric juice; they take away the appetite, without affording any nourishment to the body. In other parts of South America, on the coast of Rio de la Hacha, the Guajiros swallow lime alone, without adding any vegetable matter to it. They carry with them a little box filled with lime, as we do snuff-boxes, and as in Asia people carry a betel-box. This American custom excited the curiosity of the first Spanish navigators. Lime blackens the teeth; and in the Indian Archipelago, as among several American hordes, to blacken the teeth is to beautify them. In the cold regions of the kingdom of Quito, the natives of Tigua eat habitually from choice, and without any injurious consequences, a very fine clay, mixed with quartzose sand. This clay, suspended in water, renders it milky. We find in their huts large vessels filled with this water, which serves as a beverage, and which the Indians call agua or leche de llanka.[*]

When we reflect on these facts, we perceive that the appetite for clayey, magnesian, and calcareous earth is most common among the people of the torrid zone; that it is not always a cause

---

[*] Water or milk of clay. Llanka is a word of the general language of the Incas, signifying fine clay.

of disease; and that some tribes eat earth from choice, whilst others (as the Ottomacs in America, and the inhabitants of New Caledonia in the Pacific) eat it from want and to appease hunger. A great number of physiological phenomena prove that a temporary cessation of hunger may be produced though the substances that are submitted to the organs of digestion may not be, properly speaking, nutritive. The earth of the Ottomacs, composed of alumine and silex, furnishes probably nothing, or almost nothing, to the composition of the organs of man. These organs contain lime and magnesia in the bones, in the lymph of the thoracic duct, in the colouring matter of the blood, and in white hairs; they afford very small quantities of silex in black hair; and, according to Vauquelin, but a few atoms of alumine in the bones, though this is contained abundantly in the greater part of those vegetable substances which form part of our nourishment. It is not the same with man as with animated beings placed lower in the scale of organization. In the former, assimilation is exerted only on those substances that enter essentially into the composition of the bones, the muscles, and the medullary matter of the nerves and the brain. Plants, on the contrary, draw from the soil the salts that are found accidentally mixed in it; and their fibrous texture varies according to the nature of the earths that predominate in the spots which they inhabit. An object well worthy of research, and which has long fixed my attention, is the small number of simple substances (earthy and metallic) that enter into the composition of animated beings, and which alone appear fitted to maintain what we may call the chemical movement of vitality.

We must not confound the sensations of hunger with that vague feeling of debility which is produced by want of nutrition, and by other pathologic causes. The sensation of hunger ceases long before digestion takes place, or the chyme is converted into chyle. It ceases either by a nervous and tonic impression exerted by the aliments on the coats of the stomach; or, because the digestive apparatus is filled with substances that excite the mucous membranes to an abundant secretion of the gastric juice.

To this tonic impression on the nerves of the stomach the prompt and salutary effects of what are called nutritive medicaments may be attributed, such as chocolate, and every substance that gently stimulates and nourishes at the same time. It is the absence of a nervous stimulant that renders the solitary use of a nutritive substance (as starch, gum, or sugar) less favourable to assimilation, and to the reparation of the losses which the human body undergoes. Opium, which is not nutritive, is employed with success in Asia, in times of great scarcity; it acts as a tonic. But when the matter which fills the stomach can be regarded neither as an aliment, that is, as proper to be assimilated, nor as a tonic stimulating the nerves, the cessation of hunger is probably owing only to the secretion of the gastric juice. We here touch upon a problem of physiology which has not been sufficiently investigated. Hunger is appeased, the painful feeling of inanition ceases, when the stomach is filled. It is said that this viscus stands in need of ballast; and every language furnishes figurative expressions which convey the idea that a mechanical distension of the stomach causes an agreeable sensation. Recent works of physiology still speak of the painful contraction which the stomach experiences during hunger, the friction of its sides against one another, and the action of the gastric juice on the texture of the digestive apparatus. The observations of Bichat, and more particularly the fine experiments of Majendie, are in contradiction to these superannuated hypotheses. After twenty-four, forty-eight, or even sixty hours of abstinence, no contraction of the stomach is observed; it is only on the fourth or fifth day that this organ appears to change in a small degree its dimensions. The quantity of the gastric juice diminishes with the duration of abstinence. It is probable that this juice, far from accumulating, is digested as an alimentary substance. If a cat or dog be made to swallow a substance which is not susceptible of being digested, a pebble for instance, a mucous and acid liquid is formed abundantly in the cavity of the stomach, somewhat resembling in its composition the gastric juice of the human body. It appears to me very probable, that when the want of aliments compels the Ottomacs and the inhabitants of New Caledonia to swallow clay and steatite during a part of the year, these earths occasion a powerful secretion of the gastric and pancreatic juices in the digestive apparatus of these people. The observations which I made on the banks of the Orinoco, have been recently confirmed by the direct experiments of two distinguished young physiologists, MM. Cloquet and Breschet. After long fasting they ate as much as five ounces of a silvery green and very flexible laminar talc. Their hunger was completely satisfied, and they felt no inconvenience from a kind of food to which their organs were unaccustomed. It is known that great use is still made in the East of the bolar and sigillated earths of Lemnos, which are clay mingled with oxide of iron. In Germany the workmen employed in the quarries of sandstone worked at the mountain of Kiffhauser spread a very fine clay upon their bread, instead of butter, which they call steinbutter[*] (stone-butter).

The state of perfect health enjoyed by the Ottomacs during the time when they use little muscular exercise, and are subjected to so extraordinary a regimen, is a phenomenon difficult to be explained. It can be attributed only to a habit prolonged from generation to generation. The structure of the digestive apparatus differs much in animals that feed exclusively on flesh or on seeds; it is even probable that the gastric juice changes its nature, according as it is employed in effecting the digestion of animal or vegetable substances; yet we are able gradually to change the regimen of herbivorous and carnivorous animals, to feed the former with flesh, and the latter with vegetables. Man can accustom himself to an extraordinary abstinence and find it but little painful if he employ tonic or stimulating substances (various drugs, small quantities of opium, betel, tobacco, or leaves of coca); or if he supply his stomach,

---

[*] This steinbutter must not be confounded with the mountain butter (bergbutter) which is a saline substance, produced by a decomposition of aluminous schists.

from time to time, with earthy insipid substances that are not in themselves fit for nutrition. Like man in a savage state some animals, when pressed by hunger in winter, swallow clay or friable steatites; such are the wolves in the northeast of Europe, the reindeer and, according to the testimony of M. Patrin, the kids in Siberia. The Russian hunters, on the banks of the Yenisei and the Amour, use a clayey matter which they call rock-butter, as a bait. The animals scent this clay from afar, and are fond of the smell; as the clays of bucaro, known in Portugal and Spain by the name of odoriferous earths (tierras olorosas), have an odour agreeable to women.[*] People are fond of drinking out of these vessels on account of the smell of the clay. The women of the province of Alentejo acquire a habit of masticating the bucaro earth; and feel a great privation when they cannot indulge this vitiated taste.) Brown relates in his History of Jamaica that the crocodiles of South America swallow small stones and pieces of very hard wood, when the lakes which they inhabit are dry, or when they are in want of food. M. Bonpland and I observed in a crocodile, eleven feet long, which we dissected at Batallez, on the banks of the Rio Magdalena, that the stomach of this reptile contained half-digested fish, and rounded fragments of granite three or four inches in diameter. It is difficult to admit that the crocodiles swallow these stony masses accidentally, for they do not catch fish with their lower jaw resting on the ground at the bottom of the river. The Indians have framed the absurd hypothesis that these indolent animals like to augment their weight, that they may have less trouble in diving. I rather think that they load their stomach with large pebbles to excite an abundant secretion of the gastric juice. The experiments of Majendie render this explanation extremely probable. With respect to the habit of the granivorous birds, particularly the gallinaceae and ostriches, of swallowing sand and small pebbles, it has been hitherto attributed to an instinctive desire of

---

[*] Bucaro (vas fictile odoriferum).

accelerating the trituration of the aliments in a muscular and thick stomach.

We have mentioned that tribes of Negroes on the Gambia mingle clay with their rice. Some families of Ottomacs were perhaps formerly accustomed to cause the maize and other farinaceous seeds to rot in their poya, in order to eat earth and amylaceous matter together: possibly it was a preparation of this kind, that Father Gumilla described indistinctly in the first volume of his work when he affirms that the Guamos and the Ottomacs feed upon earth only because it is impregnated with the sustancia del maiz (substance of maize) and the fat of the cayman. I have already observed that neither the present missionary of Uruana, nor Fray Juan Gonzales, who lived long in those countries, knew anything of this mixture of animal and vegetable substances with the poya. Perhaps Father Gumilla has confounded the preparation of the earth which the natives swallow with the custom they still retain (of which M. Bonpland acquired the certainty on the spot) of burying in the ground the beans of a species of mimosacea,* to cause them to enter into decomposition so as to reduce them into a white bread, savoury, but difficult of digestion. I repeat that the balls of poya, which we took from the winter stores of the Indians, contained no trace of animal fat, or of amylaceous matter. Gumilla being one of the most credulous travellers we know, it almost perplexes us to credit facts which even he has thought fit to reject. In the second volume of his work he however gainsays a great part of what he advanced in the first; he no longer doubts that half at least (a lo menos) of the bread of the Ottomacs and the Guamos is clay. He asserts, that children and full grown persons not only eat this bread without suffering in their health, but also great pieces of pure clay (muchos terrones de pura greda.) He adds that those who feel a weight on the stomach physic themselves with the fat of the crocodile which restores their appetite and enables them to

---

* Of the genus Inga.

continue to eat pure earth.[*] It is certain that the Guamos are very fond, if not of the fat, at least of the flesh of the crocodile, which appeared to us white, and without any smell of musk. In Sennaar, according to Burckhardt, it is equally esteemed, and sold in the markets.

The little village of Uruana is more difficult to govern than most of the other missions. The Ottomacs are a restless, turbulent people, with unbridled passions. They are not only fond to excess of the fermented liquors prepared from cassava and maize, and of palm-wine, but they throw themselves into a peculiar state of intoxication, we might say of madness, by the use of the powder of niopo. They gather the long pods of a mimosacea which we have made known by the name of Acacia niopo,[†] cut them into pieces, moisten them, and cause them to ferment. When the softened seeds begin to grow black, they are kneaded like a paste; mixed with some flour of cassava and lime procured from the shell of a helix, and the whole mass is exposed to a very brisk fire, on a gridiron made of hard wood. The hardened paste takes the form of small cakes. When it is to be used, it is reduced to a fine powder, and placed on a dish five or six inches wide. The Ottomac holds this dish, which has a handle, in his right hand, while he inhales the niopo by the nose, through the forked bone of a bird, the two extremities of which are applied to the nostrils. This bone, without which the Ottomac believes that he could not take this kind of snuff, is seven inches long: it appeared to me to be the leg-bone of a large sort of plover. The niopo is so stimulating that the smallest portions of it produce violent sneezing in those who are not accustomed to

---

[*] Gumilla volume 2 page 260.

[†] It is an acacia with very delicate leaves, and not an Inga. We brought home another species of mimosacea (the chiga of the Ottomacs and the sepa of the Maypures) that yields seeds, the flour of which is eaten at Uruana like cassava. From this flour the chiga bread is prepared, which is so common at Cunariche, and on the banks of the Lower Orinoco. The chiga is a species of Inga, and I know of no other mimosacea that can supply the place of the cerealia.

its use. Father Gumilla says this diabolical powder of the Ottomacs, furnished by an arborescent tobacco-plant, intoxicates them through the nostrils (emboracha por las narices), deprives them of reason for some hours, and renders them furious in battle. However varied may be the family of the leguminous plants in the chemical and medical properties of their seeds, juices, and roots, we cannot believe, from what we know hitherto of the group of mimosaceae, that it is principally the pod of the Acacia niopo which imparts the stimulant power to the snuff of the Ottomacs. This power is owing, no doubt, to the freshly calcined lime. We have shown above that the mountaineers of the Andes of Popayan, and the Guajiros, who wander between the lake of Maracaybo and the Rio la Hacha, are also fond of swallowing lime as a stimulant, to augment the secretion of the saliva and the gastric juice.

A custom analogous to the use of the niopo just described was observed by La Condamine among the natives of the Upper Maranon. The Omaguas, whose name is rendered celebrated by the expeditions attempted in search of El Dorado, have like the Ottomacs a dish, and the hollow bone of a bird, by which they convey to their nostrils their powder of curupa. The seed that yields this powder is no doubt also a mimosacea; for the Ottomacs, according to Father Gili, designate even now, at the distance of one hundred and sixty leagues from the Amazon, the Acacia niopo by the name of curupa. Since the geographical researches which I have recently made on the scene of the exploits of Philip von Huten, and the real situation of the province of Papamene, or of the Omaguas, the probability of an ancient communication between the Ottomacs of the Orinoco and the Omaguas of the Maranon has become more interesting and more probable. The former came from the Meta, perhaps from the country between the Meta and the Guaviare; the latter assert that they descended in great numbers to the Maranon by the Rio Jupura, coming from the eastern declivity of the Andes of New Grenada. Now, it is precisely between the Guayavero (which joins the Guaviare) and the Caqueta (which takes lower down the name of Japura) that the country of the Omagua appears to be situate, of which the adventurers of Coro and Tocuyo in vain attempted the conquest. There is no doubt a striking contrast between the present barbarism of the Ottomacs and the ancient civilization of the Omaguas; but all parts of the latter nation were not perhaps alike advanced in civilization, and the example of tribes fallen into complete barbarism are unhappily but too common in the history of our species. Another point of resemblance may be remarked between the Ottomacs and the Omaguas. Both of these nations are celebrated among all the tribes of the Orinoco and the Amazon for their employment of caoutchouc in the manufacture of various articles of utility.

The real herbaceous tobacco[*] (for the missionaries have the habit of calling the niopo or curupa tree-tobacco) has been cultivated from time immemorial by all the native people of the Orinoco; and at the period of the conquest the habit of smoking was found to be alike spread over both North and South America.

The Tamanacs and the Maypures of Guiana wrap maize-leaves round their cigars, as the Mexicans did at the time of the arrival of Cortes. The Spaniards have substituted paper for the leaves of maize in imitation of them. The poor Indians of the

---

[*] The word tobacco (tabacco), like the words savannah, maize, cacique, maguey (agave), and manati, belongs to the ancient language of Haiti, or St. Domingo. It did not properly denote the herb but the tube through which the smoke was inhaled. It seems surprising that a vegetable production so universally spread should have different names among neighbouring people. The pete-ma of the Omaguas is, no doubt, the pety of the Guaranos; but the analogy between the Cabre and Algonkin (or Lenni-Lenape) words which denote tobacco may be merely accidental. The following are the synonyms in thirteen languages.

North America. Aztec or Mexican; yetl: Algonkin; sema: Huron; oyngoua.

South America. Peruvian or Quichua; sayri: Chiquito; pais. Guarany; pety: Vilela; tusup: Mbaja (west of the Paraguay), nalodagadi: Moxo (between the Rio Ucayale and the Rio Madeira); sabare. Omagua; petema. Tamanac; cavas. Maypure; jema. Cabre; scema.

forests of the Orinoco know as well as did the great nobles at the court of Montezuma that the smoke of tobacco is an excellent narcotic; and they use it not only to procure their afternoon nap, but also to put themselves into that state of quiescence, which they call dreaming with the eyes open, or day-dreaming. The use of tobacco appears to me to be now very rare in the missions; and in New Spain, to the great regret of the revenue-officers, the natives, who are almost all descended from the lowest class of the Aztec people, do not smoke at all. Father Gili affirms that the practice of chewing tobacco is unknown to the Indians of the Lower Orinoco. I rather doubt the truth of this assertion, having been told that the Sercucumas of the Erevato and the Caura, neighbours of the whitish Taparitos, swallow tobacco chopped small, and impregnated with some other very stimulant juices, to prepare themselves for battle. Of the four species of nicotiana cultivated in Europe[*] we found only two growing wild; but the Nicotiana loxensis, and the Nicotiana andicola, which I found on the back of the Andes, at the height of eighteen hundred and fifty toises (almost the height of the Peak of Teneriffe), are very similar to the N. tabacum and N. rustica. The whole genus, however, is almost exclusively American, and the greater number of the species appeared to me to belong to the mountainous and temperate region of the tropics.

It was neither from Virginia, nor from South America, but from the Mexican province of Yucatan, that Europe received the first tobacco seeds, about the year 1559.[†] The celebrated Raleigh contributed most to introduce the custom of smoking among the

---

[*] Nicotiana tabacum, N. rustica, N. paniculata, and N. glutinosa.

[†] The Spaniards became acquainted with tobacco in the West India Islands at the end of the 15th century. I have already mentioned that the cultivation of this narcotic plant preceded the cultivation of the potato in Europe more than 120 or 140 years. When Raleigh brought tobacco from Virginia to England in 1586, whole fields of it were already cultivated in Portugal. It was also previously known in France, where it was brought into fashion by Catherine de Medicis, from whom it received the name of herbe a la reine, the queen's herb.

nations of the north. As early as the end of the sixteenth century bitter complaints were made in England of this imitation of the manners of a savage people. It was feared that, by the practice of smoking tobacco, Englishmen would degenerate into a barbarous state.[*]

When the Ottomacs of Uruana, by the use of niopo (their arborescent tobacco), and of fermented liquors, have thrown themselves into a state of intoxication, which lasts several days, they kill one another without ostensibly fighting. The most vindictive among them poison the nail of their thumb with curare; and, according to the testimony of the missionary, the mere impression of this poisoned nail may become a mortal wound if the curare be very active and immediately mingle with the mass of the blood. When the Indians, after a quarrel at night, commit a murder, they throw the dead body into the river, fearing that some indications of the violence committed on the deceased may be observed. "Every time," said Father Bueno, "that I see the women fetch water from a part of the shore to which they are not accustomed to go, I suspect that a murder has been committed in my mission."

We found in the Indian huts at Uruana the vegetable substance called touchwood of ants,[†] with which we had become acquainted at the Great Cataracts, and which is employed to stop bleeding. This substance, which might less improperly be called ants' nests, is in much request in a region whose inhabitants are of so turbulent a character. A new species of ant, of a fine

---

[*] This remarkable passage of Camden is as follows, Annal. Elizabet. page 143 1585; "ex illo sane tempore [tabacum] usu cepit esse creberrimo in Anglia et magno pretio dum quamplurimi graveolentem illius fumum per tubulum testaceum hauriunt et mox e naribus efflant; adeo ut Auglornm corporum in barbarorum naturam degenerasse videantur, quum iidem ac barbari delectentur." We may see from this passage that they emitted the smoke through the nose; but at the court of Montezuma the pipe was held in one hand, while the nostrils were stopped with the other, in order that the smoke might be more easily swallowed. Life of Raleigh volume 1 page 82.

[†] Yesca de hormigas.

emerald-green (Formica spinicollis), collects for its habitation a cotton-down, of a yellowish-brown colour, and very soft to the touch, from the leaves of a melastomacea. I have no doubt that the yesca or touchwood of ants of the Upper Orinoco (the animal is found, we were assured, only south of Atures) will one day become an article of trade. This substance is very superior to the ants' nests of Cayenne, which are employed in the hospitals of Europe, but can rarely be procured.

On the 7th of June we took leave with regret of Father Ramon Bueno. Of the ten missionaries whom we had found in different parts of the vast extent of Guiana, he alone appeared to me to be earnestly attentive to all that regarded the natives. He hoped to return in a short time to Madrid, where he intended to publish the result of his researches on the figures and characters that cover the rocks of Uruana.

In the countries we had just passed through, between the Meta, the Arauca, and the Apure, there were found, at the time of the first expeditions to the Orinoco, in 1535, those mute dogs, called by the natives maios, and auries. This fact is curious in many points of view. We cannot doubt that the dog, whatever Father Gili may assert, is indigenous in South America. The different Indian languages furnish words to designate this animal, which are scarcely derived from any European tongue. To this day the word auri, mentioned three hundred years ago by Alonzo de Herrera, is found in the Maypure. The dogs we saw at the Orinoco may perhaps have descended from those that the Spaniards carried to the coast of Caracas; but it is not less certain that there existed a race of dogs before the conquest, in Peru, in New Granada, and in Guiana, resembling our shepherds' dogs. The allco of the natives of Peru, and in general all the dogs that we found in the wildest countries of South America, bark frequently. The first historians, however, all speak of mute dogs (perros mudos). They still exist in Canada; and, what appears to me worthy of attention, it was this dumb variety that was eaten in preference in Mexico,[*] and at the Orinoco. A very well informed traveller, M. Giesecke, who resided six years in Greenland, assured me that the dogs of the Esquimaux, which pass their lives in the open air and bury themselves in winter beneath the snow, do not bark, but howl like wolves.[†]

The practice of eating the flesh of dogs is now entirely unknown on the banks of the Orinoco; but as it is a Tartar custom spread through all the eastern part of Asia, it appears to me highly interesting for the history of nations to have ascertained that it existed heretofore in the hot regions of Guiana and on the table-lands of Mexico. I must observe, also, that on the confines of the province of Durango, at the northern extremity of New Spain, the Comanches have preserved the habit of loading the backs of the great dogs that accompany them in their migrations with their tents of buffalo-leather. It is well known that employing dogs as beasts of burthen and of draught is equally common near the Slave Lake and in Siberia. I dwell on these features of conformity in the manners of nations, which become of some weight when they are not solitary, and are connected with the analogies furnished by the structure of languages, the division of time, and religious creeds and institutions.

We passed the night at the island of Cucuruparu, called also Playa de la Tortuga, because the Indians of Uruana go thither to collect the turtles' eggs. It is one of the best determined points of latitude along the banks of the Orinoco. I was there fortunate enough to observe the passage of three stars over the meridian.

---

[*] See on the Mexican techichi and on the numerous difficulties that occur in the history of mute dogs and dogs destitute of hair the Views of Nature Bohn's edition page 85.

[†] They sit down in a circle, one of them begins to howl alone and the others follow in the same tone. The groups of alouate monkeys howl in the same manner, and among them the Indians distinguish the leader of the band. It was the practice at Mexico to castrate the mute dogs in order to fatten them. This operation must have contributed to alter the organ of the voice.

To the east of the island is the mouth of the Cano de la Tortuga, which descends from the mountains of Cerbatana, continually wrapped in electric clouds. On the southern bank of the Cano, between the tributary streams Parapara and Oche, lies the almost ruined mission of San Miguel de la Tortuga. The Indians assured us that the environs of this little mission abound in otters with a very fine fur, called by the Portuguese water-dogs (perritos de agua); and what is still more remarkable, in lizards (lagartos) with only two feet. The whole of this country, which is very accessible between the Rio Cuchivero and the strait of Baraguan, is worthy of being visited by a well-informed zoologist. The lagarto destitute of hinder extremities is perhaps a species of Siren, different from the Siren lacertina of Carolina. If it were a saurian, a real Bimanis (Chirotes, Cuvier), the natives would not have compared it to a lizard. Besides the arrau turtles, of which I have in a former place given a detailed account, an innumerable quantity of land tortoises also, called morocoi, are found on the banks of the Orinoco, between Uruana and Encaramada. During the great heats of summer, in the time of drought, these animals remain without taking food, hidden beneath stones, or in the holes they have dug. They issue from their shelter and begin to eat, only when the humidity of the first rains penetrates into the earth. The terekay, or tajelu turtle which lives in fresh water, has the same habits. I have already spoken of the summer-sleep of some animals of the tropics. As the natives know the holes in which the tortoises sleep amidst the dried lands, they get out a great number at once, by digging fifteen or eighteen inches deep. Father Gili says that this operation, which he had seen, is not without danger, because serpents often bury themselves in summer with the terekays.

From the island of Cucuruparu, to the capital of Guiana, commonly called Angostura, we were but nine days on the water. The distance is somewhat less than ninety-five leagues. We seldom slept on shore but the torment of the mosquitos diminished in proportion as we advanced. We landed on the 8th of June at a farm (Hato de San Rafael del Capuchino) opposite the mouth of the Rio Apure. I obtained some good observations of latitude and longitude.[*] Having two months before taken horary angles on the bank opposite Capuchino, these observations were important for determining the rate of my chronometer, and connecting the situations on the Orinoco with those on the shore of Venezuela. The situation of this farm, being at the point where the Orinoco changes its course (which had previously been from south to north), and runs from west to east, is extremely picturesque. Granite rocks rise like islets amidst vast meadows. From their tops we discerned towards the north the Llanos of Calabozo bounding the horizon. We had been so long accustomed to the aspect of forests, that this view made a powerful impression on us. The steppes after sunset assume a tint of greenish gray. The visual ray being intercepted only by the rotundity of the earth, the stars seemed to rise as from the bosom of the ocean, and the most experienced mariner would have fancied himself placed on a projecting cape of a rocky coast. Our host was a Frenchman who lived amidst his numerous herds. Though he had forgotten his native language, he seemed pleased to learn that we came from his country, which he had left forty years before; and he wished to retain us for some days at his farm. The small towns of Caycara and Cabruta were only a few miles distant from the farm; but during part of the year our host was in complete solitude. The Capuchino becomes an island by the inundations of the Apure and the Orinoco, and the communication with the neighbouring farms can be kept up only by means of a boat. The horned cattle then seek the higher grounds which extend on the south toward the chain of the mountains of Encaramada. This granitic chain is intersected by valleys which contain magnetic sands (granulary

---

[*] I had found, on the 4th of April, for the Boca del Rio Apure (on the western bank of the Orinoco), the latitude 7° 36' 30", the longitude 59° 7' 30"; on the 8th of June I found, for the Hato del Capuchino (on the eastern bank of the Orinoco), the latitude 7° 37' 45", the longitude 69° 5' 30".

oxidulated iron), owing no doubt to the decomposition of some amphibolic or chloritic strata.

On the morning of the 9th of June we met a great number of boats laden with merchandize sailing up the Orinoco, in order to enter the Apure. This is a commercial road much frequented between Angostura and the port of Torunos in the province of Varinas. Our fellow-traveller, Don Nicolas Soto, brother-in-law of the governor of Varinas, took the same course to return to his family. At the period of the high waters, several months are lost in contending with the currents of the Orinoco, the Apure, and the Rio de Santo Domingo. The boatmen are forced to carry out ropes to the trunks of trees and thus warp their canoes up. In the great sinuosities of the river whole days are sometimes passed without advancing more than two or three hundred toises. Since my return to Europe the communications between the mouth of the Orinoco and the provinces situated on the eastern slope of the mountains of Merida, Pamplona, and Santa Fe de Bogota, have become more active; and it may be hoped that steamboats will facilitate these long voyages on the Lower Orinoco, the Portuguesa, the Rio Santo Domingo, the Orivante, the Meta, and the Guaviare. Magazines of cleft wood might be formed, as on the banks of the great rivers of the United States, sheltering them under sheds. This precaution would be indispensable, as, in the country through which we passed, it is not easy to procure dry fuel fit to keep up a fire beneath the boiler of a steam-engine.

We disembarked below San Rafael del Capuchino, on the right, at the Villa de Caycara, near a cove called Puerto Sedeno. The Villa is merely a few houses grouped together. Alta Gracia, la Ciudad de la Piedra, Real Corona, Borbon, in short all the towns or villas lying between the mouth of the Apure and Angostura, are equally miserable. The presidents of the missions, and the governors of the provinces, were formerly accustomed to demand the privileges of villas and ciudades at Madrid, the moment the first foundations of a church were laid. This was a means of persuading the ministry that the colonies were augmenting rapidly in population and prosperity. Sculptured figures of the sun and moon, such as I have already mentioned, are found near Caycara, at the Cerro del Tirano.[*] It is the work of the old people (that is of our fathers), say the natives. On a rock more distant from the shore, and called Tecoma, the symbolic figures are found, it is said, at the height of a hundred feet. The Indians knew heretofore a road, that led by land from Caycara to Demerara and Essequibo.

On the northern bank of the Orinoco, opposite Caycara, is the mission of Cabruta, founded by the Jesuit Rotella, in 1740, as an advanced post against the Caribs. An Indian village, known by the name of Cabritu,[†] had existed on the same spot for several ages. At the time when this little place became a Christian settlement, it was believed to be situate in 5° latitude, or two degrees forty minutes more to the south than I found it by direct observations made at San Rafael, and at La Boca del Rio Apure. No idea was then conceived of the direction of a road that could lead by land to Nueva Valencia and Caracas, which were supposed to be at an immense distance. The merit of having first crossed the Llanos to go to Cabruta from the Villa de San Juan Baptista del Pao belongs to a woman. Father Gili relates that Dona Maria Bargas was so devoted to the Jesuits that she attempted herself to discover the way to the missions. She was seen with astonishment to arrive at Cabruta from the north. She took up her abode near the fathers of St. Ignatius, and died in their settlements on the banks of the Orinoco. Since that period the northern part of the Llanos has been considerably

---

[*] The tyrant after whom these mountains are named is not Lope de Aguirre, but probably, as the name of the neighbouring cove seems to prove, the celebrated conquistador Antonio Sedeno, who, after the expedition of Herrera, sought to penetrate by the Orinoco to the Rio Meta. He was in a state of rebellion against the audiencia of Santo Domingo. I know not how Sedeno came to Caycara; for historians relate that he was poisoned on the banks of the Rio Tisnado, one of the tributary streams of the Portuguesa.

[†] A cacique of Cabritu received Alonzo de Herrera at his dwelling, on the expedition undertaken by Herrera for ascending the Orinoco in 1535.

peopled; and the road leading from the valleys of Aragua by Calabozo to San Fernando de Apure and Cabruta is much frequented. The chief of the famous expedition of the boundaries made choice of the latter place in 1754 to establish dock-yards for building the vessels necessary for conveying his troops intended for the Upper Orinoco. The little mountain that rises northeast of Cabruta can be discerned from afar in the steppes and serves as a landmark for travellers.

We embarked in the morning at Caycara; and driving with the current of the Orinoco, we soon passed the mouth of the Rio Cuchivero, which according to ancient tradition is the country of the Aikeambenanos, or women without husbands; and we there reached the paltry village of Alta Gracia, which is called a Spanish town. It was near this place that Jose de Iturriaga founded the Pueblo de Ciudad Real, which still figures on the most modern maps, though it has not existed for fifty years past, on account of the insalubrity of its situation. Beyond the point where the Orinoco turns to the east, forests are constantly seen on the right bank, and the llanos or steppes of Venezuela on the left. The forests which border the river are not however so thick as those of the Upper Orinoco. The population, which augments perceptibly as you advance toward the capital, comprises but few Indians, and is composed chiefly of whites, negroes, and men of mixed descent. The number of the negroes is not great; but here, as everywhere else, the poverty of their masters does not tend to procure for them more humane treatment. An inhabitant of Caycara had just been condemned to four years' imprisonment, and a fine of one hundred piastres for having, in a paroxysm of rage, tied a negress by the legs to the tail of his horse, and dragged her at full gallop through the savannah till she expired. It is gratifying to record that the Audiencia was generally blamed in the country for not having punished more severely so atrocious an action. Yet some few persons, who pretended to be the most enlightened and most sagacious of the community, deemed the punishment of a white contrary to sound policy, at the moment when the blacks of St. Domingo were in complete insurrection. Since I left those countries, civil dissensions have put arms into the hands of the slaves; and fatal experience has led the inhabitants of Venezuela to regret that they refused to listen to Don Domingo Tovar, and other right-thinking men, who, as early as the year 1795, lifted up their voices in the cabildo of Caracas, to prevent the introduction of blacks, and to propose means that might ameliorate their condition.

After having slept on the 10th of June in an island in the middle of the river (I believe that called Acaru by Father Caulin), we passed the mouth of the Rio Caura. This, the Aruy and the Carony, are the largest tributary streams which the Orinoco receives on its right bank. All the Christian settlements are near the mouth of the river; and the villages of San Pedro, Aripao, Urbani, and Guaraguaraico, succeed each other at the distance of a few leagues. The first and the most populous contains only about two hundred and fifty souls. San Luis de Guaraguaraico is a colony of negroes, some freed and others fugitives from Essequibo. This colony merits the particular attention of the Spanish Government, for it can never be sufficiently recommended to endeavour to attach the slaves to the soil, and suffer them to enjoy as farmers the fruits of their agricultural labours. The land on the Caura, for the most part a virgin soil, is extremely fertile. There are pasturages for more than 15,000 beasts; but the poor inhabitants have neither horses nor horned cattle. More than five-sixths of the banks of the Caura are either desert, or occupied by independent and savage tribes. The bed of the river is twice choked up by rocks: these obstructions occasion the famous Raudales of Mura and of Para or Paru, the latter of which has a portage, because it cannot be passed by canoes. At the time of the expedition of the boundaries, a small fort was erected on the northern cataract, that of Mura; and the governor, Don Manuel Centurion, gave the name of Ciudad de San Carlos to a few houses which some families, consisting of whites and mulattos, had constructed near the fort. South of the cataract of Para, at the confluence of the

Caura and the Erevato, the mission of San Luis was then situated; and a road by land led thence to Angostura, the capital of the province. All these attempts at civilization have been fruitless. No village now exists above the Raudal of Mura; and here, as in many other parts of the colonies, the natives may be said to have reconquered the country from the Spaniards. The valley of Caura may become one day or other highly interesting from the value of its productions, and the communications which it affords with the Rio Ventuari, the Carony, and the Cuyuni. I have shown above the importance of the four tributary streams which the Orinoco receives from the mountains of Parima. Near the mouth of the Caura, between the villages of San Pedro de Alcantara and San Francisco de Aripao, a small lake of four hundred toises in diameter was formed in 1790, by the sinking of the ground, consequent on an earthquake. It was a portion of the forest of Aripao, which sunk to the depth of eighty or a hundred feet below the level of the neighbouring land. The trees remained green for several months; and some of them, it was believed, continued to push forth leaves beneath the water. This phenomenon is the more worthy of attention as the soil of these countries is probably granitic. I doubt the secondary formations of the Llanos being continued southward as far as the valley of Caura.

On the 11th of June we landed on the right bank of the Orinoco at Puerto de los Frailes, at the distance of three leagues above the Ciudad de la Piedra, to take altitudes of the sun. The longitude of this point is 67° 26' 20", or 1° 41' east of the mouth of the Apure. Farther on, between the towns of La Piedra and Muitaco, or Real Corona, are the Torno and Boca del Infierno, two points formerly dreaded by travellers. The Orinoco suddenly changes its direction; it flows first east, then north-north-west, and then again east. A little above the Cano Marapiche, which opens on the northern bank, a very long island divides the river into two branches. We passed on the south of this island without difficulty; northward, a chain of small rocks, half covered at high water, forms whirlpools and rapids. This is La Boca del

Infierno, and the Raudal de Camiseta. The first expeditions of Diego Ordaz (1531) and Alonzo de Herrera (1535) have given celebrity to this bar. The Great Cataracts of the Atures and Maypures were then unknown; and the clumsy vessels (vergantines), in which travellers persisted in going up the river, rendered the passage through the rapids extremely difficult. At present no apprehension is felt in ascending or descending the Orinoco, at any season, from its mouth as far as the confluence of the Apure and the Meta. The only falls of water in this space are those of Torno or Camiseta, Marimara, and Cariven or Carichana Vieja. Neither of these three obstacles is to be feared with experienced Indian pilots. I dwell on these hydrographic details because a great political and commercial interest is now connected with the communications between Angostura and the banks of the Meta and the Apure, two rivers that lead to the eastern side of the Cordilleras of New Grenada. The navigation from the mouth of the Lower Orinoco to the province of Varinas is difficult only on account of the current. The bed of the river nowhere presents obstacles more difficult to be surmounted than those of the Danube between Vienna and Linz. We meet with no great bars, no real cataracts, until we get above the Meta. The Upper Orinoco, therefore, with the Cassiquiare and the Rio Negro, forms a particular system of rivers, where the active industry of Angostura and the shore of Caracas will remain long unknown.

I obtained horary angles of the sun in an island in the midst of the Boca del Infierno, where we had set up our instruments. The longitude of this point according to the chronometer is 67° 10′ 31″. I attempted to determine the magnetic dip and intensity, but was prevented by a heavy storm of rain. As the sky again became serene in the afternoon, we lay down to rest that night on a vast beach, on the southern bank of the Orinoco, nearly in the meridian of the little town of Muitaco, or Real Corona. I found the latitude by three stars to be 8° 0′ 26″, and the longitude 67° 5′ 19″. When the Observantin monks in 1752 made their first entradas on the territory of the Caribs, they constructed on this spot a small fort. The proximity of the lofty mountains of Araguacais renders Muitaco one of the most healthy places on the Lower Orinoco. There Iturriaga took up his abode in 1756, to repose after thc fatigues of the expedition of the boundaries; and as he attributed his recovery to this hot rather than humid climate, the town, or more properly the village, of Real Corona took the name of Pueblo del Puerto sano. Going down the Orinoco more to the east, we left the mouth of the Rio Pao on the north, and that of the Arui on the south. The latter river, which is somewhat considerable, is often mentioned by Raleigh. The current of the Orinoco diminished in velocity as we advanced. I measured several times a base along the beach, to ascertain the time taken by floating bodies in traversing a known distance. Above Alta Gracia, near the mouth of the Rio Ujape, I had found the velocity of the Orinoco 2.3 feet in a second; between Muitaco and Borbon it was only 1.7 foot. The barometric observations made in the neighbouring steppes prove the small slope of the ground from the longitude of 69° to the eastern coast of Guiana. We found in this country, on the right bank of the Orinoco, small formations of primitive grunstein, superimposed on granite (perhaps even embedded in the rock). We saw between Muitaco and the island of Ceiba a hill entirely composed of balls with concentric layers, in which we perceived a close mixture of hornblende and feldspar, with some traces of pyrites. The grunstein resembles that in the vicinity of Caracas; but it was impossible to ascertain the position of a formation which appeared to me to be of the same age as the granite of Parima. Muitaco was the last spot where we slept in the open air on the shore of the Orinoco: we proceeded along the river two nights more before we reached Angostura, which terminated our voyage.

It would be difficult for me to express the satisfaction we felt on landing at Angostura, the capital of Spanish Guiana. The inconveniences endured at sea in small vessels are trivial in comparison with those that are suffered under a burning sky, surrounded by swarms of mosquitos, and lying stretched in a canoe, without the possibility of taking the least bodily exercise. In seventy-five days we had performed a passage of five hundred leagues (twenty to a degree) on the five great rivers, Apure, Orinoco, Atabapo, Rio Negro, and Cassiquiare; and in this vast extent we had found but a very small number of inhabited places. After the life we had led in the woods, our dress was not in the very best order, yet nevertheless M. Bonpland and I hastened to present ourselves to Don Felipe de Ynciarte, the governor of the province of Guiana. He received us in the most cordial manner, and lodged us in the house of the secretary of the Intendencia. Coming from an almost desert country, we were struck with the bustle of the town, though it contained only six thousand inhabitants. We admired the conveniences which industry and commerce furnish to civilized man. Humble dwellings appeared to us magnificent; and every person with whom we conversed seemed to be endowed with superior intelligence. Long privations give a value to the smallest enjoyments; and I cannot express the pleasure we felt when we saw for the first time wheaten bread on the governor's table. Sensations of this sort are doubtless familiar to all who have made distant voyages.

A painful circumstance obliged us to sojourn a whole month in the town of Angostura. We felt ourselves on the first days after our arrival tired and enfeebled, but in perfect health. M. Bonpland began to examine the small number of plants which he had been able to save from the influence of the damp climate; and I was occupied in settling by astronomical observations the longitude and latitude of the capital,[*] as well as the dip of the magnetic needle. These labours were soon interrupted. We were both attacked almost on the same day by a disorder which with my fellow-traveller took the character of a debilitating fever. At this period the air was in a state of the greatest salubrity at

---

[*] I found the latitude of Santo Tomas de la Nueva Guiana, commonly called Angostura, or the Strait, near the cathedral, 8° 8' 11", the longitude 66° 15' 21".

Angostura; and as the only mulatto servant we had brought from Cumana felt symptoms of the same disorder, it was suspected that we had imbibed the germs of typhus in the damp forests of Cassiquiare. It is common enough for travellers to feel no effects from miasmata till, on arriving in a purer atmosphere, they begin to enjoy repose. A certain excitement of the mental powers may suspend for some time the action of pathogenic causes. Our mulatto servant having been much more exposed to the rains than we were, his disorder increased with frightful rapidity. His prostration of strength was excessive, and on the ninth day his death was announced to us. He was however only in a state of swooning, which lasted several hours, and was followed by a salutary crisis. I was attacked at the same time with a violent fit of fever, during which I was made to take a mixture of honey and bark (the cortex Angosturae): a remedy much extolled in the country by the Capuchin missionaries. The intensity of the fever augmented but it left me on the following day. M. Bonpland remained in a very alarming state which during several weeks caused us the most serious inquietude. Fortunately he preserved sufficient self-possession to prescribe for himself; and he preferred gentler remedies better adapted to his constitution. The fever was continual and, as almost always happens within the tropics, it was accompanied by dysentery. M. Bonpland displayed that courage and mildness of character which never forsook him in the most trying situations. I was agitated by sad presages for I remembered that the botanist Loefling, a pupil of Linnaeus, died not far from Angostura, near the banks of the Carony, a victim of his zeal for the progress of natural history. We had not yet passed a year in the torrid zone and my too faithful memory conjured up everything I had read in Europe on the dangers of the atmosphere inhaled in the forests. Instead of going up the Orinoco we might have sojourned some months in the temperate and salubrious climate of the Sierra Nevada de Merida. It was I who had chosen the path of the rivers; and the danger of my fellow-traveller presented itself to my mind as the fatal consequence of this imprudent choice.

After having attained in a few days an extraordinary degree of exacerbation the fever assumed a less alarming character. The inflammation of the intestines yielded to the use of emollients obtained from malvaceous plants. The sidas and the melochias have singularly active properties in the torrid zone. The recovery of the patient however was extremely slow, as it always happens with Europeans who are not thoroughly seasoned to the climate. The period of the rains drew near; and in order to return to the coast of Cumana, it was necessary again to cross the Llanos, where, amidst half-inundated lands, it is rare to find shelter, or any other food than meat dried in the sun. To avoid exposing M. Bonpland to a dangerous relapse, we resolved to stay at Angostura till the 10th of July. We spent part of this time at a neighbouring plantation, where mango-trees and bread-fruit trees[*] were cultivated. The latter had attained in the tenth year a height of more than forty feet. We measured several leaves of the Artocarpus that were three feet long and eighteen inches broad, remarkable dimensions in a plant of the family of the dicotyledons.

END OF VOLUME 2.

---

[*] Artocarpus incisa. Father Andujar, Capuchin missionary of the province of Caracas, zealous in the pursuit of natural history, has introduced the bread-fruit tree from Spanish Guiana at Varinas, and thence into the kingdom of New Grenada. Thus the western Coasts of America, washed by the Pacific, receive from the English Settlements in the West Indies a production of the Friendly Islands.

APPENDIX:
MEASURES AND TEMPERATURE

MEASURES:

In this narrative, as well as in the Political Essay on New Spain, all the prices are reckoned in piastres, and silver reals (reales de plata). Eight of these reals are equivalent to a piastre, or one hundred and five sous, French money (4 shillings 4½ pence English). Nouv. Esp. volume 2 pages 519, 616 and 866.

The magnetic dip is always measured in this work, according to the centesimal division, if the contrary be not expressly mentioned.

One flasco contains 70 or 80 cubic inches, Paris measure.

112 English pounds = 105 French pounds; and 160 Spanish pounds = 93 French pounds.

An arpent des eaux et forets, or legal acre of France, of which 1.95 = 1 hectare. It is about 1¼ acre English.

A tablon, equal to 1849 square toises, contains nearly an acre and one-fifth: a legal acre has 1344 square toises, and 1.95 legal acre is equal one hectare.

For the sake of accuracy, the French Measures, as given by the Author, and the indications of the Centigrade Thermometer, are retained in the translation. The following tables may, therefore, be found useful.

TABLE OF LINEAR MEASURE.

1 toise[*]	6 feet 4.73 inches.
1 foot	12.78 inches.
1 metre	3 feet 3.37 inches.

---

[*] Transcriber's Note: The 'toise' was introduced by Charlemagne in 790; it originally represented the distance between the fingertips of a man with outstretched arms, and is thus the same as the British 'fathom'. During the founding of the Metric System, less than 20 years before the date of this work, the 'toise' was assigned a value of 1.949 meters, or a little over two yards. The 'foot'; actually the 'French foot', or 'pied', is defined as 1/6 of a 'toise', and is a little over an English foot.

CENTIGRADE THERMOMETER REDUCED TO FAHRENHEIT'S SCALE.

Cent.	Fahr.	Cent.	Fahr.	Cent.	Fahr.	Cent.	Fahr.
100	212	65	149	30	86	-5	23
99	210.2	64	147.2	29	84.2	-6	21.2
98	208.4	63	145.4	28	82.4	-7	19.4
97	206.6	62	143.6	27	80.6	-8	17.6
96	204.8	61	141.8	26	78.8	-9	15.8
95	203	60	140	25	77	-10	14
94	201.2	59	138.2	24	75.2	-11	12.2
93	199.4	58	136.4	23	73.4	-12	10.4
92	197.6	57	134.6	22	71.6	-13	8.6
91	195.8	56	132.8	21	69.8	-14	6.8
90	194	55	131	20	68	-15	5
89	192.2	54	129.2	19	66.2	-16	3.2
88	190.4	53	127.4	18	64.4	-17	1.4
87	188.6	52	125.6	17	62.6	-18	-0.4
86	186.8	51	123.8	16	60.8	-19	-2.2
85	185	50	122	15	59	-20	-4
84	183.2	49	120.2	14	57.2	-21	-5.8
83	181.4	48	118.4	13	55.4	-22	-7.6
82	179.6	47	116.6	12	53.6	-23	-9.4
81	177.8	46	114.8	11	51.8	-24	-11.2
80	176	45	113	10	50	-25	-13
79	174.2	44	111.2	9	48.2	-26	-14.8
78	172.4	43	109.4	8	46.4	-27	-16.6
77	170.6	42	107.6	7	44.6	-28	-18.4
76	168.8	41	105.8	6	42.8	-29	-20.2
75	167	40	104	5	41	-30	-22
74	165.2	39	102.2	4	39.2	-31	-23.8
73	163.4	38	100.4	3	37.4	-32	-25.6
72	161.6	37	98.6	2	35.6	-33	-27.4
71	159.8	36	96.8	1	33.8	-34	-29.2
70	158	35	95	0	32	-35	-31
69	156.2	34	93.2	-1	30.2	-36	-32.8
68	154.4	33	91.4	-2	28.4	-37	-34.6
67	152.6	32	89.6	-3	26.6	-38	-36.4
66	150.8	31	87.8	-4	24.8	-39	-38.2

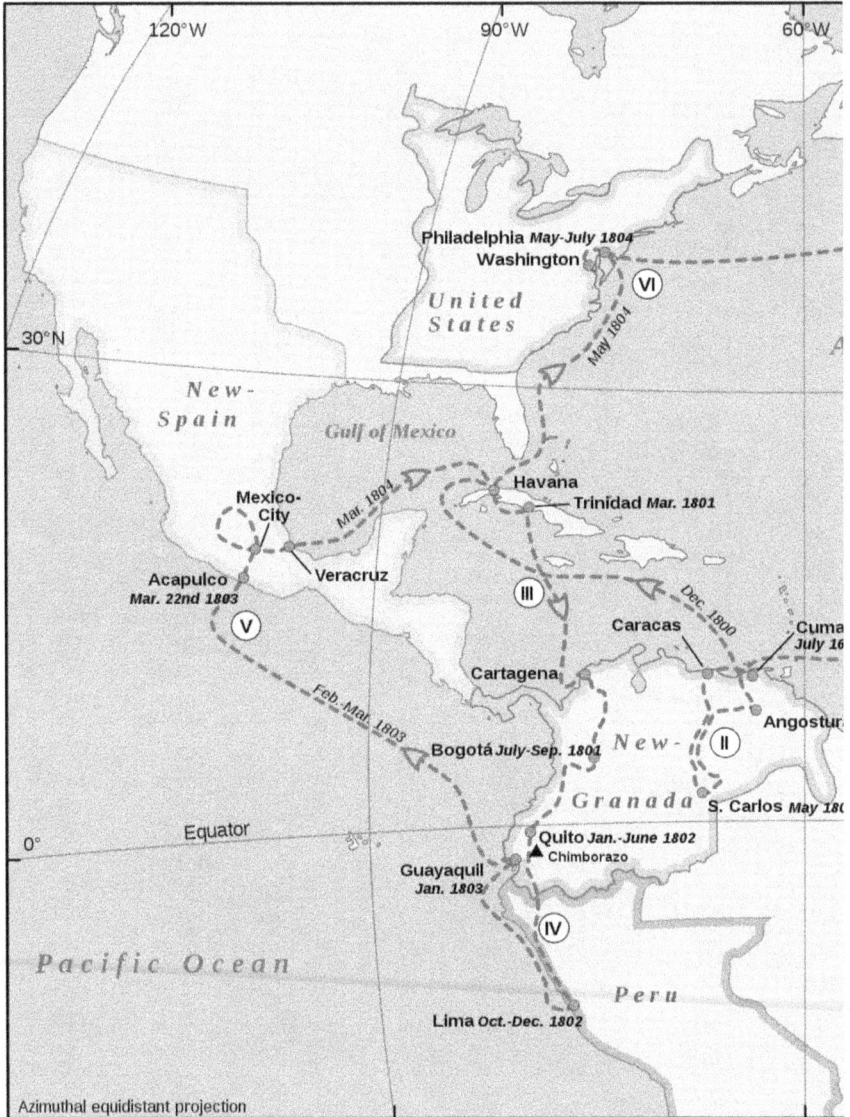

120°W   90°W   60°W

Philadelphia *May-July 1804*
Washington
VI

*United
States*

30°N

*New-
Spain*

Gulf of Mexico

Mexico-
City
Mar. 1804

Havana

Trinidad *Mar. 1801*

Acapulco
Mar. 22nd 1803
V

Veracruz

III

Dec. 1800

Caracas

Cuma
July 16

Cartagena

New-

Angostur

Feb.-Mar. 1803

Bogotá *July-Sep. 1801*

Granada

II

S. Carlos *May 180*

Equator

Quito *Jan.-June 1802*
▲ Chimborazo

Guayaquil
Jan. 1803

IV

Pacific Ocean

Peru

Lima *Oct.-Dec. 1802*

Azimuthal equidistant projection

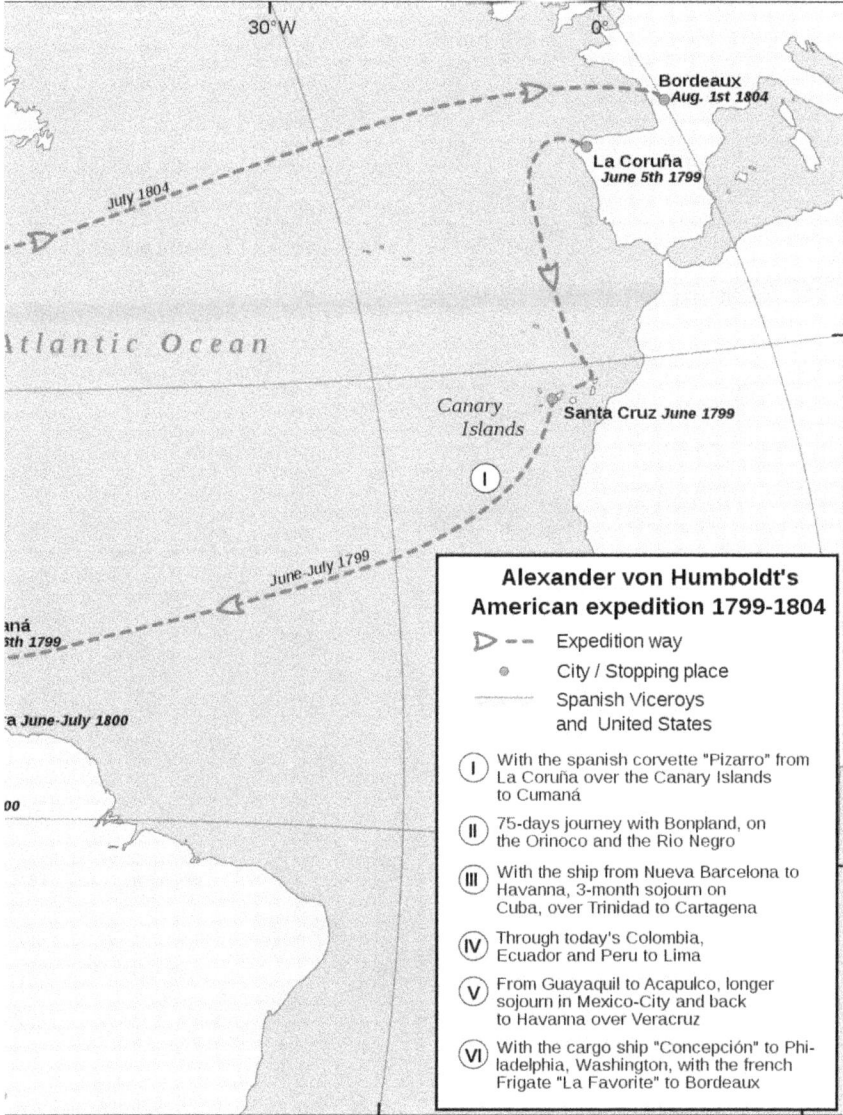

30°W   0°

Bordeaux
Aug. 1st 1804

La Coruña
June 5th 1799

July 1804

Atlantic Ocean

Santa Cruz *June 1799*

*Canary
Islands*

Ⓘ

June-July 1799

aná
5th 1799

a *June-July 1800*

00

Alexander von Humboldt's American expedition 1799-1804

▷ - -   Expedition way

⊙   City / Stopping place

―――   Spanish Viceroys
and United States

Ⓘ With the spanish corvette "Pizarro" from La Coruña over the Canary Islands to Cumaná

ⒾⒾ 75-days journey with Bonpland, on the Orinoco and the Rio Negro

ⓘⓘⓘ With the ship from Nueva Barcelona to Havanna, 3-month sojourn on Cuba, over Trinidad to Cartagena

Ⓥ Through today's Colombia, Ecuador and Peru to Lima

Ⓥ From Guayaquil to Acapulco, longer sojourn in Mexico-City and back to Havanna over Veracruz

Ⓥⓘ With the cargo ship "Concepción" to Philadelphia, Washington, with the french Frigate "La Favorite" to Bordeaux

ALEXANDER VON HUMBOLDT

Personal Narrative of Travels to the
Equinoctial Regions of America,
During the Years 1799-1804

Volume I

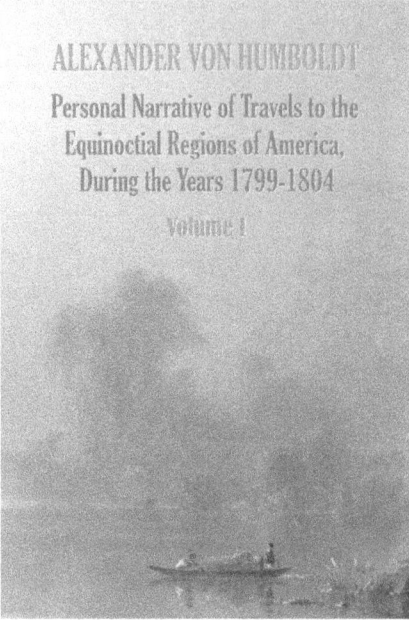

The journey begins...

Volume I covers Humboldt's
preparations, stop at Tenerife,
landfall at Cumaná and
journeys inland in what is
now Venezuela.
ISBN: 978-1-781-39330-7.

The jouney continues...

Volume III sees him
recording more information
on Venezuela, visiting
Cuba where he also writes
about local politics and
speaks out fervently against
the slave trade; he then
sails for Colombia. The
volume ends with a
comprehensive geognostic
description of the northern
part of South America.
ISBN: 978--1781-39332-1

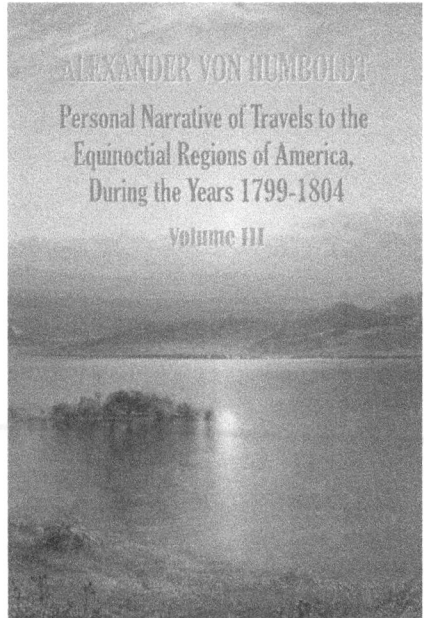

ALEXANDER VON HUMBOLDT

Personal Narrative of Travels to the
Equinoctial Regions of America,
During the Years 1799-1804

Volume III

A

Voyage

to

Terra Australis

Volume I

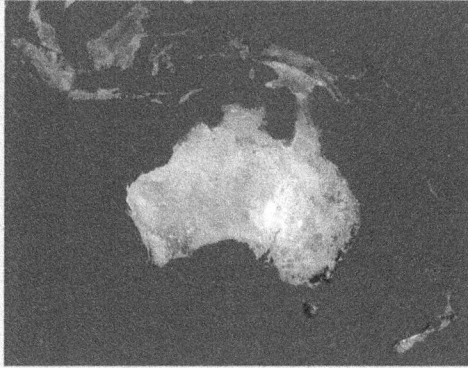

Matthew Flinders

Hardback, 2010, ISBN: 978-1-849-02565-2

When Matthew Flinders set out in 1801 to carry out a 'complete examination and survey' of the coast of New Holland, little did he know that he would be away for over ten years. Although he did not coin the term 'Australia' he keenly advocated its use, rather than the clumsy 'Terra Australis' and will always be associated with its adoption. As well as his meticulous surveys and maps, he made many observations on ship-board life, flora and fauna, and the appearance and customs of the native peoples he encountered. Volume 1 starts with a thorough review of previous exploratory voyages to the great Southern continent, and then proceeds to describe the first part of his journey, from Portsmouth to Port Jackson (Sydney).